Filling Regulatory Gaps in High Seas Fisheries

Publications on Ocean Development
Volume 75

A Series of Studies on
the International, Legal, Institutional, and Policy Aspects
of Ocean Development

The titles published in this series are listed at brill.com/pood

Filling Regulatory Gaps in High Seas Fisheries

Discrete High Seas Fish Stocks, Deep-sea Fisheries and Vulnerable Marine Ecosystems

By

Yoshinobu Takei

MARTINUS NIJHOFF PUBLISHERS

LEIDEN • BOSTON
2013

Library of Congress Cataloging-in-Publication Data

Takei, Yoshinobu.
Filling regulatory gaps in high seas fisheries : discrete high seas fish stocks, deep-sea fisheries, and vulnerable marine ecosystems / by Yoshinobu Takei.
pages cm. – (Publications on Ocean Development ; volume 75)
Includes bibliographical references and index.
ISBN 978-90-04-24859-5 (hardback : alk. paper) – ISBN 978-90-04-24860-1 (e-book)
1. Fishery law and legislation. 2. Fishery management, International. 3. Marine ecology. I. Title.

K3895.T35 2013
343.07'692—dc23

2012049187

ISSN: 0924-1922
ISBN: 978-90-04-24859-5 (hardback)
ISBN: 978-90-04-24860-1 (e-book)

This book is printed on acid-free paper.

Contents

Acknowledgements

This book is a revised and updated version of my Ph.D. thesis submitted to Utrecht University in 2008. After I defended my thesis in December 2008, there have been numerous important developments relating to the subject-matter of this book. While they attest to the continued importance of the topic of this book, this circumstance made it more difficult for me to complete the revision of the thesis. This publication would have been impossible if I had not been allowed to work on the revision and update of the thesis as part of the post-doctoral research project entitled "Adapting governance and regulation of the marine Arctic to climate change" generously funded by the Netherlands Polar Programme (NPP). I wish to express my sincere gratitude to Professor A.H.A. Soons, Dr. A.G. Oude Elferink and Professor E.J. Molenaar for their constant support and guidance throughout the Ph.D. project and thereafter.

Thanks are also due to all the other people who helped me in writing the Ph.D. thesis and revising it, in particular (in alphabetical order): Prof. Kinji Akashi, Prof. Jan Klabbers, Titia Kloos, Peter Morris, Prof. Masahito Omori, Prof. C.G. Roelofsen and Arie Trouwborst. Last but not least, my sincere gratitude goes to my family members, Yoshiyuki, Mariko and Yoko, for their unfailing support.

It goes without saying that any possible errors or omissions are my responsibility. The present manuscript is up to date as of 31 August 2012 unless otherwise indicated. Links to internet sites are accurate as of 27 September 2012.

Utrecht, September 2012

List of Abbreviations

AFZ	Australian Fishing Zone
BFSP	*British and Foreign State Papers*
CBD	Convention on Biological Diversity
CBS Convention	Convention on the Conservation and Management of Pollock Resources in the Central Bering Sea
CCAMLR	Commission for the Conservation of Antarctic Marine Living Resources
CECAF	Fishery Committee for the Eastern Central Atlantic
CEM	Conservation and Enforcement Measures
CHM	Common Heritage of Mankind
CLCS	Commission on the Limits of the Continental Shelf
COFI	Committee on Fisheries
COP	Conference of the Parties
CPPS	Permanent Commission for the South Pacific
CRFM	Caribbean Regional Fisheries Mechanism
CSC	Convention on the Continental Shelf
DHSFS	discrete high seas fish stocks
EC	European Community
EEZ	exclusive economic zone
EFZ	exclusive fishing zone
EIA	environmental impact assessment
ENB	*Earth Negotiations Bulletin*
EU	European Union
FAO	Food and Agriculture Organization of the United Nations
FSA	United Nations Agreement for the Implementation of the Provisions of the United Nations Convention on the Law of the Sea of 10 December 1982 relating to the Conservation and Management of Straddling Fish Stocks and Highly Migratory Fish Stocks
Galapagos Agreement	Framework Agreement for the Conservation of Living Marine Resources on the High Seas of the South Pacific
GFCM	General Fisheries Commission for the Mediterranean
HSC	Convention on the High Seas
HSFC	Convention on Fishing and Conservation of the Living Resources of the High Seas

IATTC	Inter-American Tropical Tuna Commission
ICES	International Council for the Exploration of the Sea
ICJ	International Court of Justice
ICP	United Nations Open-ended Informal Consultative Process on Oceans and the Law of the Sea
ICRW	International Convention for the Regulation of Whaling
ICSP	Informal Consultations of States Parties
ILA	International Law Association
ILC	International Law Commission
ILM	*International Legal Materials*
IMO	International Maritime Organization
IPOA	International Plan of Action
IPOA-IUU	International Plan of Action to Prevent, Deter and Eliminate Illegal, Unreported and Unregulated Fishing
ISA	International Seabed Authority
ISNT	Informal Single Negotiating Text
ITLOS	International Tribunal for the Law of the Sea
IUU	illegal, unreported and unregulated
IWC	International Whaling Commission
JPOI	Johannesburg Plan of Implementation
LNTS	*League of Nations Treaty Series*
LOSB	*Law of the Sea Bulletin*
LOSC	United Nations Convention on the Law of the Sea
MAR	Mid Atlantic Ridge
MoU	Memorandum of Understanding
MPA	marine protected area
MSR	marine scientific research
MSY	maximum sustainable yield
NAFO	Northwest Atlantic Fisheries Organization
NEAFC	North East Atlantic Fisheries Commission
PCIJ	Permanent Court of International Justice
PECMAS	Permanent Committee on Management and Science
REIO	regional economic integration organization
RFMO	regional fisheries management organization
RFMO/A	regional fisheries management organization or arrangement
RIAA	*Reports of International Arbitral Awards*
RILC	*Report of the International Law Commission*
SAFC	South Atlantic Fisheries Commission
SAI	significant adverse impact
SEAFO	South East Atlantic Fisheries Organization
SIODFA	Southern Indian Ocean Deepwater Fishers Association
SIOFA	Southern Indian Ocean Fisheries Agreement

SPRFMO	South Pacific Regional Fisheries Management Organization
SPRFMO Convention	Convention on the Conservation and Management of High Seas Fishery Resources in the South Pacific Ocean
SWIOFC	South West Indian Ocean Fisheries Commission
TAC	total allowable catch
UNCLOS III	Third United Nations Conference on the Law of the Sea
UNEP	United Nations Environment Programme
UNGA	United Nations General Assembly
UNTS	*United Nations Treaty Series*
UK	United Kingdom
US	United States of America
VCLT	Vienna Convention on the Law of Treaties
VME	vulnerable marine ecosystem
WCPFC	Western and Central Pacific Fisheries Commission
WECAFC	Western Central Atlantic Fishery Commission

Introduction

The state of marine capture fisheries is in crisis in many parts of the world's oceans.[1] Fisheries management has so far maintained a poor record concerning its effectiveness in the conservation and management of marine living resources. Furthermore, according to the Food and Agriculture Organization of the United Nations (FAO), "[t]he situation seems more critical for some highly migratory, straddling and other fishery resources that are exploited solely or partially in the high seas" and

> the state of straddling stocks and of other high seas fishery resources is even more problematic than for highly migratory species, with nearly two-thirds of the stocks for which the state of exploitation can be determined being classified as overexploited or depleted.[2]

Problems of high seas fisheries have been complicated by a multitude of causes. According to Agenda 21, the main problems affecting high seas fisheries are "unregulated fishing, overcapitalization, excessive fleet size, vessel reflagging to escape controls, insufficiently selective gear, unreliable databases and lack of sufficient cooperation between States".[3] These factors were again identified in the 2006 United Nations (UN) Secretary-General's Report on Oceans and the Law of the Sea when it considered the factors limiting the contribution of fisheries to sustainable development.[4]

Particularly pressing are the problems of deep-sea fisheries on the high seas and the protection of marine ecosystems from damage caused by destructive fishing practices.[5] Deep-water species include orange roughy, oreo dory,

---

1 See FAO Fisheries and Aquaculture Department, The State of World Fisheries and Aquaculture 2012, at p. 53. The overall situation did not change in recent years. See FAO Fisheries Department, The State of World Fisheries and Aquaculture 2004 (SOFIA 2004), at p. 32; FAO Fisheries and Aquaculture Department, The State of World Fisheries and Aquaculture 2006 (SOFIA 2006), at p. 29; FAO Fisheries and Aquaculture Department, The State of World Fisheries and Aquaculture 2008, at p. 7; FAO Fisheries and Aquaculture Department, The State of World Fisheries and Aquaculture 2010, at p. 35.

2 FAO Fisheries and Aquaculture Department, SOFIA 2006, at p. 33.

3 Agenda 21, Rio de Janeiro, 13 August 1992, para. 17.45.

4 Oceans and the Law of the Sea, Report of the Secretary-General, A/60/63, 4 March 2005, at pp. 57–58, paras. 210–211.

5 For example, Gjerde identified three major challenges to the legal regime for high seas fisheries, namely, declining high seas fish stocks, biodiversity concerns and IUU fishing. Regarding the first and second concerns, she specifically refers to deep-sea fisheries as examples. K.M. Gjerde,

alfonsino, toothfish, pelagic armourhead and hoki, yet they are not limited to these species: an important feature of deep-water fishes is that new discoveries continue.[6] Deep-sea fisheries first developed off New Zealand and Australia in the late-1970s and 1980s, and rapidly developed elsewhere in the 1990s.[7] The development of deep-sea fisheries has been prompted by three factors: (1) reduced fishing opportunities in shallower waters due to the depletion of species and stocks and extended maritime jurisdiction; (2) the high value of some deep-water species; and (3) advances in technology that make fishing in deep water possible and commercially viable.[8] The development of deep-sea fisheries has outpaced fisheries management and, in recent years, these problems have come under the increasing attention of the international community for the following reasons.

First, serious doubts have been cast on the sustainability of fish stocks targeted in deep-sea fisheries. Deep-water species are particularly vulnerable to overexploitation and depletion because of their slow growth and late maturity. Besides, since deep-water fisheries often exploit aggregations associated with specific features (e.g., seamounts, ocean ridges and canyons), they do not last presumably because of a localized depletion of the fishery resource.[9] In this type of operation, fishing vessels reduce a particular stock to depletion in one area, and then quickly explore new aggregations in another area and easily transfer fishing efforts there. In this way, some commercially attractive deep-sea species had already been depleted during quite a short period before any effective conservation and management measures were put in place.[10] In addition, concerns are amplified by the fear that, to the extent that demersal species are linked to pelagic species through the food chain, the degradation of the stock status of

---

"High Seas Fisheries Management under the Convention on the Law of the Sea", in D. Freestone, R. Barnes and D.M. Ong (eds.), *The Law of the Sea: Progress and Prospects* (2006), at p. 281.

6 J.-J. Maguire *et al.*, The State of World Highly Migratory, Straddling and Other High Seas Fishery Resources and Associated Species, FAO Fisheries Technical Paper No. 495, at pp. 49–55.

7 Ibid., at p. 50. It can be noted, however, that during the period between the 1960s and the 1990s, many regions of the world's oceans and areas beyond national jurisdiction were explored, fished and researched by distant water fishing nations although much of these historical data are not reported in FAO catch statistics. Report and Documentation of the Expert Consultation on Deep-sea Fisheries in the High Seas, Bangkok, Thailand, 21–23 November 2006, FAO Fisheries Report No. 838, FIEP/R838 (En), at para. 113.

8 Maguire *et al.*, The State of World Highly Migratory, Straddling and Other High Seas Fishery Resources and Associated Species, at pp. 50 and 64. See also FAO Fisheries Report No. 838, at paras. 46 and 50–51.

9 Maguire *et al.*, The State of World Highly Migratory, Straddling and Other High Seas Fishery Resources and Associated Species, at p. 49. See also Gjerde, "High Seas Fisheries Management under the Convention on the Law of the Sea", at p. 284.

10 See M. Lack *et al.*, *Managing Risk and Uncertainty in Deep-sea Fisheries: Lessons from Orange Roughy* (2003), at p. 47. See also FAO Fisheries Report No. 838, at para. 57.

demersal species may lead to a decline in the populations of pelagic species in the same ecosystem through predator-prey relationships.

Second, mounting concerns for the marine environment arose because species targeted in deep-sea fisheries often aggregate around seamounts, which are vulnerable to, *inter alia*, bottom trawl fishing.[11] Other components of vulnerable marine ecosystems (VMEs) such as cold-water corals, sponges and hydrothermal vents could also be subject to threats from deep-sea fisheries.

It has been argued that major regulatory gaps in the regime of high seas fisheries currently exist. The weakness of the existing international legal framework has been pointed out *inter alia* in relation to 'discrete high seas fish stocks' (DHSFS),[12] on which "[t]he attention of the international community has focused increasingly".[13] This category of fish stocks only occurs in the high seas. The exploitation of DHSFS is relatively recent and less is known about their biology and stock structure.[14] As deep-sea fisheries developed, target species of high seas fisheries came to include DHSFS due to, *inter alia*, their localized population distribution; most of the currently known DHSFS are deep-water species, but several others may be pelagic.[15] The United Nations Convention on the Law of the Sea (LOSC) is the basic instrument covering all activities in the oceans, including fisheries.[16] It stipulates the breadth of various marine areas under coastal state jurisdiction and the rights and jurisdiction of coastal states and other states in such areas,[17] while the rest of the oceans remains high seas where no state may exercise jurisdiction over vessels of other states unless otherwise expressly provided for.[18] While the LOSC explicitly recognizes the sovereign rights of coastal states over fisheries in areas under national jurisdiction, high seas fisheries remain subject to the principle of freedom of fishing and other general principles. Lacking precise rules for implementation, these general principles by

11 See UNGA Resolution 59/25, Sustainable Fisheries, Including through the 1995 Agreement for the Implementation of the Provisions of the United Nations Convention on the Law of the Sea of 10 December 1982 relating to the Conservation and Management of Straddling Fish Stocks and Highly Migratory Fish Stocks, and Related Instruments, 17 November 2004, paras. 66–71.

12 See also FAO Fisheries Report No. 838, at para. 53.

13 FAO Fisheries and Aquaculture Department, SOFIA 2006, at p. 125.

14 Maguire *et al.*, The State of World Highly Migratory, Straddling and Other High Seas Fishery Resources and Associated Species, at p. 5.

15 See Report Submitted in Accordance with Paragraph 17 of General Assembly Resolution 59/25 of 17 November 2004, to Assist the Review Conference to Implement Its Mandate Under Paragraph 2, Article 36 of the United Nations Fish Stocks Agreement (FSA Review Conference Preparatory Report), A/CONF.210/2006/1, 4 January 2006, at para. 104.

16 United Nations Convention on the Law of the Sea (LOSC), Montego Bay, 10 December 1982.

17 For example, ibid., Articles 2, 49 and 56.

18 See ibid., Article 92.

themselves do not necessarily ensure the effective conservation and management of the fisheries resources of the high seas.[19] The 1995 United Nations Fish Stocks Agreement (FSA),[20] which was concluded to implement the provisions of the LOSC, is aimed at the conservation and management of straddling fish stocks and highly migratory fish stocks and is not applicable to DHSFS.[21]

Institutional gaps are found in the coverage of high seas by regional fisheries management organizations or arrangements (RFMO/As). No RFMO/As with the competence to adopt conservation and management measures for DHSFS and straddling fish stocks exist in such areas as the South-West Atlantic, the Northern Indian Ocean, part of the Arctic Ocean and the Central Atlantic.[22] Under these circumstances, fisheries could become a major threat to the sustainability of the stocks concerned unless necessary conservation and management measures are taken individually by flag states.[23]

Even if appropriate conservation and management measures are in place in relation to the members of regional fisheries management organizations (RFMOs), various forms of illegal, unreported and unregulated fishing (IUU fishing) may undermine the effectiveness of the conservation and management measures of these RFMOs.[24] IUU fishing constitutes a serious impediment to achieving sustainable fisheries.[25] While a number of measures against IUU fishing are stipulated in the IPOA-IUU, the range of measures available to non-flag

19 See R.R. Churchill and A.V. Lowe, *The Law of the Sea*, 3rd edition (1999), at pp. 296–297; F. Orrego Vicuña, *The Changing International Law of High Seas Fisheries* (1999), at pp. 51–52; T. Scovazzi, "The Evolution of International Law of the Sea: New Issues, New Challenges", 286 *Recueil des Cours* (2000), at p. 146 ("A régime is needed to [...] fill the persistent legal vacuum"). See also FAO Fisheries Department, SOFIA 2004, at p. 96 (indicating that high seas conservation and management regimes under the LOSC are "limited to transboundary stocks, marine mammals and the use of driftnets").

20 Agreement for the Implementation of the Provisions of the United Nations Convention on the Law of the Sea of 10 December 1982 relating to the Conservation and Management of Straddling Fish Stocks and Highly Migratory Fish Stocks (FSA), New York, 4 August 1995.

21 See Section 2.1.4 below. But see remarks by Satya Nandan in Conference Report, Conference on the Governance of High Seas Fisheries and the UN Fish Agreement: Moving from Words to Action, St. John's, Newfoundland and Labrador, May 1–5 2005 (St. John's Conference Report), 1 June 2005, at p. 12.

22 See further Chapter 5 below.

23 Such fisheries may be regarded as unregulated fishing if "conducted in a manner inconsistent with State responsibilities for the conservation of living marine resources under international law". International Plan of Action to Prevent, Deter and Eliminate Illegal, Unreported and Unregulated Fishing (IPOA-IUU), Rome, approved by the FAO Committee on Fisheries on 2 March 2001, para. 3.3.2.

24 See ibid., para. 3.1.

25 IUU fishing may affect high seas fisheries, including DHSFS, not only because of its economically unfair practices but also because it causes ecological damage. See D.G.M. Miller, "Management and Governance Conventions and Protocols: SEAFC, WCPFC and SADC", in R. Shotton

states has been controversial. In addition, there are ambiguities in the scope of unregulated fishing against which states can take action. Certain unregulated fishing is not considered to be in violation of applicable international law and does not require the application of the measures envisaged under the IPOA-IUU.[26] Faced with ambiguities in the scope of unregulated fishing, states other than the flag state might hesitate in taking necessary measures allowed under the IPOA-IUU.

States have been addressing the problems formulated above in a variety of settings. For example, the United Nations General Assembly (UNGA) has urged responses at the global, regional and national levels, both in the short and the long term, to achieve sustainable fisheries by filling the regulatory gaps.[27] Successive UNGA Resolutions on sustainable fisheries have also continuously called for actions addressing bottom fisheries on the high seas since 2004. However, the regulatory gaps in the regime of high seas fisheries have not yet been completely filled. For example, the UNGA conducted reviews of actions taken by states and RFMOs and arrangements in response to the relevant paragraphs of UNGA Resolutions on bottom fisheries in 2009 and 2011, but found that further actions were needed and decided to conduct a further review in 2015.[28]

The international community is now faced with the urgent question of what further action should be taken to address the regulatory gaps in high seas fisheries. In search of an answer to this question, it is essential to the development of future strategies that the current state of the issue is thoroughly examined with a view to identifying the remaining regulatory gaps. On the basis of the above, this study aims to articulate the general principles applicable to high seas fisheries and their implications for new challenges in the conservation and management of the living resources of the high seas through an analysis of the actions taken by states, collectively or individually, to address the regulatory gaps in the regime of high seas fisheries at the global, regional and national levels.

Following this Introduction, this book consists of six chapters. The first two chapters mainly discuss the general principles of high seas fisheries whereas the

---

(ed.), *Deep Sea 2003: Conference on the Governance and Management of Deep-sea Fisheries: Part 1: Conference Reports. Queenstown, New Zealand, 1–5 December 2003* (2005), at p. 630.

26 IPOA-IUU, para. 3.4. A similar phrase was contained in some of the drafts for the FAO Port State Measures Agreement. See, e.g., Chairperson's Draft Agreement on Port State Measures to Prevent, Deter and Eliminate Illegal, Unreported and Unregulated Fishing, Chairperson's text as at 5 February 2009, Articles 1(1)(e)(iii.2) and 9(1)(c).

27 See, e.g., UNGA Resolution 66/68, Sustainable Fisheries, Including through the 1995 Agreement for the Implementation of the Provisions of the United Nations Convention on the Law of the Sea of 10 December 1982 relating to the Conservation and Management of Straddling Fish Stocks and Highly Migratory Fish Stocks, and Related Instruments, 6 December 2011, para. 94.

28 See further Chapter 3 below.

subsequent three chapters (Chapters 3–5) consider actions taken by states to address regulatory gaps, focusing on DHSFS, deep-sea fisheries and the protection of vulnerable marine ecosystems. These analyses are related to each other and should therefore be taken together. Chapter 6 concludes with some final remarks, synthesizing the analyses in this book and suggesting action for the international community to address regulatory gaps in high seas fisheries. The following paragraphs describe the main discussions in this book in more detail.

Chapter 1 considers the historical development of the general principles of high seas fisheries up to the adoption of the LOSC. The general principles of high seas fisheries reflected in the LOSC are not new; the LOSC is the culmination of decades of efforts to codify and develop the law of the sea. During that process, the general principles of high seas fisheries were devised, agreed upon and elaborated. It is because of this historical context that the study first needs to look at the evolution of the general principles. Putting these principles into historical perspective contributes to revealing the objectives underlying these principles and to better understanding what is meant by them.

Chapter 2 analyzes the general principles of high seas fisheries under the LOSC and other relevant international instruments. A point of departure for the investigation of the general principles of high seas fisheries under the contemporary international law of the sea is the LOSC. It is the basic instrument dealing with law of the sea issues and underlies the current international legal framework for the conservation and management of all marine living resources in the high seas. The LOSC presents the general principles of high seas fisheries, including the freedom of fishing on the high seas, the duty of cooperation between states and the conservation of marine living resources. While the existence of (at least some of) these principles was recognized, *inter alia*, by the judgments of the International Court of Justice (ICJ) in the *Fisheries Jurisdiction* cases, the LOSC is the first (almost) universally accepted multilateral treaty articulating these principles. The study needs to establish the content of the general principles of high seas fisheries under the LOSC in the light of the negotiating history of the LOSC and contemporary developments in the law of the sea and other fields of international law. In particular, one new aspect that emerged in the 1970s is the protection of the marine environment, now constituting an important element of the law of the sea.[29] The content of the general principles could be affected by this aspect of the development of the law.

---

29 Another important factor that has affected the evolution of the high seas fisheries regime is the 'economic' factor, including the legality of trade restrictions for the conservation of marine living resources under international trade law and the economic aspect as one of the three pillars of the concept 'sustainable development'. The present study deals with the development and relevance of sustainable development, rather than international economic law as such, in a high seas fisheries context.

The analysis in this chapter cannot be complete without examining developments that took place after the adoption of the LOSC in 1982. The LOSC is not a static instrument: the general principles contained in the LOSC have been further elaborated during the period following the adoption of the LOSC through treaties and a variety of other normative instruments. The most notable of those are the FSA, the FAO Compliance Agreement and UNGA Resolutions. Their impacts on the general principles of high seas fisheries in the LOSC are examined in this chapter.

Other important questions include: whether the general principles of high seas fisheries articulated under the LOSC are applicable to sedentary species beyond the outer limit of the continental shelf and whether these principles need to be applied to DHSFS in the same manner as to straddling fish stocks.

It is against this background that the practice of states addressing, collectively or individually, new challenges in high seas fisheries needs to be examined in this study. States have been addressing the regulatory gaps in high seas fisheries at the global, regional and national levels, although actions are linked to each other. For example, while action at the regional level is aimed at implementing commitments at the global level, a uniform trend in practice at the regional level may contribute to the formation of rules of customary international law at the global level.

Chapter 3 considers the actions of states at the global level addressing the regulatory gaps in high seas fisheries in three different ways. First, in view of the inapplicability of the provisions of the FSA to DHSFS, states decided to recognize that principles of the FSA should be applicable to DHSFS. Is it now generally agreed that (at least, some of) the principles of the FSA should be applicable to DHSFS? If that is so, which principles are considered to be applicable? Second, states have attempted to regulate deep-sea fisheries, especially bottom trawling, by prescribing actions to be taken by states and RFMO/As. What actions are called for, and how and to what extent are they implemented? In addition, it should be noted that the boom and bust cycle of deep-sea fisheries has caused some scientists to characterize them as mining activities, vastly different from harvesting a renewable resource.[30] If sustainable fisheries are feasible in deep-sea fisheries, the following question could be raised: does the same legal regime apply to deep-sea fisheries as other high seas fisheries, or are deep-sea fisheries governed by a different regime due to their special characteristics? Third, area-based management tools such as high seas marine protected areas (MPAs) have been advocated as useful tools to address the conservation of marine biodiversity in general and problems of high seas fisheries in particular. But, doubts have

30 See Gjerde, "High Seas Fisheries Management under the Convention on the Law of the Sea", at p. 304.

also been cast on the use of area-based management tools, notably high seas MPAs. Are they compatible with international law? Is the establishment of MPAs for fisheries purposes influenced by other considerations?

Chapter 4 examines the work of the five RFMOs with the competence to manage fisheries for DHSFS and straddling fish stocks. Cooperative mechanisms at the regional level are one of the options for cooperation explicitly mentioned in the LOSC, and subsequent international fisheries instruments stipulate a major role to be played by RFMOs. These five RFMOs appear to have the competence to regulate straddling fish stocks and DHSFS. The analysis of the constitutive instruments and practice of these RFMOs allows a measurement to be made of the extent to which modern approaches to fisheries management elaborating the general principles are incorporated. Are these approaches sufficiently reflected in the constitutive instruments and practice of RFMOs? How are they implemented? The five RFMOs have taken conservation and management measures, addressing deep-sea fisheries and/or the protection of VMEs such as seamounts and corals. The standards applied in adopting these measures could be used to take stock of the effectiveness of addressing regulatory gaps at the regional level.

Chapter 5 considers other developments at the regional level. As indicated earlier, not all oceans are covered by operational RFMO/As. In some areas, initiatives have been launched to establish RFMO/As, while in other areas no such formal initiatives exist. The texts of the three conventions concluded to establish RFMO/As in such areas are useful in considering what principles were employed in these negotiations. The consideration of interim measures adopted in these processes to implement commitments made at the global level also provides insights into the acceptance of global commitments at the regional level. It is also of interest to know the current situation in areas where no initiatives are launched, for example, whether negotiations for establishing a new mechanism are likely to take place in the near future.

Despite the primary role played by regional cooperative mechanisms, may national legislation affect the regime of high seas fisheries? One way is for flag states (and an entity to which flag states have transferred certain competence regarding fisheries) to enact domestic laws and regulations governing high seas fisheries by their vessels (e.g., the European Union (EU)). Another way is to enact laws and regulations that have impacts on the activities of third state vessels and nationals (e.g., the United States of America (US)). The present study investigates the content of the EU regulations in the context of high seas fisheries in the South-West Atlantic.[31]

[31] In Chapter 6, a brief reference is made to the Magnuson-Stevens Reauthorization Act of the US, which will potentially impact on high seas fisheries by third state vessels.

The study addresses the research questions explained above, among other things, by articulating rights granted to, and duties assumed by, states under international law; by observing the processes in which states develop, invoke, interpret and comply with (or violate) rules of international law in a variety of settings; and, in some cases, by exposing the adequacy or inadequacy of the current international legal framework in addressing new challenges.

Materials enumerated in Article 38(1) of the Statute of the ICJ, such as treaties and custom, constitute useful (yet non-exhaustive) sources to look at. Other materials are also important in the sense that the behaviour of states could be influenced by them. The recommendations of international organizations such as UNGA Resolutions, declarations recording political commitments obtained at international conferences, memoranda of understanding, international plans of action and technical guidelines are among those materials potentially constituting a basis for rules of international law. Article 119(1) of the LOSC stipulates that generally recommended international minimum standards shall be taken into account in determining conservation and management measures. Among other instruments, the FAO Code of Conduct is the result of the efforts to produce such standards. Although they are not legally binding, the content of the Code as well as instruments made within its framework should be taken into account in interpreting the provisions of the LOSC.[32] In addition, the subsequent agreement and practice of states parties to a treaty are factors to be taken into account in interpreting a treaty. From this perspective, the practice of the UNGA concerning high seas fisheries, most notably its annual Resolutions, could be a relevant factor, given the fact that all parties to the LOSC (except for the EU) are members of the UN. The critical role of the UNGA in the implementation of the LOSC has been widely recognized while the review function of the meeting of states parties to the LOSC has proven to be controversial: the latter's role has been mainly confined to procedural and administrative aspects.[33]

In the case of the present study, there are difficulties in taking the same approach as the study of multilateral environmental regimes since the LOSC cannot be described as having the typical characteristics of the framework treaties often used in multilateral environmental agreements. For one thing, the LOSC does not specify which institution(s) shall implement provisions concerning high seas fisheries except for the implementation through cooperation between

32 See Chapter 2.

33 See, generally, A.G. Oude Elferink, "Reviewing the Implementation of the LOS Convention: The Role of the United Nations General Assembly and the Meeting of States Parties", in A.G. Oude Elferink and D.R. Rothwell (eds.), *Oceans Management in the 21st Century: Institutional Frameworks and Responses* (2004), at pp. 295–312; T. Treves, "The General Assembly and the Meeting of States Parties in the Implementation of the LOS Convention", in A.G. Oude Elferink (ed.), *Stability and Change in the Law of the Sea: The Role of the LOS Convention* (2005), at pp. 55–74.

states.[34] In fact, the regime of high seas fisheries after the adoption of the LOSC developed in a multitude of ways, including multilateral treaty-making at the global and regional levels, both within and outside the UN system, recommendations and other normative instruments adopted at international conferences and unilateral state action. Thus, the processes to be analyzed include discussions in international organizations, fora established under multilateral treaties, bilateral and multilateral negotiations, international litigation, unilateral action taken by states individually or jointly through national legislation and other governmental actions as well as reactions by other states against such unilateral action.

The term 'discrete high seas fish stock' is used throughout the study. It is not a term of art, although this term is often found in recent international fisheries instruments. Other terms used in international fisheries instruments to denote the same or similar concepts include 'purely high seas stock',[35] 'other high seas fish stocks',[36] and 'high-seas discrete stock'.[37]

The term 'discrete high seas fish stock' has begun to be used recently as interest in the conservation and management of this type of fish stocks has grown. Conceptually, a fish stock in the high seas that does not fall within the categories of anadromous stocks, catadromous species, marine mammals or highly migratory species defined in the LOSC is classified either as a straddling fish stock or a discrete high seas fish stock. In other words, if such a fish stock is not a straddling fish stock, it is a discrete high seas fish stock.

Some ambiguities are inherent in this definition, however. Although the term 'straddling fish stock' is generally used to refer to stocks that occur both in the exclusive economic zone (EEZ) and an area of the high seas adjacent to and beyond the zone,[38] the term is neither used in Article 63(2) or any other provision of the LOSC, nor defined in the FSA.

---

34 For example, Article 319(2) provides for, among other things, reporting on issues of a general nature that has arisen with respect to the LOSC and the convening of necessary meetings of states parties in accordance with the LOSC, but it is silent on the potential supervisory role with respect to the LOSC.

35 FAO Fisheries Department, World Review of Highly Migratory Species and Straddling Stocks, FAO Fisheries Technical Paper No. 337.

36 See FSA Review Conference Preparatory Report, at para. 14. See also Maguire *et al.*, The State of World Highly Migratory, Straddling and Other High Seas Fishery Resources and Associated Species.

37 See, e.g., Report of the Review Conference on the Agreement for the Implementation of the Provisions of the United Nations Convention on the Law of the Sea of 10 December 1982 relating to the Conservation and Management of Straddling Fish Stocks and Highly Migratory Fish Stocks (FSA Review Conference Report), A/CONF.210/2006/15, 5 July 2006.

38 Orrego Vicuña, *The Changing International Law of High Seas Fisheries*, at p. 41. But, for a more restrictive use, see E.D. Brown, *The International Law of the Sea*, vol. I (1994), at p. 320; C.C. Joyner and P.N. De Cola, "Chile's Presential Sea Proposal: Implications for Straddling Stocks

The phrase 'occur both within the exclusive economic zone and in an area beyond and adjacent to the zone' is subject to different interpretations. What does the word 'occur' mean? In fact, discussions took place when the FSA was negotiated, and delegations made proposals for a definition. For example, Russia proposed to classify fish stocks into five categories, and it advised the UN Fish Stocks Conference to exclude issues of both (i) stocks which basically are constituted in and inhabit high seas areas and migrate into the EEZ only at a specific period of their life-cycle and (ii) stocks which for the duration of their entire life-cycle inhabit only areas of the high seas. It appears that Russia considered that the conservation and management of the former type of stocks should not form part of the discussion on straddling fish stocks at that conference ("this issue can become the subject of consideration by future forums").[39] Faced with complex definitions as found in this Russian proposal, efforts to define the term 'straddling fish stock' were soon abandoned because of the potential difficulties to reach agreement.[40] One FAO paper argues that usage seems to indicate that as long as there is some directed fishing effort exerted to catch the stock on either side of the outer limit of the EEZ, it is considered to be straddling.[41] This way of classification could imply that DHSFS include fish stocks for which no fisheries exist in the EEZ even if small amounts of biomass exist within the EEZ.

Another factor that could influence discussions on the scope of the terms is what is included in 'fish' stocks. As explained in Chapters 2, 4 and 5, the constitutive instruments of RFMOs deal with marine living resources widely and the legislation of states indicates that the term 'fish' could be interpreted broadly, including invertebrates and corals and sponges. For this reason, this study considers the conservation and management of not only finfish but also invertebrates and coral and sponges unless otherwise indicated. In the same vein, the term 'fishery' is used to mean the 'harvesting' of finfish as well as invertebrates, corals and sponges.

Throughout the present study, the terms 'deep-sea fisheries' and 'bottom fisheries' are used interchangeably unless the context indicates otherwise. Certainly, there is a difference in the connotation of these terms. On the one hand, the term

---

and the International Law of Fisheries", 24 *Ocean Development and International Law* (1993), at p. 103.

39 Conceptual Approach to the Conservation of Straddling Fish Stocks by Improving Their Management (Submitted by the Delegation of the Russian Federation), A/CONF.164/L.38, 2 March 1994, at p. 2.

40 See Orrego Vicuña, *The Changing International Law of High Seas Fisheries*, at p. 140.

41 Maguire *et al.*, The State of World Highly Migratory, Straddling and Other High Seas Fishery Resources and Associated Species, at p. 4. For discussions, see E.J. Molenaar, "The South Tasman Rise Arrangement of 2000 and Other Initiatives on Management and Conservation of Orange Roughy", 16 *International Journal of Marine and Coastal Law* (2001), at pp. 86–87.

'deep-sea fisheries' has often been used in connection with certain depths.[42] On the other hand, the term 'bottom fisheries' concerns contact with the seabed, as seen in the definition of the term in the relevant EU Regulation.[43] Nevertheless, such trends have by no means resulted in a rigid distinction: the FAO Technical Consultation on the International Guidelines for the management of *deep-sea fisheries* in the high seas agreed to use the criterion of likely contact with the seafloor during the normal course of fishing operations.[44] For this reason, the present study does not distinguish between the two terms unless circumstances dictate otherwise.

42 See Section 3.2 below.
43 See Section 5.4.3 below.
44 See Section 3.2 below.

Chapter One

The Evolution of the General Principles of High Seas Fisheries

This chapter examines the evolution of the general principles of high seas fisheries from the late nineteenth century to the adoption of the LOSC in 1982. The chapter is divided into three sections. The first section describes the background and developments up to the 1940s, looking at instances of multilateral regulatory regimes in which high seas fishing states chose to regulate the fishing activities of their own vessels as well as coastal states' attempts to challenge the dominant principle, the freedom of fishing, through the extension of fisheries jurisdiction to adjacent high seas areas. This is followed by an account of developments between 1945 and 1960 with special emphasis on an analysis of the codification process under the auspices of the United Nations between 1950 and 1958. During this period, general principles of high seas fisheries were established. The concept of freedom of fishing was elaborated and other general principles were framed. The third section focuses on the competing claims concerning the proposed legal framework for fisheries beyond the limits of national jurisdiction as part of the process of establishing a new legal order for the oceans during the period from the aftermath of the 1960 Geneva Conference on the Law of the Sea to the conclusion of the LOSC in 1982.

The development of the general principles of high seas fisheries during this period mainly concerns the area adjacent to the territorial sea, which is included in the EEZ/exclusive fishing zone (EFZ) under the contemporary international law of the sea. In this sense, the discussions on the regime of high seas fisheries during this period did not primarily concern the area now called the high seas. Nevertheless, it is necessary to examine the development of the general principles of high seas fisheries in the area which was considered to be the 'high seas' at that time because the general principles developed in relation to that area also influenced the general principles which were applicable in the area now considered to be the high seas.

1.1 *Freedom of Fishing on the High Seas Challenged: The Late 19th Century to World War II*

Freedom of the high seas, including the freedom to fish on the high seas, is an established principle in international law.[1] Freedom of fishing on the high seas consists of, *inter alia*, the recognition that all states have the right to engage in fishing operations on the high seas, that is, free access to fishery resources.[2] In addition, freedom of the high seas includes the principle that the flag state has exclusive jurisdiction over ships on the high seas.[3]

Technological developments gave rise to conflicts and cooperation between states over fisheries in some parts of the world in the nineteenth century. There were two strands in the attempts to regulate fisheries on the high seas: (1) multilateral regulatory regimes established by high seas fishing states; (2) unilateral measures by coastal states in areas adjacent to their territorial sea.

First of all, a corollary of freedom of fishing on the high seas was the proposition that the regulations of each state were binding on its own nationals. The joint regulation of high seas fisheries could be pursued by agreement among the interested states; if a state refused to participate in the resulting agreement, regulations agreed among participants in the agreement were not applicable to the nationals of the non-participating state.[4]

Among the earliest attempts to provide for a regulation on fisheries on the high seas through multilateral conventions was the 1882 North Sea Policing Convention.[5] The object of the Convention was formulated as regulating the policing

1 See, e.g., "Mémorandum présenté par le Secrétariat, Regime of the High Seas", *Yearbook of the International Law Commission 1950*, at p. 69; A.P. Daggett, "The Regulation of Maritime Fisheries by Treaty", 28 *American Journal of International Law* (1934), at p. 704; G. Gidel, *Le Droit International Public de la Mer: Le Temps de Paix*, vol. I (1932), at pp. 197–200; R.G. Rayfuse, *Non-Flag State Enforcement in High Seas Fisheries* (2004), at p. 20.

2 On this and other elements of the principle, see, e.g., A.W. Koers, *International Regulation of Marine Fisheries: A Study of Regional Fisheries Organizations* (1973), at p. 19.

3 See *The Case of the S.S. "Lotus"*, Judgment, 7 September 1927, *PCIJ, Recueil des Arrêts, Série A, No. 10*, p. 3, at p. 25.

4 See "Mémorandum présenté par le Secrétariat, Regime of the High Seas", at p. 85.

5 Convention Between Belgium, Denmark, France, Germany, Great Britain and the Netherlands, for Regulating the Police of the North Sea Fisheries, The Hague, 6 May 1882. For the history of conflicts in this region and negotiations, see T.W. Fulton, *The Sovereignty of the Sea: An Historical Account of the Claims of England to the Dominion of the British Seas, and of the Evolution of the Territorial Waters: With Special Reference to the Rights of Fishing and the Naval Salute* (1911), at p. 630 *et seq*. For a recent account of the Convention, see, e.g., K. Bangert, "The Effective Enforcement of High Seas Fishing Regimes: The Case of the Convention for the Regulation of the Policing of the North Sea Fisheries of 6 May 1882", in G.S. Goodwin-Gill and S. Talmon (eds.), *The Reality of International Law: Essays in Honour of Ian Brownlie* (1999), at pp. 1–20.

of fisheries in the North Sea outside territorial waters.[6] This Convention and modern fisheries management mechanisms differ in two important respects. First, its narrow focus on policing shows that there was no need to attempt any conservation of fisheries resources as such even in an area where considerable fishing efforts were concentrated, such as the North Sea.[7] Second, the concept of conservation was not elaborated in the Convention.

Another multilateral regulatory regime for high seas fisheries was the whaling regime. Although the whaling regime did not explicitly state its objectives, economic considerations were the prevalent factor.[8] In addition, the Agreements did not place any limits on the total catch number, but rather prohibited the taking or killing of certain species and small individual specimens and restricted operations in certain areas.[9] The whaling regime established by these Agreements did not, however, result in effective protection for whales.[10] The elements embodied in the regime could not be regarded as reflecting existing or emerging trends at that time.

The unilateral extension of coastal state fisheries jurisdiction to adjacent high seas areas was attempted, among others, in the dispute between the United States and Great Britain over the protection of fur seals in the Bering Sea.[11] The United States' attempt to regulate sealing outside its territorial sea in the Bering Sea in order to protect herds frequenting its islands gave rise to a dispute with Great Britain, on behalf of Canada, which eventually led to arbitration.[12] The arbitral tribunal rejected the American claims, and ruled that freedom of

6 Convention Between Belgium, Denmark, France, Germany, Great Britain and the Netherlands, for Regulating the Police of the North Sea Fisheries, Article I.

7 The French delegation even proposed a clause providing that "[i]n the extra-territorial part of the North Sea, fishery shall be free at all seasons, and with all kinds of implements, without any sort of distinction". Fulton, *The Sovereignty of the Sea*, at pp. 636–637. See also Gidel, *Le Droit International Public de la Mer*, vol. I, at p. 438.

8 See, e.g., International Agreement for the Regulation of Whaling, London, 8 June 1937, preamble.

9 Ibid., Articles 4–9. See also Convention for the Regulation of Whaling, Geneva, 24 September 1931, Articles 4–5.

10 Reasons include the non-participation of major players such as Japan in the 1937 Agreement as well as the failure to agree on the total allowable catch due to the conflict between traditional whaling states and new entrants. See Koers, *International Regulation of Marine Fisheries*, at p. 87; M.S. McDougal and W.T. Burke, *The Public Order of the Oceans: A Contemporary International Law of the Sea* (1962), at pp. 950 and 965.

11 On other major instances, see, e.g., Juda, *International Law and Ocean Use Management* (1996).

12 *Bering Sea Fur-Seals Case*, Award of the Tribunal of Arbitration Constituted under the Treaty Concluded at Washington, the 29th of February 1892, Between the United States of America and Her Majesty the Queen of the United Kingdom of Great Britain and Ireland 15 August 1893, *International Environmental Law Reports*, vol. 1 (Cairo A.R. Robb ed.), p. 67, at pp. 67–77. For a description of the facts, see pp. 43–62.

fishing is applicable on the high seas and the United States had neither exclusive jurisdiction nor rights in the area concerned, and neither did it have the right of protection or property for stock found outside of its territorial sea.[13]

Some argue that the tribunal did admit the need to prevent overexploitation and recognized restraints on the exercise of freedom of fishing and thus set a precedent for later conservation measures.[14] However, the circumstances surrounding the dispute make this case exceptional.[15] These features were not seen in other contemporary fisheries so that it can be said that the vulnerability of seals to the fishery concerned in this case was exceptional.

1.2 *General Principles Established: Codification Efforts and State Practice during the Period 1945–1960*

During the period from 1945 to 1960, ever increasing efforts in offshore fisheries caused by, among other things, advancements in technology and the rising demand for food supplies gave rise to the need for regulation in many parts of the world's oceans. In response to these developments, a number of coastal states successively challenged freedom of fishing on the high seas in various ways.

Codification efforts started once again under the auspices of the United Nations. The International Law Commission (ILC) chose the regime of the high seas as one of the priority topics for codification in 1949 and appointed J.P.A. François as its Special Rapporteur.[16] The existing position of international law with regard to high seas fisheries was summarized by the ILC as follows: first, regulations issued by a state were only binding upon the nationals of that state; second, if two or more states agreed upon regulations affecting a particular area, these regulations were only binding upon the nationals of the states concerned; third, in treaties concluded by states for the joint regulation of fisheries, the authority created for the purpose had been, as a rule, entrusted merely with the power to make recommendations.[17] In light of the recognition that the existing law on the subject provided no adequate protection of marine fauna against extermination, the Commission attempted to go beyond the codification of existing law.[18] While some of the innovative ideas adopted at the early

13 Ibid., at pp. 69–70. Upon the request of the parties, the tribunal presented a model treaty for conservation measures. Ibid., at pp. 70–73.

14 See P.W. Birnie *et al.*, *International Law and the Environment*, 3rd edition (2009), at pp. 707–708. See also Orrego Vicuña, *The Changing International Law of High Seas Fisheries*, at pp. 14–15.

15 See *International Environmental Law Reports*, vol. 1, at pp. 51 and 55.

16 *Yearbook of the International Law Commission 1949*, at p. 238.

17 See, e.g., commentary to draft articles on fisheries in *Yearbook of the International Law Commission 1953*, vol. II, at p. 218, para. 95.

18 For the task of the ILC, see Statute of the International Law Commission, Article 1(1).

stages were later abandoned, general principles formulated by the ILC during this period constituted a basis for subsequent discussions on this issue at the 1958 and 1960 Geneva Conferences and beyond.

The work of the ILC on fisheries was heavily influenced by an international technical conference held in Rome in 1955 to study the problem of the international conservation of the living resources of the sea and to make appropriate scientific and technical recommendations.[19] The result of the Conference covered not only technical issues but also such important issues as the extent of coastal states' rights and responsibility, problems involved in reaching agreement on conservation measures and procedures, and new entrants into a fishery already subject to conservation measures.[20]

The United Nations Conference on the Law of the Sea held in Geneva in 1958 discussed the ILC draft articles, which were adopted in 1956, and adopted four conventions concerning the law of the sea, three of which, namely the Convention on Fishing and Conservation of the Living Resources of the High Seas (HSFC), the Convention on the High Seas (HSC) and the Convention on the Continental Shelf (CSC), were concerned with high seas fisheries.[21]

Two years after the adoption of the four Geneva Conventions, the Second United Nations Conference on the Law of the Sea was held in 1960 with a view to the determination of the breadth of the territorial sea, on which states had not been able to agree at the 1958 Conference.[22] The 1960 Conference again failed to reach any agreement on the breadth of the territorial sea despite support for a compromise proposal combining the 6-mile territorial sea and the 6-mile exclusive fishing zone contiguous to the territorial sea by an almost two-thirds majority. It was evident that many states were considering a departure from the regime which adopted a dichotomy between the narrow territorial sea and the high seas fishing freedom combined with the special interests of coastal states in adjacent high seas areas. The high seas fisheries regime based on the jurisdictional framework established under the Geneva Conventions was not widely accepted

19 UNGA Resolution 900(IX), International technical conference on the conservation of the living resources of the sea, 14 December 1954, para. 1.

20 See Report of the International Technical Conference on the Conservation of the Living Resources of the Sea, 18 April to 10 May 1955, Rome (1955 Rome Conference Report). For a detailed account of the Conference and its preparatory work as well as its implications for subsequent discussions at the ILC, see F.V. Garcia-Amador, *The Exploitation and Conservation of the Resources of the Sea: A Study of Contemporary International Law* (Second and enlarged edition, 1963), at pp. 134–200.

21 Convention on Fishing and Conservation of the Living Resources of the High Seas (HSFC), Geneva, 29 April 1958; Convention on the High Seas (HSC), Geneva, 29 April 1958; Convention on the Continental Shelf, Geneva, 29 April 1958; Convention on the Territorial Sea and Contiguous Zone, Geneva, 29 April 1958.

22 See, generally, McDougal and Burke, *The Public Order of the Oceans*, at pp. 962–964.

by the international community.[23] Due to this, many authors question the significance of the HSFC.[24]

Notwithstanding the small number of parties to the HSFC, it is during this period that the general principles of high seas fisheries were framed and developed.[25] In fact, the ICJ ruled in 1974 that some principles formulated during this process of codification were rules of general international law.[26] To identify general principles established during this period, it is indispensable to look into the codification process in the 1950s.[27] Thus, the current section closely examines the rules of high seas fisheries as codified in this process, taking account of the contemporaneous practice of states and international organizations.

Before proceeding to an examination of the development of the general principles, a warning should be given with regard to the limited implications of the discussion during this period for the regime of fisheries for DHSFS. The following comments made by Special Rapporteur François exemplify where the concerns of the members of the ILC lay at the time when the high seas fisheries regime was discussed in the Commission:

> [T]he whole question of protection of the products of the sea (marine resources) was closely bound up with the problems of the continental shelf and contiguous zones. If the Commission could reach conclusions on those two problems, the whole question of protection of marine resources would be settled too.[28]

As illustrated by the above comments, when it discussed the 'Protection of the Products of the Sea' as part of the regime of the high seas, the Commission was principally, if not exclusively, interested in the protection of marine living resources in areas adjacent to the territorial sea.

---

23 The HSFC attracted only 38 parties. Information on the status of the HSFC is available at <http://treaties.un.org/pages/ViewDetails.aspx?src=TREATY&mtdsg_no=XXI-3&chapter=21&lang=en>.

24 See, e.g., Churchill and Lowe, *The Law of the Sea*, at p. 287; Birnie *et al.*, *International Law and the Environment*, at p. 709; W.T. Burke, *The New International Law of Fisheries: UNCLOS 1982 and Beyond* (1994), at pp. 13–14.

25 See also Kaye, *International Fisheries Management*, at pp. 66 and 74; Orrego Vicuña, *The Changing International Law of High Seas Fisheries*, at p. 19.

26 *Fisheries Jurisdiction Case* (*United Kingdom of Great Britain and Northern Ireland v. Iceland*), Judgment (Merits), 25 July 1974, *I.C.J. Reports 1974*, p. 3, at p. 24, para. 54 *et seq.*; *Fisheries Jurisdiction Case* (*Federal Republic of Germany v. Iceland*), Judgment (Merits), 25 July 1974, *I.C.J. Reports 1974*, p. 175, at p. 193, para. 46 *et seq.* On the history of the cases, see *I.C.J. Reports 1974*, at pp. 4–22; *I.C.J. Reports 1974*, at pp. 176–191. The analysis of the relevant part of the judgments of the Court will be conducted in the next chapter.

27 See R. Bernhardt, "Custom and Treaty in the Law of the Sea", 205 *Recueil des Cours* (1987-V), at p. 275.

28 *Yearbook of the International Law Commission 1950*, vol. I, at p. 204, para. 103. A similar view was implicit in the secretariat study prepared for the Commission. See "Mémorandum présenté par le Secrétariat, Regime of the High Seas", at pp. 77–78 and 82–87.

In fact, during this period, most high seas fisheries took place within areas adjacent to the territorial sea and most international fisheries commissions incorporated the interests of coastal states in the region. Some commissions whose mandate covered the high seas were established without the participation of distant water fishing states; notably, the commission in the South-East Pacific was established to strengthen the assertion of coastal state interests in adjacent high seas areas.[29] In contrast, the International Whaling Commission (IWC) was competent to regulate whaling in all of the world's oceans; no particular state or group of states were considered to be coastal state(s) under the International Convention for the Regulation of Whaling (ICRW). In fact, there is no specific provision for coastal states in the ICRW. Even for whaling, however, claims to coastal state interests gave rise to disputes between coastal states and high seas fishing states.[30]

As the main purpose of the present section is to explore the general principles of high seas fisheries, the special status of coastal states with regard to fisheries in the adjacent high seas area as well as those for sedentary species and fisheries employing fishing gear embedded in the seafloor is not dealt with in this chapter.[31] The remainder of this section attempts to analyze general principles which are applicable to high seas fisheries in general, whether or not adjacent to the territorial sea.

1.2.1 *Freedom of Fishing on the High Seas*

Throughout the period under consideration, freedom of fishing remained a basic principle of high seas fisheries. The HSC confirmed the freedom of the high seas; among the four enumerated freedoms was freedom of fishing.[32] The concept of freedom of fishing was elaborated during the codification process, and the HSFC articulated the right of fishing on the high seas, as seen below.

The comment on Article 2 of the 1955 ILC Draft on the Regime of the High Seas contained a statement that states were bound to refrain from any acts which might adversely affect the use of the high seas by nationals of other states.

---

29 See Koers, *International Regulation of Marine Fisheries*, at pp. 100–101.

30 See especially the conflicts between high seas whaling states and Chile and Peru. In fact, it has been pointed out that the desire to gain increased control over profits from whaling in offshore areas was one of the reasons for Chile, Ecuador and Peru to establish their 200-mile maritime zones. See McDougal and Burke, *The Public Order of the Oceans*, at p. 951.

31 On these issues, there are a number of monographs and articles, including: Garcia-Amador, *The Exploitation and Conservation of the Resources of the Sea: A Study of Contemporary International Law*; D.M. Johnston, *The International Law of Fisheries: A Framework for Policy-Oriented Inquiries* (1965); McDougal and Burke, *The Public Order of the Oceans*; S. Oda, *International Control of Sea Resources* (1963).

32 HSC, Article 2.

Moreover, it stated that "the law of the high seas contains certain rules, most of them already recognized in positive international law, which are designed, not to limit or restrict the freedom of the high seas but to safeguard its exercise in the interests of the entire international community", enumerating, among other things, the right of states concerning the conservation of the living resources of the high seas.[33] In 1956, the Special Rapporteur sought to introduce the reasonableness test for scientific research and the testing of new weapons.[34] Although intense discussions took place within the Commission with regard to atomic bomb tests,[35] the above-mentioned statements on the freedom of the high seas were retained in the Commentary to Article 27 of the 1956 Draft.[36]

At the 1958 Geneva Conference, discussions on the legality of the weapons tests led to proposals to explicitly recognize limitations on the freedom of the high seas.[37] However, some states were reluctant to include such a clause in the HSC. Incorporating the British and Mexican proposals,[38] the HSC provides that the freedom shall be exercised under the conditions laid down by these articles and by the other rules of international law and shall be exercised by all states with reasonable regard to the interests of other states in their exercise of the freedom of the high seas.[39] In other words, Article 2 of the HSC was formulated to balance the exercise of the freedom of the high seas, whether in fisheries by two or more states or between different uses of the high seas, despite the original intention to restrict the use of the high seas for nuclear tests.[40]

Restrictions on the freedom of the high seas were also discussed for each freedom, including the freedom of fishing, during this period. In fact, among the most important contributions of the ILC at the early stage of its work was the explicit recognition that, contrary to the concept of absolute freedom, freedom of fishing on the high seas had limitations under general international law. The comment on Article 3 in the 1953 draft referred to the concept of the abuse of rights in the context of the competence of the proposed international organization. The commentary stated that when a state arbitrarily and without good reason, in rigid reliance upon the principle of the freedom of the seas, declines to play its part in the reasonably necessary conservation measures, it is considered to abuse a right conferred upon it by international law.[41] This clear articulation

33 *Yearbook of the International Law Commission 1955*, vol. II, at pp. 21–22 (Comment on Article 2).

34 See *Yearbook of the International Law Commission 1956*, vol. II, at p. 10.

35 Ibid., vol. I, at p. 11 *et seq.*

36 Ibid., at p. 278 (Commentary on Article 27).

37 A/CONF.13/C.2/L.29 (Poland), Article 27(3).

38 A/CONF.13/C.2/L.3.

39 HSC, Article 2.

40 In fact, the ICJ applied the reasonableness test to high seas fisheries in the *Fisheries Jurisdiction* cases. For the details of the Court's statements on this issue, see Chapter 2.

41 *Yearbook of the International Law Commission 1953*, vol. II, at pp. 218–219, para. 100.

of the inherent restriction on the freedom of fishing on the high seas underlies subsequent general provisions in the HSFC such as the obligation to take conservation measures.

Freedom of fishing was elaborated by its clarification as a 'right' belonging to states. Article 1 of the HSFC formulates the right of fishing in line with the previous ILC draft.[42] This formulation as a 'right' necessarily implies that it is restrained by corresponding obligations. The right was made subject to several restrictions as provided for in Article 1 of the HSFC. In line with the ILC draft, the right is subject to treaty obligations and other articles of the HSFC. Moreover, the words 'the interests and rights of the coastal states' were added during the Conference as a new restriction.[43] One commentator observes that the insertion of that phrase in Article 1 has weakened or even altered the original intent of the authors of this provision, which was, according to him, an enunciation of the fundamental principle, namely, the freedom of nationals of all states to fish in the high seas.[44]

As states have the right 'for their nationals', the scope of the right depends on the definition of 'nationals'. On the one hand, the draft articles of the ILC dealing with fishing consistently refer to 'nationals', rather than 'fishing vessels' or 'ships'. In other words, members of the ILC were thinking of the behaviour of fishermen, i.e., of crews.[45] On the other hand, it has been suggested that when members of the ILC discussed the draft articles on the nationality of 'ships', they usually had merchant vessels in mind and thought of shipowners and operators.[46] Only in 1956 did the ILC seek to diminish the lack of clarity in the formula and to employ the same objective; in its comment on article 49 (Right to fish) it stated that the term 'nationals' means fishing boats having the nationality of the state concerned, irrespective of the nationality of members of their crews.[47] It then became clear that articles concerning high seas fisheries were to regulate the behaviour of fishing vessels flying the flag of parties to the HSFC, rather than that of crews on board.[48]

---

42 See ILC Draft Articles (1956), Article 49.

43 A/CONF.13/C.3/L.49; HSFC, Article 1(b). It seems that the phrase was intended to emphasize the importance of the interests and rights of coastal states. See, e.g., comments by the representative of Mexico. *United Nations Conference on the Law of the Sea, Geneva, 24 February–27 April 1958, Official Records* (*UNCLOS I Official Records*), vol. V, at p. 45, para. 27.

44 D. Nelson, "The Development of the Legal Regime of High Seas Fisheries", in A. Boyle and D. Freestone (eds.), *International Law and Sustainable Development: Past Achievements and Future Challenges* (1999), at pp. 117–118.

45 H. Meyers, *The Nationality of Ships* (1967), at pp. 287–288.

46 Ibid., at p. 287.

47 *Yearbook of the International Law Commission 1956*, vol. II, at p. 286.

48 HSFC, Article 14. On the minor changes in this article, see *UNCLOS I Official Records*, vol. II, at p. 46. See also Meyers, *The Nationality of Ships*, at pp. 287–289.

1.2.2 *Cooperation between States*

The discussions in the ILC on the general framework for a high seas fisheries regime revolved around two notions: the concept of exclusive flag-state jurisdiction[49] as well as the special position of coastal states, recognizing their special interests in high seas areas adjacent to their territorial sea.[50] The ILC recognized the need for the involvement of coastal states.[51] Throughout its deliberations during the period between 1951 and 1956, it recognized that coastal states were entitled to take part *on an equal footing* in regulation even though their nationals were not engaged in the fishery concerned, while the area was not to be closed to nationals of other states.[52] Article 6(1) of the HSFC repeats the concept of the special interest of all coastal states provided in Article 54(1) of the 1956 ILC draft verbatim. In short, the regime formulated during the process of codification between 1950 and 1958 was based on the principle of cooperation between high seas fishing states and coastal states or among high seas fishing states, as the case may be.

A fundamental question was left unanswered: who finally determines conservation measures if no agreement has been reached? Up to 1953, the ILC took the position that regulatory gaps caused by the lack of agreement between the states concerned should be filled by an institution serving the interests of the international community as a whole. The prohibition on the abuse of rights was, according to its commentary, to provide a satisfactory legal basis for the prescription of a binding conservation measure by an international organization to be established under the auspices of the United Nations.[53] In addition to research on fisheries, the

49 This notion was reflected in the ILC's draft articles. See ILC Draft (1951), draft articles on the continental shelf and related subjects, Part II (Related subjects), Article 1; ILC Draft (1953), draft articles on fisheries, Article 1; ILC Draft (1955), Draft articles relating to the conservation of the living resources of the sea, Article 2(1); ILC Draft (1956), Articles concerning the law of the sea, Article 52(1). It was eventually incorporated as Articles 3 and 4(1) in the HSFC. For the theoretical basis underlying the duty of cooperation in the context of shared natural resources, see also E. Hey, *The Regime for the Exploitation of Transboundary Marine Fisheries Resources: The United Nations Law of the Sea Convention Cooperation between States* (1989), at p. 35; R. Rayfuse, "Countermeasures and High Seas Fisheries Enforcement", LI *Netherlands International Law Review* (2004), at p. 54.

50 See, e.g., second report of the *Special Rapporteur*, Mr J.P.A. François (1951), at para. 81.

51 *Yearbook of the International Law Commission 1951*, vol. I, at p. 309.

52 Draft articles on the continental shelf and related subjects, Part II, Article 1 (1951); ILC Draft (1953), draft articles on fisheries, Article 2; 1955 Draft, Article 29(1); 1956 Draft, Article 29(1). The distance of 100 miles was specified in the 1951 and 1953 Drafts, but the figure was eliminated in the 1955 Draft. According to Garcia-Amador, "the sub-committee had abandoned the limit of 100 miles adopted by the Commission [...] because in some cases it was inadequate and in others excessive". *Yearbook of the International Law Commission 1955*, vol. I, at p. 105.

53 *Yearbook of the International Law Commission 1953*, vol. II, at pp. 218–219, para. 100.

organization was to be empowered to adopt regulations in cases where the states concerned did not agree on a necessary measure to be taken.[54]

The idea to establish an international organization under the auspices of the United Nations which was competent to deal with conservation measures, however, was not well received by governments: "some were not prepared to delegate their powers on fishery regulation to an international body, while others were sincerely convinced that conservation problems, being regional and even local in character, called for an *ad hoc* system".[55] After the Rome Conference, the idea of a universal organization was abandoned. For the high seas area adjacent to the territorial sea of coastal states, it was replaced by the recognition of unilateral measures of coastal states in the area combined with the safeguard that such unilateral measures are subject to third party dispute-settlement procedures;[56] where the fishery took place beyond that area, any of the fishing states concerned were entitled to have recourse to compulsory dispute settlement procedures.[57]

In addition to the special interests of coastal states, it was proposed during the discussion on the Special Rapporteur's draft articles in the ILC in 1955 to recognize the special interests of non-coastal states whose nationals were not engaged in the fishery concerned. Scelle proposed an amendment to the draft article dealing with the special interests of coastal states by adding an equal right of any other state by analogy.[58] François sought to revise Scelle's proposal by differentiating between the positions of coastal and non-coastal states and granting the non-coastal state only the right to take part in (or to initiate) any system of regulation if it could be proved that the extermination of a given species in that area would affect its interests elsewhere.[59] The commentary to Article 30 in the 1955 draft stated that "[t]his case may arise, for example, if the exhaustion of the resources of the sea in the area would affect the results of fishing in another area in which the nationals of the State concerned do engage in fishing".[60]

It is doubtful whether there was any common understanding in the Third Committee of the Geneva Conference on the situations covered by ILC Draft Article 56. Some states even argued that there was no situation covered by this article.[61] Some participants still held the position that the article covered

---

54 ILC Draft (1951), draft articles on the continental shelf and related subjects, Part II (Related subjects), Article 2; ILC Draft (1953), draft articles on fisheries, Article 3.

55 *Yearbook of the International Law Commission 1955*, vol. I, at p. 78.

56 See ILC Draft 1955, Articles 31–33. See also HSFC, Article 7.

57 See ILC Draft 1955, Article 26(2). See also HSFC, Article 4(2).

58 *Yearbook of the International Law Commission 1955*, vol. I, at p. 107, para. 7.

59 See ibid., at pp. 107–111.

60 Ibid., vol. II, at p. 30.

61 See *UNCLOS I, Official Records*, vol. V, at pp. 51–54.

species in the high seas migrating to the territorial sea or internal waters,[62] but, in view of the amendment to Draft Article 52, Draft Article 56 no longer appeared necessary to address the impacts of one fishery on another fishery for the same stock in a different part of the high seas.[63] The United Kingdom (UK) suggested that Draft Article 56 was meant to relate to consumer states whose nationals were not engaged in the fishery concerned. In response to this suggestion, Canada considered it somewhat exaggerated to extend the article to consumer interests since nearly all states were consumers of marine resources.[64]

A proposal to entitle non-coastal and non-fishing states having a special interest to participate in the regulatory system on an equal footing, like coastal states,[65] was rejected and, in the end, the ILC Draft Article was adopted with only minor amendments.[66] Article 8 of the HSFC provides for the special interest of non-coastal states not engaged in the harvesting of a stock in question: that state may request fishing states to take necessary conservation measures.

In summary, the duty of cooperation was recognized in two ways. On the one hand, in the high seas area adjacent to the territorial sea, the fishing states needed to cooperate with the coastal state if the coastal state so wished even though its nationals were not engaged in the fishery concerned; only if an agreement was not achieved among the states concerned was the coastal state entitled to take unilateral conservation measures. On the other hand, in areas outside the area of the high seas adjacent to the territorial sea, it was implicitly recognized that the duty of cooperation only concerned high seas fishing states *inter se*. Non-coastal and non-fishing states having a special interest were not entitled to participate in the regulatory system on an equal footing, but were allowed to request fishing states to take necessary conservation measures.

1.2.3 *Conservation*

Pre-World War II international fisheries agreements tended to have general and vague objectives.[67] In contrast, international fisheries commissions established

62 E.g., J.H.W. Verzijl, "The United Nations Conference on the Law of the Sea, Geneva, 1958 II", 6 *Nederlands Tijdschrift voor Internationaal Recht* (1959), at p. 126. See also J.L. Brierly, *The Law of Nations: An Introduction to the International Law of Peace*, 6th edition (1963), at p. 316; Oda, *International Control of Sea Resources*, at p. 118.

63 *UNCLOS I, Official Records*, vol. V, at p. 52.

64 Ibid.

65 A/CONF.13/C.3/L.30.

66 *UNCLOS I, Official Records*, vol. V, at p. 54.

67 Koers, *International Regulation of Marine Fisheries*, at p. 179. One notable exception was the 1911 Fur Seals Convention whose objective was to gather the optimum economic yield once effective control was established. Convention between Great Britain, the United States, Japan, and Russia, respecting Measures for the Preservation and Protection of the Fur Seals in the North Pacific Ocean, Washington, 7 July 1911.

after the Second World War were given explicit management goals.[68] On the one hand, the ICRW made the economic proficiency of whaling operations as the primary objective without expressing biological or scientific conservation criteria.[69] On the other hand, the International Convention for the Northwest Atlantic Fisheries pursued the objective of achieving the maximum sustainable yield (MSY).[70] This objective became prevalent in the constitutive instrument of fisheries commissions during the period until 1958.[71]

At the early stages of the work of the ILC, the meaning of 'conservation' was not discussed in detail.[72] In contrast, the Rome Conference contributed to the elaboration and clarification of the concept of conservation in a number of ways, and the subsequent work of the ILC and the Geneva Conference followed the framework adopted at the Rome Conference.

First, the Rome Conference confirmed that conservation is "essential in the development of a rational exploitation of the living resources of the seas" and stated that "conservation measures should be applied when scientific evidence shows that fishing activity adversely affects the magnitude and composition of the resources or that such effects are likely".[73] At the 1958 Geneva Conference, the duty to take conservation measures was embodied in Articles 1(2) (with the qualifying wording "as may be necessary for the conservation of the living resources of the high seas"), 3 and 4(1) of the HSFC on the basis of 1956 ILC draft articles.[74]

---

68 However, the 1946 North-East Atlantic Fishing Convention did not provide management goals at all in terms of stock biomass. See Convention for the Regulation of the Meshes of Fishing Nets and the Size Limits of Fish, London, 5 April 1946.

69 International Convention for the Regulation of Whaling (ICRW), Washington, 2 December 1946, preambular paragraphs 2, 4 and 7. See also Article V(2)(a) and (d). But this does not mean that scientific information was not considered important. For example, the Convention stipulates that the amendment of the Schedule shall be based on scientific findings. ICRW, Article V(2)(b).

70 International Convention for the Northwest Atlantic Fisheries, Washington, 8 February 1949, preamble.

71 Koers, *International Regulation of Marine Fisheries*, at p. 180; McDougal and Burke, *The Public Order of the Oceans*, at pp. 945–946. However, the 1957 Fur Seals Convention went beyond this by providing that the goal was maximum sustainable productivity with due regard to the relation of the fur seals to the productivity of the other living marine resources of the area. Interim Convention on Conservation of North Pacific Fur Seals, Washington, 9 February 1957, Preamble.

72 See, e.g., ILC Draft (1951), Part II, Article 1.

73 1955 Rome Conference Report, at para. 16. Despite general agreement that conservation of the living resources of the high seas was necessary, the ILC did not explicitly set out conservation as a 'duty' of fishing states in general terms by 1956. See, e.g., *Yearbook of the International Law Commission 1955*, vol. I, at pp. 103–105.

74 The general duty to take conservation measures in Article 1(2) of the HSFC was inserted based on the proposal by the UK at the 1958 Geneva Conference. A/CONF.13/C.3/L.72.

Second, the Rome Conference articulated that effective conservation requires scientific information. The Conference considered it essential that "any nation engaging in sea fishing collect adequate statistical records of fishing and catch" and should conduct pertinent biological and other investigations.[75] While the HSFC does not expressly require states to collect scientific information (and neither did the ILC draft articles), it is obvious that states have to collect sufficient scientific information since the special commission to be established under the HSFC is to employ appropriate scientific findings as a basis for its determination of the legitimacy of requests and the validity of measures.[76]

Third, taking into account not only scientific and technical aspects but also socio-economic aspects, the Rome Conference formulated the objectives of fishery conservation in the following way:

> The immediate aim of conservation of living marine resources is to conduct fishing activities so as to increase, or at least to maintain, the average sustainable yield of products in desirable form. [...] [T]he principal objective of conservation of the living resources of the seas is to obtain the optimum sustainable yield so as to secure a maximum supply of food and other marine products.[77]

This definition reflected the economic consideration to maximize the yield to be harvested by fisheries. The use of this reference point marked a major departure from the previously prevailing negative objectives such as preventing extermination, and higher levels of stock biomass were set to be maintained or restored. In addition, the biological wealth of living resources was considered as a means to meet the end of maximizing the supply of food and other marine products. Put differently, the utilization aspect of marine living resources was the principal element of the concept of conservation. This concept of conservation was also adopted in the ILC draft articles.

At the Geneva Conference, some coastal states proposed to introduce the recognition of the special interests of coastal states in the definition of conservation.[78] However, other states opposed the mixture of scientific criteria and the recognition of coastal state interests within the definition of conservation. Article 2

[75] 1955 Rome Conference Report, at paras. 19–22.

[76] Kaye, *International Fisheries Management*, at p. 73. Articles in the HSFC not related to dispute settlement procedures do not make reference to scientific advice. As pointed out by Kaye, "[n]o guidance is given how those measures are to be formulated consistent with the Convention's objective, except in the event of a dispute or where a coastal State attempts to unilaterally prescribe measures in the absence of agreement between parties" (pp. 70–71).

[77] 1955 Rome Conference Report, at paras. 17–18. On the relationship between MSY and the term 'optimal sustainable yield', see Section 2.1.3.1 below. See also Kaye, *International Fisheries Management*, at p. 69.

[78] See *UNCLOS I, Official Records*, vol. V, at pp. 38–40.

of the HSFC reiterates the ILC Draft for the most part.[79] In addition, reference to the conservation of the living resources 'affected' in Articles 3 and 4 of the HSFC broadened the scope of conservation. By virtue of this phrase, fishing states also have to take into account resources not exploited by their nationals but belonging to the same ecosystem.[80]

1.3 *Towards a New Regime for High Seas Fisheries under the LOSC*

By the mid-1960s, 12-mile exclusive fishing rights were generally accepted even by major maritime powers.[81] During the period between 1968 and 1982, competing claims were made to establish a new law of the sea: distant water fishing nations sought to maintain the freedom of high seas fishing while coastal states attempted to extend their control over marine resources seawards. Agreement on a new general framework of high seas fisheries as part of the new law of the sea was not secured either in the Committee on the Peaceful Uses of the Sea-Bed and the Ocean Floor beyond the Limits of National Jurisdiction (the Seabed Committee) or during the early sessions of the Third United Nations Conference on the Law of the Sea (UNCLOS III). During that time, it was suggested that a new treaty should not refer to the freedom of fishing or should not even use the term 'high seas'. For example, in the Seabed Committee, various proposals used such terms as 'international ocean space', 'international sea area' and 'international seas';[82] these proposals did not list freedom of fishing as one of the freedoms exercised in such areas.[83] During this transitional period, the ICJ delivered

79 As pointed out by Kindt and Wintheiser, the "definition clearly emphasizes utilization and leaves little room for aesthetic and moral values". J.W. Kindt and C.J. Wintheiser, "The Conservation and Protection of Marine Mammals", 7 *University of Hawaii Law Review* (1985), at p. 358. The only change in the definition of conservation was the amendment based on a Swedish proposal. See A/CONF.13/C./L.8.

80 See the comments by Chile, El Salvador and Peru and the proposal by the United Kingdom accommodating the views of these states. See *UNCLOS I, Official Records*, vol. V, at pp. 44–46. See also Oda, *International Control of Sea Resources*, at pp. 112–113. On the other hand, some commentators consider that the HSFC does not apply to species caught incidentally. Kindt and Wintheiser, "The Conservation and Protection of Marine Mammals", at p. 358.

81 See Juda, *International Law and Ocean Use Management*, at p. 176. See also Kaye, *International Fisheries Management*, at p. 86. This was, however, not a uniform trend among maritime states. For example, it was only in 1971 that Japan accepted the 12-mile limit for fisheries jurisdiction in its proposal for the 12-mile territorial sea. See S. Oda, "Proposals Regarding a 12-mile Limit for the Territorial Sea by the United States in 1970 and Japan in 1971: Implications and Consequences", 22 *Ocean Development and International Law* (1991), at pp. 192–193.

82 See S.N. Nandan and S. Rosenne (eds.), *United Nations Convention on the Law of the Sea 1982: A Commentary*, vol. III (1995), at p. 75.

83 Ibid., at pp. 75–76.

two important judgments on fisheries disputes. It confirmed some principles of high seas fisheries established during the preceding period.

The rest of this section focuses on the analysis of various proposals relating to the management of fisheries beyond the territorial sea, including those beyond the 200-mile zone, during the period 1968–1982.

1.3.1 *The Seabed Committee and Contemporary Developments: 1968–1973*

Following the gradual acceptance of the idea to establish an economic zone between the territorial sea and the high seas in this period, more focus was put on fisheries beyond the 200-mile zone by those advocating the exclusive economic zone. Several proposals argued that fisheries beyond the 200-mile zone should be regulated.[84] For example, China proposed regulation by a unified international fishery organization. Pending its establishment, it was proposed, states of a given sea area may set up a regional committee to elaborate appropriate rules and regulations for the regulation of fishing and the conservation of marine living resources, while vessels of other states may enter the region for fishing activities only when complying with the relevant rules and regulations of the region.[85] Some states proposed that the common heritage of mankind (CHM) principle should be applied to waters lying over the seabed beyond the limits of national jurisdiction.[86]

Many proposals linked the need to regulate high seas fisheries to their impact on resources within the 200–mile zone by pronouncing special interests of coastal states in the productivity of resources of the high seas adjacent to their jurisdictional zone.[87] These proposals were followed by more assertive claims which attempted to grant to coastal states preferential rights to resources beyond the 200-mile zone.[88]

In sum, it was reflected in the Report of Sub-Committee II of the Seabed Committee that the interest of the international community beyond the 200-mile zone shall be safeguarded by way of regulating high seas fisheries, while the

---

84 A/AC.138/SC.II/L.21 (Colombia, Mexico and Venezuela), Articles 16–17; A/AC.138/SC.II/L.27 (Ecuador, Panama and Peru), Articles 20–22; A/AC.138/SC.II/L.38 (Canada, India, Kenya and Sri Lanka), Articles 8–12; A/AC.138/SC.II/L.45 (China); A/AC.138/SC.II/L.54 (Ecuador, Panama and Peru), Articles G-M.

85 A/AC.138/SC.II/L.45, para. 6. See also A/AC.138/SC.II/L.38 (Canada, India, Kenya, Sri Lanka).

86 Orrego Vicuña, *The Changing International Law of High Seas Fisheries*, at p. 10 note 26 and its accompanying text; J.E. Carroz, "Institutional Aspects of Fishery Management under the New Regime of the Oceans", 21 *San Diego Law Review* (1984), at pp. 516–517.

87 See, e.g., A/AC.138/SC.II/L.21 (Colombia, Mexico, Venezuela); A/AC.138/SC.II/L.27 (Ecuador, Panama, Peru). See also the 1973 OAU Declaration.

88 A/AC.138/SC.II/L.38 (Canada, India, Kenya, Sri Lanka); A/AC.138/SC.II/L.54 supplementing A/AC.138/SC.II/L.27.

special interests of coastal states in straddling stocks and highly migratory and anadromous species were also emphasized.[89]

1.3.2 *UNCLOS III: 1973–1982*

As the concept of sovereign rights over living resources in the EEZ was accepted at UNCLOS III by the end of the 1975 session, the new regime governing high seas fisheries gradually became clear.[90] While coastal states gained exclusive rights over resources in the vast area adjacent to their territorial sea, the freedom of fishing remained a general principle of high seas fisheries beyond that area. While various formulations had still existed in the Main Trends Working Paper,[91] the freedom of fishing, subject to the conditions laid down in section 2, appeared as part of the freedoms of the high seas in Article 75 of the ISNT.[92]

The Main Trends Working Paper provided two basic paragraphs for the exploitation of the living resources of the high seas in Provision 138.[93] One of them, i.e., Formula B, provided for the right of all states to engage in fishing on the high seas subject to conservation measures.[94] Formula B and other rules of conservation and management later developed as common provisions for all the high seas fisheries (except for marine mammals) and developed into Articles 116–119 of the LOSC. Fisheries for stocks occurring solely on the high seas are subject to these provisions in Part VII, section 2, while fisheries for straddling stocks are also covered by Article 63(2). Thus, initiatives to safeguard the interests of the international community distinct from those of coastal and fishing states through, *inter alia*, a unified international fishery organization were not included in the LOSC.

---

89 Nandan and Rosenne (eds.), *United Nations Convention on the Law of the Sea 1982: A Commentary*, vol. III, at p. 283.

90 While the Main Trends Working Paper (1974) incorporated various proposals with regard to the EEZ, the text of the Informal Single Negotiating Text (ISNT) struck the balance among the competing interests within the EEZ and the main elements of the ISNT/Part II (Articles 45–61) were retained in subsequent texts and appear in Part V of the LOSC. See, generally, S.N. Nandan and S. Rosenne (eds.), *United Nations Convention on the Law of the Sea 1982: A Commentary*, vol. II (1993), at pp. 498–501.

91 Provision 154.

92 ISNT, Article 75(1)(e).

93 See Nandan and Rosenne (eds.), *United Nations Convention on the Law of the Sea 1982: A Commentary*, vol. III, at pp. 283–284. Note that the term 'high seas' was not yet uniformly used at this stage to denote the area beyond national jurisdiction.

94 Formula A stipulated the preferential rights of coastal states to a certain portion of resources of the high seas adjacent to their EEZ. Some other provisions of the Main Trends Working Paper also referred to the special status of coastal states with regard to the marine living resources of the high seas adjacent to their maritime zone. See Provision 155 and Provision 156, Formula A, paragraph 2.

1.4 *Concluding Remarks*

The principle of the freedom of fishing was solidly established by the 19th century. Before 1945, states could, outside the narrow belt of the territorial sea, adopt conservation measures only for their nationals and no regulation could take place unless the states concerned agreed on measures applicable among themselves. In other words, the freedom of fishing on the high seas was the single operating principle under general international law. International regulations ensued only for some species of marine mammals and for some regions if states agreed thereon. The instances reviewed at the beginning of this chapter illustrate the dominance of the freedom of fishing on the high seas.

As the concerns of overexploitation caused by the further advance of technology grew and the attempts of coastal states to gain greater control over the living resources in the adjacent waters for economic development and food supply increased, a conceptual change took place after the Second World War. Several general principles appear in the work of the ILC and, eventually, in the Geneva Conventions: freedom of high seas fishing as restrained by the requirement of due regard, the duty of cooperation and the duty of conservation. During the process leading up to the adoption of the Geneva Conventions, it was generally recognized that the duties of due regard for the interests of other states relegated the concept of unfettered freedom to history, as subsequently confirmed in the judgments of the ICJ in the *Fisheries Jurisdiction* cases. The duty to cooperate between high seas fishing states was clearly recognized, while coastal states were also entitled to participate in the regulatory system if the fishery took place in the high seas area adjacent to their territorial sea. Other states such as market states were not granted participatory rights, apart from the right to request fishing states to take conservation measures. The concept of conservation which was framed and elaborated during this period still forms the basis of the current concept of conservation. The concept introduced both biological and scientific considerations as well as social and economic considerations. The general interests of states were thus formulated as attaining optimum sustainable yield so as to maximize food supply and other marine products. The utilization aspect dominated the concept of conservation.

While the special interests of coastal states gradually gained universal acceptance in the following period, it took the form of an extension of coastal state jurisdiction seawards by establishing a new jurisdictional zone, rather than a change in the general principles of high seas fisheries. The adoption of the LOSC completely transformed the jurisdictional framework of the oceans: fisheries in the area up to 200 miles were separated from the high seas regime, and high seas fisheries for certain stocks and species were subjected to specific regimes. Fishing for stocks occurring solely in the high seas is to be regulated by, *inter alia*, the provisions of Part VII, section 2 of the LOSC.

It is noted that the legal regime of high seas fisheries during this period developed along the lines of tensions between coastal states and distant water fishing nations at both global and regional levels. At the global level, the progressive development of the international law of high seas fisheries was attempted in the ILC by incorporating the interests of coastal states in fisheries in high seas areas adjacent to the territorial sea. At the regional level, several commissions covered vast areas of the high seas, including those parts that were not considered as adjacent high seas areas, but their practice was mostly concentrated on adjacent areas.

In summary, despite the formulation of the rules governing high seas fisheries in general at the global level, practice at the regional level was scarce with regard to fisheries outside the high seas areas adjacent to the territorial sea. One might consider that there was no international law of high seas fisheries *per se* that reflected the actual practice of states while there was international law governing fisheries in high seas areas adjacent to the territorial sea. It is thus difficult to argue that existing state practice with regard to fisheries for DHSFS at that time was reflected in the provisions of Part VII, section 2 of the LOSC. It could rather be argued that these provisions had the *potential* to be applied to fisheries for DHSFS and to provide a legal framework to address such fisheries once they took place after the adoption of the LOSC. The next chapter examines this framework as elaborated by developments that took place after the adoption of the LOSC.

Chapter Two

The Legal Framework for High Seas Fisheries under the LOSC

This chapter examines the legal framework for high seas fisheries provided by the LOSC. The LOSC provides a general legal framework for all kinds of activities in the seas and oceans. Fisheries on the high seas are regulated by rules, among others, contained in Part VII, Section 2. Its importance is exemplified by the quasi-universal participation in the LOSC by states.[1] In addition, the importance of the LOSC as the legal framework for the conservation and sustainable use of marine living resources and biological diversity has been emphasized in a number of recent international fora.[2]

The chapter presents an overview of the general principles of high seas fisheries under the LOSC and explores the implications arising from those general principles. It examines various interpretations of the provisions of the LOSC regarding high seas fisheries in the light of the text of the Convention, its drafting history, judgments of the ICJ, the International Tribunal for the Law of the Sea (ITLOS) and regional courts and arbitral awards by international tribunals as well as state practice subsequent to the adoption of the LOSC in 1982. In particular, the section ultimately attempts to articulate the likely interpretation of the contentious provisions of the LOSC in the light of the subsequent practice of states and international organizations.

Some commentators argue that the concept of high seas fishing rights was fundamentally altered by Article 116 of the LOSC,[3] while others do not agree with

---

1 As of 24 September 2012, there are as many as 164 parties to the LOSC. Information is available on the website of the United Nations Treaty Section <http://treaties.un.org/Pages/ViewDetailsIII.aspx?&src=TREATY&mtdsg_no=XXI~6&chapter=21&Temp=mtdsg3&lang=en>.

2 See, for example, Report of the Ad Hoc Open-ended Informal Working Group to study issues relating to the conservation and sustainable use of marine biological diversity beyond areas of national jurisdiction, A/61/65, 20 March 2006, at p. 7, para. 22.

3 Nandan and Rosenne (eds.), *United Nations Convention on the Law of the Sea 1982: A Commentary*, vol. III, at pp. 29 and 287. See also Orrego Vicuña, *The Changing International Law of High Seas Fisheries*, at p. 25.

this proposition for one reason or another.[4] This section examines whether this proposition holds true for high seas fisheries in general.

Before analyzing the general principles of high seas fisheries under the LOSC, it is necessary to elaborate the relationship between the LOSC and subsequent international instruments. In accordance with the law of treaties, the provisions of the LOSC are interpreted in the light of the subsequent agreements and practice.[5] A question is whether international fisheries-related instruments adopted subsequent to the adoption of the LOSC may be considered to be subsequent agreements in the sense of the VCLT. The question is particularly relevant with regard to the FSA, whose title indicates that the Agreement is intended for the implementation of the provisions of the LOSC. It has been argued that the FSA is to be qualified as a subsequent agreement in this sense.[6] Apart from its potential as a subsequent agreement, many commentators point to the importance of the FSA in interpreting or elaborating the provisions of the LOSC.[7]

Furthermore, subsequent instruments, including non-binding instruments, may have legal significance if their provisions are considered to have passed into customary international law. In that case, the maxim *lex posterior derogat legi priori* could apply.[8]

---

[4] C.A. Fleischer, "The New Régime of Maritime Fisheries", 209 *Recueil des Cours* (1988-II), at p. 172; Hey, *The Regime for the Exploitation of Transboundary Marine Fisheries Resources*, at pp. 5 and 119; S. Oda, "Fisheries under the United Nations Convention on the Law of the Sea", 77 *American Journal of International Law* (1983), at p. 749.

[5] Vienna Convention on the Law of Treaties (VCLT), Vienna, 23 May 1969, Article 31(3). On the implications of the characterization of the LOSC as a 'Constitution for the Oceans', see, e.g., S. Scott, "The LOS Convention as a Constitutional Regime for the Oceans", in A.G. Oude Elferink (ed.), *Stability and Change in the Law of the Sea: The Role of the LOS Convention* (2005), at pp. 9–38.

[6] D.H. Anderson, "The Straddling Stocks Agreement of 1995: An Initial Assessment", 45 *International and Comparative Law Quarterly* (1996), at p. 468; P.G.G. Davies and C. Redgwell, "The International Legal Regulation of Straddling Fish Stocks", 67 *British Year Book of International Law* (1996), at p. 259; T. Henriksen *et al.*, *Law and Politics in Ocean Governance: The UN Fish Stocks Agreement and Regional Fisheries Management Regimes* (2006), at p. 15. But, see M. Gavouneli, *Functional Jurisdiction in the Law of the Sea* (2007), at p. 129. Also note the difference between the FSA and the 1994 Implementation Agreement with regard to the relationship with the LOSC. Article 2(1) of the 1994 Agreement stipulates that "[t]he provisions of [the 1994] Agreement and Part XI shall be interpreted and applied together *as a single instrument*" (emphasis added) and, in the event of any inconsistency between the 1994 Agreement and Part XI of the LOSC, "the provisions of [the 1994] Agreement shall prevail". The FSA is not linked to the LOSC as firmly as the 1994 Agreement, stipulating that "[n]othing in [the FSA] shall prejudice the rights, jurisdiction and duties of States under the [LOSC]" and the FSA "shall be interpreted and applied in the context of and in a manner consistent with" the LOSC in its Article 4.

[7] E.g., Orrego Vicuña, *The Changing International Law of High Seas Fisheries*, at p. 214.

[8] Bernhardt, "Custom and Treaty in the Law of the Sea", at p. 276.

2.1 *Freedom of Fishing on the High Seas*

Freedom of fishing remains as one of the freedoms of the high seas explicitly stated in the list in Article 87(1) of the LOSC. Freedom of fishing shall be exercised with due regard for the interests of other states in their exercise of the freedom of the high seas and for the rights under the Convention with respect to activities in the Area and is subject to conditions laid down in section 2 of Part VII.[9]

The high seas fishing right belonging to all states, an essential element of the freedom of fishing, is maintained in Article 116, which stipulates that the right to fish on the high seas shall be subject to the treaty obligations, rights, duties and interests of coastal states, and provisions of Part VII, Section 2, i.e., Articles 116–120. In comparison with Article 1(1) of the HSFC, what is innovative is that freedom of fishing is directly linked to the restrictions on the right to fish on the high seas. While states may ratify the HSC without ratifying the HSFC, the LOSC does not allow reservations to any of its provisions unless they are expressly allowed.[10] Therefore, there is a major difference between the LOSC and the 1958 Geneva Conventions in that the right of high seas fishing is automatically subject to the restrictions of Part VII, Section 2 under the LOSC framework.

The following sub-sections examine the content of the freedom/right of fishing on the high seas, focusing on five issues: restraints of a general nature on freedom of the high seas, the scope of 'fishing' in Articles 87 and 116, paragraphs (a)–(c) of Article 116, stocks and species covered by Article 116 and the scope of 'nationals' in Articles 116–118.

2.1.1 *Restraints of a General Nature*

Freedom of fishing may be exercised with due regard for the interests of other states and also with due regard for the rights under the LOSC with respect to activities in the Area.[11] The requirement of due regard in the LOSC, as in the HSC, is formulated so broadly that it could restrain the exercise of this freedom by all states. It has been argued that the standard of due regard requires all states to be aware of and consider the interests of other states in using the high seas, and to refrain from activities that interfere with other states exercising the freedom of the high seas.[12] In other words, it is the balance of interests among states that is required under this requirement.

---

9 LOSC, Article 87(2).

10 Ibid., Article 309.

11 Ibid., Article 87(2).

12 Nandan and Rosenne (eds.), *United Nations Convention on the Law of the Sea 1982: A Commentary*, vol. III, at p. 86. This provision may also have implications for the setting of an allowable

The requirement may play an important role in resolving a dispute over interests in fisheries between two states, as is exemplified by the *Fisheries Jurisdiction* cases. In these cases, the ICJ declared that the freedom of fishing on the high seas was not absolute, indicating that the regime of high seas fisheries moved from being *laissez-faire* to the recognition of a duty to have due regard to the rights of other states and of the needs of conservation for the benefit of all.[13] The Court considered that the Icelandic action constituted an infringement of the principle enshrined in Article 2 of the HSC concerning reasonable regard for the interests of other states to be had in exercising the freedom of the high seas.[14] This indicates that the Court considered Article 2 of the HSC to be a rule of general international law, given the fact that the Federal Republic of Germany and Iceland were not parties to the HSC.[15] Furthermore, the Court considered that not only the right of fishing of all states but also preferential rights recognized only for certain states shall be exercised with due regard for the rights of other states to fish on the high seas.

As far as due regard for the interests of other states is concerned, the importance attached to this provision in the LOSC appears to be less significant than in the HSC, given the fact that the freedom is no longer considered to be absolute under general international law as noted above and fishing is regulated in more detail in Part VII, section 2.[16] In fact, it is not clear to what extent the requirement has significance in a multilateral context, especially for states not engaged in fishing.[17] It should be noted, however, that whenever a new issue emerges, the requirement of due regard could be invoked.

On the contrary, the requirement of due regard with regard to the activities of the Area[18] is new and could be regarded as a significant addition. Articles 87(2) and 147 stipulate how fishing and mining activities shall be reconciled as conflicts between two or more activities arise. First of all, Article 147(2)(b) provides

catch and the determination of fishing quotas. Division for Ocean Affairs and the Law of the Sea, *The Regime for High-Seas Fisheries: Status and Prospects* (1992), at pp. 8–9.

13 *I.C.J. Reports 1974*, at p. 31, para. 72; *I.C.J. Reports 1974*, at pp. 200–201, para. 64.

14 *I.C.J. Reports 1974*, at p. 29, para. 67; *I.C.J. Reports 1974*, at p. 198, para. 60.

15 The Federal Republic of Germany and the UK, invoking their historic fishing rights in the waters concerned, argued that "reasonable regard must be had to such traditional rights by the coastal State, in accordance with the generally recognized principles embodied in Article 2" of the HSC. *I.C.J. Reports 1974*, at p. 24, para. 54; *I.C.J. Reports 1974*, at p. 193, para. 46.

16 For a similar view on the due regard clause in relation to the interests of other states, see C.A. Fleischer, "Fisheries and Biological Resources", in R.-J. Dupuy and D. Vignes (eds.), *A Handbook on the New Law of the Sea* (1991), at p. 1113.

17 See D.P. O'Connell, *The International Law of the Sea*, vol. I (1982), at p. 58 (the concept of reasonable use is relativistic and hence susceptible of subjective evaluation and not capable of resolving specific questions).

18 The term 'Activities in the Area' is defined as "all activities of exploration for, and exploitation of, the resources of the Area", for the purposes of the LOSC. LOSC, Article 1(1)(3).

that installations may not be established in areas of intense fishing activity. It could be argued, *a contrario*, that where fishing activities are not 'intense', installations may be established and mining activities initiated. Second, Article 147 does not give precedence to mining over fishing: read together, paragraphs 1 and 3 of Article 147 as well as Article 87(2) indicate that neither activity has precedence. In reality, various considerations need to be carefully taken into account, including the scale of these activities in spatial and economic terms, the potential impacts mining may have on fishing in large areas, and other possible environmental impacts. Third, like installations on the continental shelf, it is not clear whether regulations in safety zones around them may cover fishing activities as Article 147(2)(c) is silent on this point. It could be argued that although Article 147 does not specify that fishing is subject to safety zones around installations, the due regard requirement in Article 87 as well as the reasonable regard requirement in Article 147(3)[19] suggest that high seas fisheries need to be conducted in a manner so as not to interfere with such installations or safety zones around them.

General principles contained in the LOSC, including the concepts of good faith and the prohibition on the abuse of rights, also restrict the exercise of freedom of fishing. First, states are required to fulfil their obligations under the LOSC in good faith.[20] As opposed to the obligation to negotiate in good faith recognized in the *Fisheries Jurisdiction* cases, the duty in Article 300 of the LOSC covers all types of obligations to be performed in good faith. Second, states shall exercise rights, jurisdiction and freedoms under the LOSC in a manner that does not constitute an abuse of rights.[21] Notwithstanding the legislative history of this provision indicating the original intent to restrain the discretionary power of coastal states, Article 300 as a general provision applies to the Convention as a whole and is not solely related to an abuse of rights by coastal states.[22]

---

19 "Other activities in the marine environment shall be conducted with reasonable regard for activities in the Area."

20 LOSC, Article 300. In the context of fisheries, see *The North Atlantic Coast Fisheries Case between Great Britain and the United States*, Award of the Tribunal, 7 September 1910, *The Hague Court Reports*, vol. I, p. 141, at p. 171. On the principle of good faith in general, see *Nuclear Tests Case (Australia v. France)*, Judgment, 20 December 1974, *I.C.J. Reports 1974*, p. 253, at p. 268, para. 46; *Nuclear Tests Case (New Zealand v. France)*, Judgment, 20 December 1974, *I.C.J. Reports 1974*, p. 457, at p. 473, para. 49.

21 LOSC, Article 300.

22 See F. Orrego Vicuña, *The Exclusive Economic Zone: Regime and Legal Nature under International Law* (1989), at p. 132; S. Rosenne and L.B. Sohn (eds.), *United Nations Convention on the Law of the Sea 1982: A Commentary*, vol. V (1989), at pp. 150–151; M.T. Infante, "The Settlement of Disputes regarding the Law of the Sea and its Bearing on the Legal Nature of the Exclusive Economic Zone", in F. Orrego Vicuña (ed.), *The Exclusive Economic Zone: A Latin American Perspective* (1984), at pp. 166–167.

Article 300 of the LOSC may have major implications in high seas fisheries disputes. In the *Southern Bluefin Tuna* case, the arbitral tribunal did not exclude the possibility that a conduct "would be so egregious, and risk consequences of such gravity, that a Tribunal might find that the obligations of [the LOSC] provide a basis for jurisdiction, having particular regard to the provisions of Article 300 of [the LOSC]", even if a fisheries treaty implementing the LOSC excludes the application of the compulsory dispute settlement procedure in respect of an alleged breach of the obligations under Articles 64 and 116–119 of the LOSC.[23]

2.1.2 *Scope of 'Fishing'*

The term 'fishing' is not defined in the LOSC. What activities fall under the term 'fishing'? If an activity may not be regarded as fishing, what rules are applicable? Some activities related to the exploitation of living resources are not always considered as 'fishing' in the traditional sense of the word. First, there have been controversies over whether activities associated with or in support of harvesting are considered to be 'fishing'. Second, the purpose of the harvesting activities might be relevant in determining whether the activity concerned is fishing or not. In particular, this is the case for fisheries for scientific purposes.[24] Some RFMO/As explicitly include the harvesting of fisheries resources for scientific research purposes in 'fishing'.[25] As regards domestic legislation, some laws and

23 *Southern Bluefin Tuna Case*, Award on Jurisdiction and Admissibility, 4 August 2000, at para. 64. In that case, the Tribunal noted that while Australia and New Zealand invoked Article 300 in the proceedings before ITLOS, they made it clear that they did not consider Japan to be responsible for any independent breach of an obligation to act in good faith. See also the transcripts of the daily sessions of the hearing on jurisdiction, first round presentation of Australia and New Zealand, May 8, 2000, available at <http://icsid.worldbank.org/ICSID/FrontServlet?requestType=CasesRH&actionVal=OpenPage&PageType=AnnouncementsFrame&FromPage=NewsReleases&pageName=Archive_%20Announcement7>.

24 In addition, fishing for recreational purposes occasionally conducted by crew members during navigation on the high seas does not appear to be qualified as commercial fishing. In other words, it seems that there is a *de minimis* requirement in terms of the scale of the activity concerned.

25 E.g., Convention on the Conservation and Management of Fishery Resources in the South East Atlantic Ocean (SEAFO Convention), Windhoek, 20 April 2001, Article 1(h)(ii) and (k); Southern Indian Ocean Fisheries Agreement (SIOFA), Rome, 7 July 2006, Article 1(g)(ii). The definition of 'fishing' in one convention includes "engaging in any activity which can reasonably be expected to result in the locating, catching, taking or harvesting of fishery resources for any purpose". Convention on the Conservation and Management of High Seas Fishery Resources in the South Pacific Ocean (SPRFMO Convention), Auckland, New Zealand, 14 November 2009, Article 1(1)(g). The new NAFO Convention agreed in 2007 does not refer to fisheries for scientific research purposes in the definition of fishing, but provides that the Commission "shall in collaboration with the Scientific Council, [...] develop guidelines for the conduct of fishing activities for scientific purposes" and "shall, in relation to the Regulatory Area adopt [...] measures for the conduct

regulations extend the requirement of permits to fisheries for scientific or experimental purposes, either explicitly[26] or implicitly.[27] Others appear to exclude such fisheries by limiting the permit requirement to commercial fisheries.[28] In this context, it is recalled that it is not always easy to find a clear distinction between so-called 'applied' research as part of marine scientific research (MSR) and fishery research as part of resource exploration.[29] The former is regulated by Part XIII of the LOSC while the latter, if conducted in the high seas, is subject to Part VII, section 2 of the LOSC. Third, it is not always clear whether the exploitation of species other than invertebrates or finfish, such as flora belonging to sedentary species, is included in 'fishing'.[30]

As regards the scope of activities covered by the term 'fishing', a possible argument would be that, in view of the modernization of fishing techniques, the activities of support vessels such as refuelling vessels, processing vessels, refrigeration vessels and transport vessels may be considered to be fishing, in addition to the capturing of fish. The following two cases will shed some light on this issue.

In the *La Bretagne* case,[31] the arbitral tribunal took a cautious approach to the question of the rights of the coastal state under the 1972 Agreement between Canada and France in fishery matters. Finding that the coastal state should exercise its rights in a reasonable manner, the tribunal concluded that "the regulation of filleting at sea cannot *a priori* be justified by coastal state powers under the new law of the sea rules".[32] In other words, it confined the sovereign rights regarding the management of living resources to the conservation of resources (to the exclusion of environmental, economic and social factors).[33] Notwithstanding this, the tribunal did not deny coastal state jurisdiction altogether, indicating that

---

of fishing for scientific purposes as referred to in subparagraph 6(d)". Convention on Cooperation in the Northwest Atlantic Fisheries (New NAFO Convention), Lisbon, 28 September 2007, Article VI(6)(d) and (8)(d).

26 Australia, Fisheries Management Act, Sections 105A and 105B; Canada, Fishery (General) Regulations, Section 51.

27 See South Africa, Marine Living Resources Act, Section 83.

28 Namibia, Marine Resources Act, Articles 1 and 32; New Zealand, Fisheries Act, Section 113D (for sale); United States, High Seas Fishing Compliance Act of 1995, Pub. L. 104–43, title I, section 103; 16 US Code 5502(4)(B).

29 See, e.g., A.H.A. Soons, *Marine Scientific Research and the Law of the Sea* (1982), at p. 7.

30 In addition, it is worth noting that it is controversial what legal regime applies to bioprospecting for marine genetic resources in areas beyond national jurisdiction, but this issue goes beyond the scope of this study.

31 *Case concerning filleting within the Gulf of St. Lawrence between Canada and France* (*La Bretagne case*), Decision, 17 July 1986, *International Law Reports*, vol. 82, p. 590.

32 Ibid., at para. 52.

33 Ibid., at para. 50. See D. Anderson, "The Regulation of Fishing and Related Activities in Exclusive Economic Zones", in E. Franckx and P. Gautier (eds.), *The Exclusive Economic Zone and the*

the coastal state could regulate fish processing in the EEZ through the power to require the landing of the catch in Article 62(4)(h) and regulations could cover the 'types', 'sizes' and 'numbers' of fishing vessels.[34] On the other hand, the tribunal identified that "the natural and ordinary sense of the phrase "fishing regulations" consists in designating the legislative or regulatory prescriptions [...], which fix the conditions to which all fish-catching activities are subject [...]".[35]

In the *Saiga* case, although ITLOS did not comment on the legal regime of bunkering in the EEZ generally,[36] the parties to the dispute as well as some judges expressed divergent views on this issue.[37]

To sum up, first of all, despite the fact that these cases are concerned with the exercise of sovereign rights in the EEZ and might be argued to have limited relevance, they could shed some light on the interpretation of the term 'fishing' in the LOSC in relation to high seas fisheries as well. As indicated above, it is difficult to draw any definitive conclusion on the meaning of 'fishing' given the fact that the opinions of or before tribunals are not uniform, in particular with regard to bunkering vessels. Some academic literature seems to suggest that transshipment vessels can be included in fishing vessels while bunkering vessels cannot be included.[38]

Recent state practice at the global and regional levels indicates a trend where states are increasingly required to take necessary conservation measures not only for fish-catching vessels but also for other fishing vessels associated with fishing operations. At the global level, transport vessels and support vessels are referred to in the context of combating IUU fishing, and the at-sea transshipment of catch harvested in IUU fishing operations is subject to regulation.[39]

United Nations Convention on the Law of the Sea, 1982–2000: A Preliminary Assessment of State Practice (2003), at p. 37.

34 *La Bretagne case*, at para. 53.

35 Ibid., at para. 38.

36 *The M/V "Saiga" (No. 2) Case (Saint Vincent and the Grenadines v. Guinea)*, Judgment, 1 July 1999, Case No. 2, at para. 138.

37 Ibid., at para. 137. See the Separate Opinion of Judge Zhao, at paras. 3–4; Dissenting Opinion of Judge Warioba, at para. 91; Separate Opinion of Judge Nelson, at p. 7; Separate Opinion of Judge Vukas, at para. 17. See also Separate Opinion of Judge Anderson, at p. 5. Note that Guinea appears to maintain a similar position. See the *Luchegorsk* incident of 21 August 2006, cited in R. Lagoni, "Offshore Bunkering in the Exclusive Economic Zone", in T.M. Ndiaye and R. Wolfrum (eds.), *Law of the Sea, Environmental Law and Settlement of Disputes: Liber Amicorum Judge Thomas A. Mensah* (2007), at p. 615 note 6 and p. 626.

38 T. Treves, "Some International Law Aspects of the Use of Vessel Monitoring Systems for Preventing Illegal Unreported Unregulated Fishing", in T.M. Ndiaye and R. Wolfrum (eds.), *Law of the Sea, Environmental Law and Settlement of Disputes: Liber Amicorum Judge Thomas A. Mensah* (2007), at pp. 816–817; Lagoni, "Offshore Bunkering in the Exclusive Economic Zone", at p. 617.

39 E.g., UNGA Resolution 63/112, Sustainable Fisheries, Including through the 1995 Agreement for the Implementation of the Provisions of the United Nations Convention on the Law of

Several RFMO/As have determined or are to determine the nature and extent of participation in fisheries, including the allocation of fishing opportunities, and regulate processing, transshipment, transporting and refuelling (the last one not always explicitly) as part of fishing and include these vessels in the definition of fishing.[40] Legislation in a number of countries whose vessels are engaged in high seas fisheries, including deep-sea fisheries, requires permits not only for catch vessels but also for vessels engaging in support operations to various extents.[41]

Even if one takes a view that these activities would not be regarded as fishing, this does not mean that there is a legal vacuum, as some of the principles of the LOSC would be applicable. First, since the list of freedoms in Article 87 is not exhaustive, other activities may be conducted as part of the exercise of the freedom of the high seas.[42] Second, as in the case of fishing, such freedoms are subject to conditions laid down in the LOSC and other rules of international law as well as the requirement of due regard for the exercise of the high seas freedom by other states and for the rights of other states with regard to activities in the Area. In this regard, the activities are subject to *inter alia* the provisions of Part XII.

In addition, some of the articles of section 2 of Part VII are relevant to the regime governing these activities. Attention could be drawn to the subtle difference in the wording of Articles 116 to 119. First, Article 117 is concerned with the conservation and management of living resources of the high seas in general, not

---

the Sea of 10 December 1982 relating to the Conservation and Management of Straddling Fish Stocks and Highly Migratory Fish Stocks, and Related Instruments, 5 December 2008, paras. 43 and 56.

40 New NAFO Convention, Articles I(g)–(h) and VI(6)(d); North East Atlantic Fisheries Commission (NEAFC), Scheme of Control and Enforcement 2011, Articles 1(e) and (h) and 4; SEAFO Convention, Article 1(h) and (j); SIOFA, Article 1(g) and (i). See also SPRFMO Convention, Articles 1(1)(g), 8(b) and 27(1)(c). The FAO Port State Measures Agreement defines 'fishing related activities' including the landing, packaging, processing, transshipment or transporting of fish that have not been previously landed at a port as well as the provision of personnel, fuel, gear and other supplies at sea. Agreement on Port State Measures to Prevent, Deter and Eliminate Illegal, Unreported and Unregulated Fishing (FAO Port State Measures Agreement), Rome, 22 November 2009, Article 1(d).

41 Australia, Fisheries Management Act, Section 4(1); Canada, Fishery (General) Regulations, Section 65; South Korea, Fisheries Act, Article 2(1); Namibia, Marine Resources Act, Article 50(1); New Zealand, Fisheries Act, Section 113D(2); South Africa, Marine Living Resources Act, Sections 1 and 40; United States, High Seas Fishing Compliance Act of 1995, Pub. L. 104–43, title I, section 103; 16 US Code 5502(4)(C). For Japan, see Report on the state of Japan's Implementation of UNFSA, distributed at the Review Conference on the FSA in May 2006 (on file with the author), at p. 5 (in principle, transshipment at sea is prohibited).

42 Churchill and Lowe observe that freedoms cannot be exhaustively listed "because new ocean technology is constantly developing" and that disputes over the status of a particular activity constituting freedoms should be resolved on the principle of compatibility with the status of the high seas. Churchill and Lowe, *The Law of the Sea*, at p. 205.

being limited to fishing or other exploration and exploitation. Second, Article 118, while its first sentence appears to be broader in scope, is primarily concerned with cooperation between states whose nationals *exploit* resources on the high seas. Thus, there may be room for Articles 117 and 118 to be invoked with regard to other activities involving the living resources of the high seas. All in all, it can be argued that the LOSC provides a sufficient legal framework to address new challenges concerning the marine living resources of the high seas, including activities related to fishing.

2.1.3 *Paragraphs (a)–(c) of Article 116*

Two questions arise with regard to the content of paragraphs (a)–(c) of Article 116. The first question is what is meant by the formulation that the right is 'subject to' paragraphs (a)–(c) since the term 'subject to' may have different meanings.[43] There have been a number of works expressing different views on the implications of the interests of coastal states in the context of Article 116(b). Some commentators have argued for the possibility of coastal state unilateral conservation measures.[44] Others do not recognize any preferential position of coastal states in taking conservation measures.[45] Another approach is to limit a special interest of coastal states to participation in the regulatory system of adjacent high seas fisheries.[46]

In addition, possible consequences resulting from a violation of Article 116 are subject to different interpretations. Do states retain the right to fish as such, while being held responsible for their internationally wrongful acts? Or, does this provision imply that a state is disqualified as a right holder if it does not abide by

43 Fleischer, "The New Régime of Maritime Fisheries", at p. 161.

44 See, e.g., Burke, *The New International Law of Fisheries*, at p. 134; Orrego Vicuña, *The Changing International Law of High Seas Fisheries*, at p. 42; Scovazzi, "The Evolution of International Law of the Sea", at pp. 134–137.

45 Brown, *The International Law of the Sea*, at p. 228; Davies and Redgwell, "The International Legal Regulation of Straddling Fish Stocks", at pp. 238–239 and 241; Fleischer, "Fisheries and Biological Resources", at p. 1116; Joyner and De Cola, "Chile's Presential Sea Proposal", at pp. 106–107; Kaye, *International Fisheries Management*, at pp. 160–161; J.L. Meseguer, "Le Régime Juridique de l'Exploitation de Stocks Communs de Poissons au-delà des 200 Milles", XXVIII *Annuaire Français de Droit International* (1982), at pp. 898–899; Nelson, "The Development of the Legal Regime of High Seas Fisheries", at p. 123; Oda, "Fisheries under the United Nations Convention on the Law of the Sea", at p. 750; K. Yonezawa, "Some Thoughts on the Straddling Stock Problem in the Pacific Ocean", in T. Kuribayashi and E.L. Miles (eds.), *The Law of the Sea in the 1990s: A Framework for Further International Cooperation* (1992), at pp. 133–134.

46 Hey, *The Regime for the Exploitation of Transboundary Marine Fisheries Resources*, at p. 54 (but see p. 123).

the obligations referred to in paragraphs (a)–(c)?[47] In other words, is the exercise of the right conditional upon compliance with the obligations stipulated in paragraphs (a)–(c)? It might be argued that if the breach of the obligations under Articles 117–119 by a state is very serious, this could lead to forfeiting the right to fish on the high seas. But such a far-reaching conclusion needs to be examined in the light of state practice (see Sections 2.2.2 and 2.3.2 below).

Second, Article 116(b) subjects the right of fishing on the high seas to the interests,[48] rights (not 'sovereign rights') and duties of coastal states, *inter alia*, in Article 63(2) and Articles 64–67. First and foremost, this provision relates to transboundary fish stocks such as straddling stocks under Article 63(2).[49] It should be noted that Article 63(2) is not only concerned with 'the same stock' (namely, a straddling stock) but also 'stocks of associated species'. The latter phrase could be interpreted to the effect that a state fishing for a stock solely occurring in the high seas whose predator species straddles the EEZ and the high seas is under an obligation to seek agreement on necessary conservation and management measures, irrespective of the engagement of the coastal state in the high seas fishery concerned. If this interpretation is adopted, certain DHSFS are subject to both Article 63(2) and the provisions of Part VII, section 2.[50]

---

47 See, in the context of straddling stocks, Joyner and De Cola, "Chile's Presential Sea Proposal", at p. 111. See also, e.g., Fleischer, "Fisheries and Biological Resources", at p. 1116; L.D.M. Nelson, "Declarations, Statements and 'Disguised Reservations' with Respect to the Convention on the Law of the Sea", 50 *International and Comparative Law Quarterly* (2001), at p. 123.

48 The nature and extent of the 'interests' of coastal states appears to relate to both management and conservation. In other words, coastal states have an interest in maintaining the long-term sustainability of and exploiting resources, and in the coordination of conservation measures adopted for the outer continental shelf and those for the high seas. See also Nandan and Rosenne (eds.), *United Nations Convention on the Law of the Sea 1982: A Commentary*, vol. III, at p. 288.

49 The situation covered in Article 116 includes circumstances where it is not clear whether a given fish population comprises two fish stocks including a discrete high seas stock or one straddling fish stock. On this issue, see the management of orange roughy fisheries in the high seas areas of the South Tasman Rise by Australia and New Zealand in Section 5.2.1 below.

50 It is not clear whether this type of associated stocks was in the drafters' minds at UNCLOS III. The 1980 NEAFC Convention and the 1978 NAFO Convention which were negotiated in parallel with UNCLOS III appear to include this situation in their scope. See Convention on Future Multilateral Co-operation in North-East Atlantic Fisheries, London (1980 NEAFC Convention), 18 November 1980, Article 5(2); Convention on Future Multilateral Cooperation in the Northwest Atlantic Fisheries (1978 NAFO Convention), Ottawa, 24 October 1978, Article XI(3). The existing literature almost equates the scope of Article 63(2) to straddling stocks. See, e.g., Nandan and Rosenne (eds.), *United Nations Convention on the Law of the Sea 1982: A Commentary*, vol. II, at p. 647. A notable exception is Hey, who argues that by this provision "the consequence that fisheries management should in fact be multi-species management is accepted". Hey, *The Regime for the Exploitation of Transboundary Marine Fisheries Resources*, at p. 54. See also L. Miovski, "Solutions in the Convention on the Law of the Sea to the Problem of Overfishing

The wording '*inter alia*' in Article 116(b) seems to indicate that the enumerated articles are not exhaustive and the interests, rights and duties of coastal states are provided for elsewhere in the LOSC. Other interests of coastal states could be invoked, among others, on the following bases.[51] First, coastal states have sovereign rights to explore and exploit natural resources on the continental shelf, including that beyond 200 miles.[52] The issue was discussed in connection with the regime for the continental shelf even before the ILC recognized that coastal states have sovereign rights over sedentary species. For example, Mouton argued that "[t]rawling on [...] banks, even with the exclusive object to catch bottom-fish, will certainly be objected to by the coastal State as being destructive for the oysters or other "sedentary" species".[53]

The exercise by coastal states of rights on the continental shelf is restricted by the provision of Article 78(2), which provides that the exercise must not infringe or result in any unjustifiable interference with navigation and other rights and freedoms of other states.[54] On the one hand, one might argue that this provision prevents coastal states from prescribing conservation measures unilaterally

---

in the Central Bering Sea: Analysis of the Convention, Highlighting the Provisions Concerning Fisheries and Enclosed and Semi-Enclosed Seas", 26 *San Diego Law Review* (1989), at p. 538 (arguing that the inter alia clause refers to instances "where an EEZ species was related to, or dependent on, a species of the adjacent high seas area, or where fishing in the high seas otherwise has a direct effect on stocks in the EEZ").

51 In addition, it has been argued that Article 123 is also relevant. See Miovski, "Central Bering Sea Overfishing".

52 See LOSC, Articles 76(1) and 77(1). The argument developed here also applies to the continental shelf beyond the outer limit of the territorial sea if a state has not yet proclaimed an EEZ or EFZ. The latter situation is particularly relevant to the Mediterranean, where coastal states have so far refrained from extending their maritime zone to 200 miles. Note that the sovereign rights of coastal states on the continental shelf exist regardless of the proclamation by the coastal states concerned. In the words of the ICJ, "the rights of the coastal State in respect of the area of continental shelf [...] exist *ipso facto* and *ab initio*, by virtue of its sovereignty over the land, as an extension of it in an exercise of sovereign rights for the purpose of exploring the seabed and exploiting its natural resources". *North Sea Continental Shelf (Federal Republic of Germany/Denmark; Federal Republic of Germany/Netherlands)*, Judgment, 20 February 1969, *I.C.J. Reports 1969*, p. 3, at p. 22, para. 19. See also *Case concerning the Continental Shelf (Libyan Arab Jamahiriya/Malta)*, Judgment, 3 June 1985, *I.C.J. Reports 1985*, p. 13, at p. 33, para. 33.

53 M.W. Mouton, *The Continental Shelf* (1952), at p. 151. O'Connell, who defended coastal state sovereign rights over continental shelf resources, also argued for coastal state jurisdiction in this regard. D.P. O'Connell, "Sedentary Fisheries and the Australian Continental Shelf", 49 *American Journal of International Law* (1955), at pp. 205–206 (the doctrine of abuse of rights might be resorted to as a solution to the interesting legal problem of a hypothetical collision of two incompatible rights, and the coastal state is competent to require foreign nationals to employ suitable rollers on the foot-wire of the trawl).

54 For various proposals during the negotiations in UNCLOS III, see Nandan and Rosenne (eds.), *United Nations Convention on the Law of the Sea 1982: A Commentary*, vol. II, at pp. 901–907.

and that coastal states shall cooperate with states engaged in high seas fisheries. On the other hand, it is possible to argue that since coastal states' regulation in this regard is in any case limited to measures aimed at avoiding harm to the exploration and exploitation of natural resources on the continental shelf, Article 78(2) does not deprive coastal states of the possibility to take unilateral measures, but merely sets limitations thereon. Given the weight of wording such as 'sovereign rights' and 'exclusive right', the second argument appears to be more plausible.[55]

Second, provisions in Part XII for the protection and preservation of the marine environment may be considered to be included in articles implied in the '*inter alia*' clause.[56] Whether or not the components of marine ecosystems on the continental shelf are harvestable, the duty to protect and preserve the marine environment demands that all states, including high seas fishing states, do not cause harm to the environment of the continental shelf.[57]

Third, coastal states may regulate MSR activities on the continental shelf. If fishing activities on the high seas superjacent to the outer continental shelf are preceded by or involve MSR projects, high seas fishing states are obliged to meet the requirements in Part XIII of the LOSC. Article 246(2) of the LOSC provides that MSR on the continental shelf shall be conducted with the consent of the coastal state. On the continental shelf beyond 200 miles, coastal states may withhold consent to MSR projects in accordance with paragraphs 5 and 6 of Article 246. Under Article 248, high seas fishing states shall inform the coastal state concerned of the intended project at least six months before their vessels begin MSR activities.

Although it seems that fisheries in the superjacent water column in the high seas are subject to the rights, duties and interests of coastal states concerning the continental shelf on the basis of the above provisions,[58] state practice with regard to the regulation of high seas fisheries in the water column superjacent to the continental shelf is limited.[59] The Canadian Senate suggested that the

---

55 It is noted that the LOSC added the word 'infringe' in this provision. According to Attard, the word was inserted because "the concept of 'unjustifiable interference' was not felt to be sufficient to safeguard the international community's interests in view of the new extensive EEZ powers allocated to the coastal State". D.J. Attard, *The Exclusive Economic Zone in International Law* (1987), at p. 144.

56 Nandan and Rosenne (eds.), *United Nations Convention on the Law of the Sea 1982: A Commentary*, vol. III, at pp. 287–288.

57 See LOSC, Article 192.

58 It should of course be noted that the exercise of sovereign rights by coastal states over the continental shelf does not affect the legal status of the superjacent water. Ibid., Article 78(1).

59 O'Connell wrote in a book published in 1982 that "[...] no incompatibility between fishing for pelagic and sedentary species has been recorded". O'Connell, *The International Law of the Sea*, vol. I, at p. 500.

government should examine the possibility of action concerning the regulation of bottom trawl fishing over Canada's outer continental shelf.[60] The Chilean 'presential sea' doctrine could also be considered as an expression of coastal state interests over the high seas area superjacent to the outer continental shelf as it extends to the outermost limit of the asserted outer continental shelf of the Chilean islands in the South Pacific.[61] Opposition to the exercise of such jurisdiction is found in a European Commission paper, especially in relation to the Grand Banks off the Canadian coast and the South West Atlantic.[62] Some commentators explicitly advocate the exercise of coastal state jurisdiction over high seas areas superjacent to the continental shelf.[63]

In this context, it is of interest to note that during the first meeting of the open-ended working group to study the issues of conservation and sustainable use of biodiversity beyond areas of national jurisdiction established by the UNGA, some delegations stated that "the coastal State was fully entitled to adopt any conservation and management measures it deemed necessary to protect its sedentary species on the continental shelf. Those may include the possibility of imposing restrictive measures on fishing activities in the high seas over its continental shelf, including on fishing practices that were deemed to have

60 Standing Senate Committee on Fisheries and Oceans of the Senate of Canada, The Management of Atlantic Fish Stocks: Beyond the 200–Mile Limit, February 2007, at pp. 51–52.

61 The Preliminary Report of the International Law Association (ILA) Outer Continental Shelf Committee notes "the progressive encroachment by certain states of their fisheries jurisdiction within the superjacent waters of their outer continental shelf" and refers to Canada and Chile as examples of that type of state practice. Preliminary Report, Committee on Legal Issues of the Outer Continental Shelf, New Delhi Conference (2002), 15 January 2002, at p. 15.

62 European Commission, Background Paper No. 3 on Exclusive Economic Zones, Underwater Resources (Including Fisheries Resources, Continental Shelves, Law of the Sea), at p. 15.

63 See, e.g., Fleischer, "Fisheries and Biological Resources", at p. 1105; M. Hayashi, "Global Governance of Deep-Sea Fisheries", 19 *International Journal of Marine and Coastal Law* (2004), at p. 293; Miovski, "Central Bering Sea Overfishing", at pp. 555–556; E.J. Molenaar, "Unregulated Deep-Sea Fisheries: A Need for a Multi-Level Approach", 19 *International Journal of Marine and Coastal Law* (2004), at pp. 243–246; J. Mossop, "Protecting Marine Biodiversity on the Continental Shelf Beyond 200 Nautical Miles", 38 *Ocean Development and International Law* (2007), at pp. 297–299; Owen, *The Powers of the OSPAR Commission and Coastal State Parties to the OSPAR Convention to Manage Marine Protected Areas on the Seabed Beyond 200 nm From the Baseline: A Report for WWF Germany*, at pp. 40–41; A. Serdy, "Schrödinger's TAC: Superposition of Alternative Catch Limits from 2003 to 2006 under the South Tasman Rise Orange Roughy Arrangement between Australia and New Zealand", in R. Shotton (ed.), *Deep Sea 2003: Conference on the Governance and Management of Deep-sea Fisheries: Part 1: Conference Reports. Queenstown, New Zealand, 1–5 December 2003* (2005). See also J. Mossop, "Regulating Uses of Marine Biodiversity on the Outer Continental Shelf", in D. Vidas (ed.), *Law, Technology and Science for Oceans in Globalisation: IUU Fishing, Oil Pollution, Bioprospecting, Outer Continental Shelf* (2010), at pp. 319–337 and the literature cited therein.

a negative impact on sedentary species".[64] According to the Co-Chairpersons, "[i]t was acknowledged that under the Convention a coastal State had the right to regulate the activities that had negative impacts on the sedentary species of its continental shelf and the right to adopt the necessary measures, including restrictive measures, to protect those resources".[65]

In addition to the above discussions, the issue is further complicated by the fact that many coastal states have not yet established the outer limit of their outer continental shelf. It should be noted that until the outer limit of the outer continental shelf is determined on the basis of recommendations by the Commission on the Limits of the Continental Shelf (CLCS), states might hesitate in taking action regarding activities in the superjacent water column of their continental shelf.[66] The LOSC does not specifically provide how and to what extent coastal states may exercise sovereign rights and jurisdiction with regard to the outer continental shelf pending the establishment of the outer limit of the continental shelf by coastal states on the basis of the recommendations by the CLCS. While the rights of coastal states in respect of the continental shelf exists *ipso facto* and *ab initio*, the exercise of such rights by one coastal state may be an infringement of the competence of the ISA if the area concerned eventually becomes outside the continental shelf of that state following the recommendations from the CLCS.

An analogy can be suggested from the exercise of sovereign rights and jurisdiction with regard to the continental shelf where overlapping claims exist between opposite or adjacent coastal states. In the *Aegean Sea Continental Shelf* case, the

---

64 A/61/65, at p. 7, para. 22.

65 Ibid., at p. 23 (Summary of trends prepared by the Co-Chairpersons, para. 9). But note that, in discussing the draft Co-Chairs' summary of trends, the US objected to and suggested the deletion of the text on the rights of coastal states that do not accurately reflect the provisions in the LOSC. *Earth Negotiations Bulletin* (*ENB*), vol. 25 No. 25, at p. 4. *ENB* is available at the website of the International Institute for Sustainable Development <http://www.iisd.ca/>. See also a view in E.J. Molenaar, "Marine Biodiversity in Areas Beyond National Jurisdiction", 22 *International Journal of Marine and Coastal Law* (2007), at p. 118. In the second meeting of the Working Group, it was reiterated that any activities relating to marine biodiversity beyond areas of national jurisdiction should be conducted in a manner consistent with the principles of international law, including the rights of coastal states over their extended continental shelves. Joint statement of the Co-Chairpersons of the Ad Hoc Open-ended Informal Working Group to study issues relating to the conservation and sustainable use of marine biological diversity beyond areas of national jurisdiction, reproduced as Letter dated 15 May 2008 from the Co-Chairpersons of the Ad Hoc Open-ended Informal Working Group to study issues relating to the conservation and sustainable use of marine biological diversity beyond areas of national jurisdiction addressed to the President of the General Assembly, A/63/79, 16 May 2008, Annex, at p. 3, para. 9.

66 See also Owen, *The Powers of the OSPAR Commission and Coastal State Parties to the OSPAR Convention to Manage Marine Protected Areas on the Seabed Beyond 200 nm From the Baseline: A Report for WWF Germany*, at p. 45.

ICJ declined to indicate the provisional measures requested by Greece to order Turkey to refrain from all exploration activity or any research in the disputed area of the continental shelf.[67] In examining the Greek request for provisional measures, the ICJ considered the seismic exploration conducted by Turkey.[68] The ICJ stated that

> [...] seismic exploration of the natural resources of the continental shelf without the consent of the coastal state might, no doubt, raise a question of infringement of the latter's exclusive right of exploration; [...] accordingly, in the event that the Court should uphold Greece's claims on the merits, Turkey's activity in seismic exploration might then be considered as such an infringement and invoked as a possible cause of prejudice to the exclusive rights of Greece in areas then found to appertain to Greece.[69]

Although the ICJ later found no jurisdiction to entertain the case,[70] some implications for the present study can be inferred from the order on provisional measures. First, although Turkey had not embarked upon any operations involving the actual appropriation or other use of the natural resources of the areas of the continental shelf in dispute, the Turkish activities were found liable to entail the infringement of the rights of Greece if the areas were found to be the Greek continental shelf. In the context of the present study, it is submitted that similar activities may be considered to constitute an infringement of the interests of the international community as a whole and the competence of the ISA in particular.[71] Second, the exploration activities concerned were not considered irreparable and therefore these activities were found insufficient to require interim measures to prohibit such activities. The Court further found that the prejudice to the rights caused by the activity concerned was capable of reparation by appropriate means. This should not be taken as excluding the possibility of provisional measures. In its consideration, the Court confirmed that there was no complaint that such exploration activities were considered to involve a risk of physical damage to the seabed or subsoil or to their natural resources. In view of this, if high seas fishing activities or environmental impact assessment activities involve a risk of physical damage to the seabed or subsoil or to their natural resources, the provisional measures will remain possible.

---

67 *Aegean Sea Continental Shelf (Greece v. Turkey)*, Request for the Indication of Interim Measures of Protection, Order of 11 September 1976, *ICJ Reports 1976*, p. 3.

68 Ibid., at pp. 10–11, paras. 30–33.

69 Ibid., at pp. 10–11, para. 31.

70 *Aegean Sea Continental Shelf (Greece v. Turkey)*, Judgment (Jurisdiction) of 19 December 1978, *ICJ Reports 1978*, p. 3, at p. 45, para. 109.

71 If the area concerned is found to be another state's shelf, it is the infringement of the rights of that state.

2.1.4 *Scope of Fishery Resources*

The discussions in the previous two sub-sections give rise to the question of which general principles will apply to the harvesting of stocks and species not explicitly mentioned in Articles 63(2) and 64–67. Articles 87 and 116 are couched in an all-inclusive manner: they do not directly exclude any species of the marine living resources of the high seas from their scope. The following paragraphs examine the possibility that separate regimes apply to high seas fisheries in two additional circumstances: the exploitation of organisms belonging to sedentary species beyond the outer limit of the continental shelf and fisheries for DHSFS.[72]

Organisms belonging to sedentary species may be found both within and beyond the outer limit of the continental shelf determined in accordance with Article 76 of the LOSC. Is there any distinction in law between the harvesting of organisms belonging to sedentary species and that of other marine living resources in areas beyond the outer limit of the continental shelf? Is the former regulated by the regime of the Area?

At the outset, it should be noted that the term 'sedentary species' is not used in any context outside the continental shelf in the LOSC; it is not used in either Part VII or Part XI.[73] It is not even clear, in the context of the Area, whether it is adequate to use the term 'sedentary species' to refer to organisms which, at the harvestable stage, either are immobile on or under the seabed or are unable to move except in constant physical contact with the seabed or the subsoil.[74] Nevertheless, for the sake of convenience, the term 'sedentary species' is employed in this section to mean such organisms, whether within or beyond the outer limit of the continental shelf.

The scope of species covered by the term 'sedentary species' is not without controversy. While the term is defined in the CSC and the LOSC, disputes have arisen as to whether a particular species falls under the biological character

---

72 In theory, one may consider the possibility of a distinct regime for fisheries covered by Article 13 of the HSFC. But it is doubtful that the type of fisheries contemplated in that article has been maintained and conducted by the coastal state's nationals on the high seas even if the area covered by Article 13 is interpreted as high seas areas adjacent to the EEZ. For this reason, the present study does not look at this issue.

73 The concept of sedentary species has been primarily developed in association with the regime of the continental shelf. See also T. Scovazzi, "Mining, Protection of the Environment, Scientific Research and Bioprospecting: Some Considerations on the Role of the International Sea-Bed Authority", 19 *International Journal of Marine and Coastal Law* (2004), at pp. 400–401.

74 A.G. Oude Elferink, "The Regime of the Area: Delineating the Scope of Application of the Common Heritage Principle and Freedom of the High Seas", 22 *International Journal of Marine and Coastal Law* (2007), at p. 151; Scovazzi, "Mining, Protection of the Environment, Scientific Research and Bioprospecting", at p. 401.

defined in the treaties.[75] Moreover, recent interest in bioprospecting for marine genetic resources has further complicated the classification of marine living resources. In this regard, it is interesting to note the following comments made by the Secretary-General: "[t]he extent to which the definition of sedentary species under article 77 covers the complex and symbiotic web of life of deep-sea ecosystems may need to be addressed in order to clarify whether such ecosystems and organisms belong to the regime of the continental shelf or of the water column above it".[76]

The Area is defined as the seabed and ocean floor and the subsoil thereof beyond the limits of national jurisdiction,[77] while the term 'resources of the Area' is defined as mineral resources for the purposes of Part XI.[78] Many of the provisions of Part XI are concerned with the exploration and exploitation of mineral resources, and only Article 145 has the potential to be considered to explicitly refer to living resources. The LOSC does not define which living resources belong to the Area or to the high seas. The term 'seabed' is not defined in the LOSC, either.

Many authors argue that the provisions of Part XI or the CHM concept are not applicable to living resources,[79] while some suggest the applicability as *lege ferenda*.[80] The principal argument to deny the applicability of Part XI to living resources is that Article 133 defines resources as mineral resources, thus excluding living resources.

75 Disputes include those concerning king crab between the United States and Japan, lobster between Brazil and France, and Icelandic scallops between Canada and the United States. See E.H. Buck, U.N. Convention on the Law of the Sea: Living Resources Provisions, CRS Report for Congress, Updated January 7, 2004,

76 Oceans and the Law of the Sea, Report of the Secretary-General, A/62/66, 12 March 2007, at para. 193.

77 LOSC, Article 1(1)(1).

78 Ibid., Article 133(a).

79 E.g., Craig H. Allen, "Protecting the Oceanic Gardens of Eden: International Law Issues in Deep-Sea Vent Resource Conservation and Management", 13 *Georgetown International Environmental Law Review* (2001), at pp. 628–638; E. Holmila, "Common Heritage of Mankind in the Law of the Sea", 1 *Acta Societatis Martensis* (2005), at pp. 193–194; Orrego Vicuña, *The Changing International Law of High Seas Fisheries*, at pp. 10–11. See also R. Wolfrum and N. Matz, "The Interplay of the United Nations Convention on the Law of the Sea and the Convention on Biological Diversity", 4 *Max Planck Yearbook of United Nations Law* (2000), at p. 455.

80 S. Oda, "Sharing of Ocean Resources: Unresolved Issues in the Law of the Sea", 3 *New York Law School Journal of International and Comparative Law* (1981), at p. 13 (the CHM concept has been introduced for the exploitation of mineral resources in areas beyond the continental shelf or EEZ, although "in the future, almost without doubt, this concept [CHM] will also apply to fishery resources or even to protection of the marine environment"); K. Baslar, *The Concept of the Common Heritage of Mankind in International Law* (1998), at pp. 235 and 242.

Whether or not recognizing the applicability of the regime of the Area to living resources, most commentators argue that sedentary species of the Area are not governed by the regime of the Area, but by the regime of the high seas.[81] A notable exception is Armas Pfirter, who argues that "it would not be logical to consider living resources that are in constant contact with the seabed in the Area subject to the high seas legal regime", among other reasons, because "[t]hose which inhabit the Area have a greater dependency on the seabed and subsoil than those living in the continental shelf [and the] particular status of the Area will exert an influence on any regulatory framework dealing with biological communities associated with mineral resources in the Area".[82] Another commentator, having argued that corals are mineral resources, concludes that as bottom trawling takes advantage of corals, "it thereby should be considered exploitation"; thus it would fall under the Part XI system.[83]

State practice supports the majority proposition. Grigg notes that the US Western Pacific Fishery Council has requested the State Department to enter into multilateral arrangements with Japan and Taiwan for jointly managing the precious coral fisheries in international waters off Hawaii.[84] The constitutive instruments of RFMO/As appear to support the majority proposition: none of the RFMO/As examined in Chapters 4 and 5 exclude sedentary species beyond the limit of the continental shelf from their definition of resources to be regulated by them.[85] Many domestic laws and regulations widely define the scope of fish or

81 Allen appears to consider that the regime applicable to sedentary species in areas beyond the limits of national jurisdiction is no different from that which is applicable to non-sedentary species beyond the limits of the EEZ, namely, freedoms of the high seas. See Allen, "Protecting the Oceanic Gardens of Eden", at p. 628. Molenaar reaches the same conclusion by way of an analysis of the constitutive instruments of the relevant RFMOs. See also E.J. Molenaar, "Current Legal and Institutional Issues Relating to the Conservation and Management of High Seas Deep-Sea Fisheries", *Report and Documentation of the Expert Consultation on Deep-sea Fisheries in the High Seas, Bangkok, Thailand, 21–23 November 2006*, FAO Fisheries Report, No. 838 (2007), at p. 131.

82 F.M. Armas Pfirter, "The Management of Seabed Living Resources in 'the Area' under UNCLOS", 11 *Revista Electrónica de Estudios Internacionales* (2006), at p. 21.

83 P. Prows, "Tough Love: The Dramatic Birth and Looming Demise of UNCLOS Property Law (and What is to be Done about it)", 42 *Texas International Law Journal* (2007), at pp. 295–307. See also R.W. Grigg, "Precious Coral Fisheries of Hawaii and the U.S. Pacific Islands", 55 *Marine Fisheries Review* (1993).

84 Grigg, "Precious Coral Fisheries of Hawaii and the U.S. Pacific Islands". He refers to the CHM principle and measures to conserve living resources within and beyond the EEZ in this regard. It is not clear whether these arguments were actually made and whether any action has been taken by the State Department. It appears that vessels from Taiwan continued to harvest corals in the Northwest Pacific Ocean.

85 Note that the SEAFO Convention explicitly uses the term 'sedentary species' to denote those beyond the limit of the continental shelf. In addition, one treaty includes 'any aquatic plant' in the definition of 'fish'. South African Development Community Protocol on Fisheries (SADC

fishery resources which may not be harvested without permits.[86] Some refer to molluscs and/or crustaceans (Australia, Canada, Namibia, South Africa, Norway and the United States), while some laws appear to include corals and/or sponges (South Africa and the United States).

Assuming that the regime of the high seas, including the freedom of fishing, is applicable to sedentary species beyond the outer limit of the continental shelf, the 'inter alia' clause in Article 116 subjects, in any case, the high seas fishing right to coastal states' rights, duties and interests with regard to sedentary species occurring both within and beyond the outer limit of the continental shelf. There may be various ways to give effect to coastal states' rights and interests over sedentary species on the continental shelf.[87] It is thus necessary to confirm, through examining the practice of RFMO/As, whether there is any difference between sedentary species and non-sedentary species in giving effect to the rights and interests of coastal states. At any rate, it is noted that the issue of straddling sedentary species is marginal precisely because sedentary species do not tend to move as actively as finfish do.

The other important issue in this sub-section is the applicable legal regime for fisheries for DHSFS, in particular the application of the general principles of high seas fisheries in the LOSC. Provisions of the treaty "shall be interpreted in good faith in accordance with the ordinary meaning to be given to the terms of the treaty in their context and in the light of its object and purpose".[88] The term 'living resources' is generic and wide enough to justify the inclusion of DHSFS. Even if the drafters were not particularly concerned about the issue of DHSFS at the time of drafting high seas fisheries provisions, this does not prevent the applicability of the LOSC to DHSFS: as unequivocally clarified by the World Court, "[t]he mere fact that, at the time when the Convention [...] was concluded, certain facts or situations, which the terms of the Convention in their ordinary

Protocol), Maputo, 14 August 2001, Article 1(2). See also FAO Port State Measures Agreement, Article 1(b).

86 Australia, Fisheries Management Act, Section 4(1); Canada, Fisheries Act, Section 2; Namibia, Marine Resources Act, Article 1; New Zealand, Fisheries Act, Section 2; Norway, Act of 3 June 1983 relating to Sea-water Fisheries, Section 2; South Africa, Marine Living Resources Act, Section 1. US law lists corals and sponges under 'continental shelf fishery resource' whereas 'fish' is defined as "finfish, mollusks, crustaceans, and all other forms of marine animal and plant life other than marine mammals and birds". See 16 U.S. Code 1802. In addition, the US government has understood the exploitation of corals as falling under the regime established by Part VII of the LOSC. See Grigg, "Precious Coral Fisheries of Hawaii and the U.S. Pacific Islands".

87 Note that the FSA does not cover sedentary species of the continental shelf, although such organisms beyond the continental shelf are not explicitly excluded from the scope of Article 1(1)(c). Therefore, the compatibility requirement in Article 7 does not apply to sedentary species as opposed to finfish, to which Article 7 applies by virtue of Article 3(1).

88 VCLT, Article 31.

meaning are wide enough to cover, were not thought of, does not justify interpreting those of its provisions which are general in scope otherwise than in accordance with their terms".[89] In addition, nothing in the LOSC, subsequent agreements or state practice appears to exclude DHSFS from the scope of the provisions of Part VII, section 2 of the LOSC.

There is widespread agreement in legal writing that the LOSC is applicable to fisheries for DHSFS. For example, Munro and others state that "[i]n legal terms, [discrete high seas stocks] are covered by [LOSC] Articles 116(a), 116(c), 117–120".[90] The FAO, in its SOFIA 2006, stated that the legal framework for the conservation and management of discrete high seas stocks is provided by Part VII of the LOSC.[91] A number of publicists recognize that the provisions of Part VII, section 2 of the LOSC are applicable to DHSFS.[92]

A question remains as to whether all articles are applicable and whether there should be any difference in the implications of the relevant articles between DHSFS and straddling fish stocks. Notably, two important issues would be whether the right to fish on the high seas is subject to the rights, duties and interests of coastal states under Article 116(b) and whether the obligation and the right of states to cooperate with other states extends to coastal states not engaged in high seas fishing in the context of DHSFS.

On the one hand, the above-mentioned FAO Technical Paper by Munro and others excludes Article 116(b) from its list of articles.[93] On the other hand, writing in 1989, Miles and Burke suggested that the obligations are owed to at least

89 *Interpretation of the Convention of 1919 concerning Employment of Women during the Night*, Advisory Opinion, 15 November 1932, *PCIJ, Série A/B No. 50*, p. 365, at p. 377.

90 G. Munro *et al.*, The Conservation and Management of Shared Fish Stocks: Legal and Economic Aspects, FAO Fisheries Technical Paper No. 465, at p. 55. Notably, they continue by observing that in the LOSC, "[n]o mechanism for cooperation is suggested" in relation to these stocks, and also observes that one has no justification for "assuming that these articles will prove to be any more adequate for the conservation of discrete high seas stocks" than for the conservation of straddling and highly migratory stocks, for which "[t]hese articles proved to be inadequate".

91 FAO Fisheries and Agriculture Department, SOFIA 2006, at p. 125.

92 See also Burke, *The New International Law of Fisheries*, at pp. 93 and 99; St. John's Conference Report, at p. 12 (comments by Satya Nandan); Orrego Vicuña, *The Changing International Law of High Seas Fisheries*, at p. 51; Joyner and De Cola, "Chile's Presential Sea Proposal", at p. 119 note 45; E.L. Miles and W.T. Burke, 'Pressures on the United Nations Convention on the Law of the Sea of 1982 Arising From New Fisheries Conflicts: The Problem of Straddling Stocks', 20 *Ocean Development and International Law* (1989), at pp. 350–351. Others implicitly consider that the LOSC is applicable to DHSFS by observing that the LOSC is applicable to all living resources of the high seas. E.g., A. Tahindro, 'Conservation and Management of Transboundary Fish Stocks: Comments in Light of the Adoption of the 1995 Agreement for the Conservation and Management of Straddling Fish Stocks and Highly Migratory Fish Stocks', 28 *Ocean Development and International Law* (1997), at p. 3.

93 Munro *et al.*, The Conservation and Management of Shared Fish Stocks: Legal and Economic Aspects, at p. 55.

three recipients, namely, "another state or states fishing on the high seas, the general community of states [...], and coastal states adjoining on the high seas area" and that it is conceivable that "states of a region or subregion might be the beneficiaries of the obligation as well".[94] While the article was written in the context of straddling stocks, the same argument might be advanced with regard to DHSFS. Even if this argument is accepted, it is not clear how and to what extent coastal states whose vessels are not engaged in a particular fishery for DHSFS are allowed to participate in fisheries management for that fishery. These ambiguities over the interpretation of the provisions of Part VII of the LOSC generate the need to examine state practice in relation to DHSFS.

Another issue that might be of importance in the context of DHSFS is whether the provisions of the LOSC concerning marine living resources are considered to be part of customary international law in relation to DHSFS. True, the LOSC has attracted quasi-universal participation from states and its general principles have been argued to reflect customary international law in relation to high seas fisheries.[95] But, without state practice supporting the proposition that the general principles of high seas fisheries as reflected in the LOSC apply to DHSFS, this question may not be conclusively answered.

2.1.5 *Scope of 'Nationals'*

Under Article 116, the high seas fishing right of all states is formulated as the right *for their nationals to engage in fishing* on the high seas. The scope of the right to fish on the high seas under Article 116 remains ambiguous in respect of the meaning of the term 'nationals'. There is no article defining the term 'nationals' comparable to Article 14 of the HSFC.[96] In this respect, an argument has been put forward that this omission may be interpreted to the effect that non-flag states whose national is a member of the crew of a third state vessel must exercise jurisdiction over fishing activities by such a vessel and that they must assume responsibility to prescribe and enforce conservation measures.[97] In this regard, it is of interest to note a provision of the LOSC concerning fisheries in the EEZ. Article 62(4) starts with the following sentence: "Nationals of other States fishing in the exclusive economic zone shall comply with the conservation measures and with the other terms

94 Miles and Burke, "Pressures on the United Nations Convention on the Law of the Sea of 1982 Arising From New Fisheries Conflicts: The Problem of Straddling Stocks", at p. 351.

95 E.g., Sands, *Principles of International Environmental Law*, at p. 569; Miovski, "Central Bering Sea Overfishing", at pp. 525–574.

96 Note that the French version of the LOSC uses 'ressortissants' as opposed to 'nationaux' employed in the HSFC.

97 See, e.g., Kaye, *International Fisheries Management*, at p. 106 note 57; Fleischer, "Fisheries and Biological Resources", at p. 989; Molenaar, "Management and Conservation of Orange Roughy", at p. 106 note 121 and its accompanying text.

and conditions established in the laws and regulations of the coastal State". The article continues by stipulating that "[t]hese laws and regulations [...] may relate, *inter alia*, to the following: (a) licensing of fishermen, fishing vessels [...]". Taken together, it could be argued that both natural persons and vessels are included in the meaning of the term 'nationals'.

Difficulties arising from this interpretation are as follows. First, if the term is interpreted in this way, it is too demanding for such non-flag states. For example, not only the duty to take conservation measures under Article 117 but also the duty of negotiation under Article 118 will be imposed on such states. Second, the drafting history indicates the contrary. Almost all proposals for articles concerning the conservation and management of the living resources of the high seas seem to consistently employ 'nationals' in the same sense as the HSFC. As regards the right to fish, an exception is found in Article III(1) of the 1971 US draft articles on territorial sea, straits, and fisheries submitted to the UN Seabed Committee, which provided that the coastal state and any other state whose nationals *or vessels* exploit or desire to exploit a regulated species have an equal right. Formulated in this way, 'nationals' seem to include natural persons as opposed to 'vessels'. This formulation was not adopted in the final text of the LOSC, and both subsequent proposals and the text of the LOSC retain phrases similar to the HSFC. Thus, it is safe to say that the term 'nationals' as employed in the LOSC is used in the same sense as the HSFC, and it means vessels flying the flag of that state.[98]

Although the term 'nationals' has increasingly been used in the sense of natural and juridical persons in international and national fisheries instruments,[99]

---

98 See also Nandan and Rosenne (eds.), *United Nations Convention on the Law of the Sea 1982: A Commentary*, vol. III, at pp. 286 and 294.

99 E.g., UNGA Resolution 66/68, paras. 44 and 46; SEAFO Convention, Article 13(6)(a); South East Atlantic Fisheries Organization (SEAFO), Conservation Measure 07/06 relating to Interim Measures to Amend the Interim Arrangement of the SEAFO Convention, para. 6(e)–(f); SIOFA, Article 10(3)–(4); New NAFO Convention, Article X(1)(g)–(h); Convention on the Conservation and Management of Pollock Resources in the Central Bering Sea (CBS Convention), Washington, 16 June 1994, Articles XI(1), XII(1) and (3), XVI(4); SPRFMO Convention, Article 1(1)(k); Australia, Fisheries Management Act, Section 105E–105F; Canada's National Plan of Action to Prevent, Deter and Eliminate Illegal, Unregulated and Unreported Fishing, at p. 19; National Plan of Action to Prevent, Deter and Eliminate Illegal, Unregulated and Unreported Fishing/ Republic of Korea, at p. 11 (measures to prevent IUU fishing through fishing licence systems and to prohibit Korean nationals from being involved in IUU fishing by the Fisheries Law); New Zealand, Fisheries Act, Section 113E. See also Republic of Namibia, National Plan of Action to Prevent, Deter and Eliminate Illegal, Unreported and Unregulated Fishing, Part B, Section 2.2.2. For the United States, including the Lacey Act, which enables the US authorities to prosecute US nationals who engage in certain forms of IUU fishing where there is some nexus between the activity in question and the United States, see National Plan of Action of the United States of America to Prevent, Deter, and Eliminate Illegal, Unregulated, and Unreported Fishing, at

terms are not used in a uniform way in, among others, the FSA. The FSA uses various phrases such as 'States whose nationals fish for such stocks in the region' (Article 7(1)), 'States fishing on the high seas' (Article 8(1)) and '[a] State whose vessels fish on the high seas' (Article 18(1)). Despite different phrases in these articles, the FSA does not appear to intend to refer to different states by virtue of these articles. This view may be supported, for example, by the declarations of the European Community deposited on signature and ratification as follows:

> The European Community and its Member States understand that the term "States whose nationals fish on the high seas" shall not provide any new grounds for jurisdiction based on the nationality of persons involved in fishing on the high seas rather than on the principle of flag State jurisdiction.[100]

It appears that the declarations focus on the rejection of any potential for a new jurisdictional basis. Furthermore, it could be argued that the EC and its member states are not willing to assign a meaning to 'nationals' other than vessels in the context of high seas fisheries.

2.2 *Cooperation*

2.2.1 *Provisions of the LOSC*

Cooperation between states is one of the essential elements in the legal regime of high seas fisheries under the LOSC.[101] Article 117 of the LOSC enunciates the duty of *all* states to cooperate in taking necessary conservation measures. It has been pointed out that the duty to cooperate in taking measures cannot be avoided by individually taking measures; this view is reinforced by Article 118.[102]

---

pp. 5–6. See also Miller, "Management and Governance Conventions and Protocols: SEAFC, WCPFC and SADC", at pp. 606–607 note 13.

100 Interpretative declarations, para. 3. All the declarations regarding the FSA are available at the UN Treaty Collection database. See <http://treaties.un.org/Pages/ViewDetails.aspx?src=TREATY&mtdsg_no=XXI-7&chapter=21&lang=en>.

101 The need for cooperation at the global and regional levels is also underlined in the context of marine environmental protection. See LOSC, Article 197. On the relevance of this provision in the context of deep-sea fisheries on the high seas, see Report and Documentation of the Expert Consultation on Deep-sea Fisheries in the High Seas, Bangkok, Thailand, 21–23 November 2006, at p. 12, para. 64.

102 Nandan and Rosenne (eds.), *United Nations Convention on the Law of the Sea 1982: A Commentary*, vol. III, at p. 294; Hey, *The Regime for the Exploitation of Transboundary Marine Fisheries Resources*, at pp. 34–35. But see Fleischer, "Fisheries and Biological Resources", at p. 1114.

The provisions of the LOSC leave some ambiguities with regard to the content of the duty of cooperation.[103] It is not clear: (1) which states are required to cooperate; (2) which states are entitled to require cooperation from other states; (3) what form cooperation needs to take.

As regards the first issue, the scope of states required to cooperate under the LOSC appears to be broad: in fact, as suggested by the second sentence of Article 118, even if nationals of two states are not engaged in fishing for the same stock in the same area of the high seas, they are still under a duty to cooperate when targeted stocks are biologically related or two states are engaged in the fishery for the same stock in different areas of the high seas.[104] It has been argued that the reference to *all* states aims to include both fishing states and coastal states.[105] It is unlikely that states other than flag, coastal, port or market states have the duty to cooperate.

Regarding the second issue, it could be argued that states which do not have the duty still have the right of cooperation if they have an interest in the fishery resource concerned. In other words, arguments could be made that states not engaged in high seas fisheries, including coastal states, are entitled to require 'some' cooperation from high seas fishing states, such as allowing participation in benefit-sharing. Joyner and DeCola write: "[t]rue, all states may in principle enjoy the same rights to exploit living resources of the high seas. But, it is believed, only states having distant-water fishing fleets will be able to harvest them without having to share benefits with other nations".[106] If the coastal state in the region cannot afford to invest in costly high seas fishing fleets, the recognition of the special interests of developing states through, say, the favourable allocation of fishing opportunities will not be meaningful unless the entitlement is transferrable to other states. The increased participation in high seas fisheries through support activities might be one solution particularly where the stock is fully exploited and the states should not develop their fishing capacities.

With regard to the third issue, the relevant articles do not command particular types of cooperation to be pursued by the states concerned, nor do these articles

---

103 It has been pointed out that "[a] broader issue undermining the sustainability of high seas fisheries is the absence of a consensus on the nature of the duty to cooperate under international law for the conservation and management of high seas fisheries". A/60/63, at p. 58, para. 211. The Report continues by observing the problem associated with fishing by third party vessels without regard for RFMO conservation measures.

104 See also Davies and Redgwell, "The International Legal Regulation of Straddling Fish Stocks", at p. 229.

105 See also Nandan and Rosenne (eds.), *United Nations Convention on the Law of the Sea 1982: A Commentary*, vol. III, at p. 294.

106 Joyner and De Cola, "Chile's Presential Sea Proposal", at p. 109.

explicitly state the consequences of a failure to cooperate.[107] Cooperation may take various forms and need not necessarily be pursued in a single form at any given time.[108] It has been argued that cooperation may have objectives other than conservation such as the orderly conduct of fishing operations.[109]

Articles 116–119 do not explicitly require cooperation through international organizations. The third sentence of Article 118 stipulates that fishing states shall, as appropriate, cooperate to establish subregional or regional fisheries organizations to this end. This suggests that cooperation through international fisheries organizations, particularly regional or subregional, is generally considered to be desirable in the LOSC.[110] But, under the LOSC, there is no explicit obligation for high seas fishing states to join the relevant RFMO.[111] Thus, it is not impossible for existing fishing states to continue fishing without joining regional fisheries organizations. Such organizations do not have to be granted a management mandate; they can be advisory bodies.[112] It is up to the states concerned to decide whether or not to collectively adopt conservation and management measures.[113]

Only in three situations are particular actions required under the LOSC. First, the second sentence of Article 118 stipulates the duty to enter into negotiations with a view to taking the appropriate conservation measures when their nationals exploit identical living resources or different resources in the same area of the high seas. It has been argued that the duty to enter into negotiations means the duty to negotiate in good faith. This duty has been examined in the

107 Some commentators attach significant consequences to the lack of cooperation. See, e.g., Miovski, "Central Bering Sea Overfishing", at p. 536.

108 Burke, *The New International Law of Fisheries*, at pp. 122–124. Miles and Burke identify the obligation of contributing and exchanging available scientific information, catch and fishing effort statistics and other data relevant to the conservation of fish stocks as the most basic obligation. Miles and Burke, "Pressures on the United Nations Convention on the Law of the Sea of 1982 Arising From New Fisheries Conflicts: The Problem of Straddling Stocks", at p. 351. The ICJ in the *Fisheries Jurisdiction* cases required some forms of cooperation such as a joint examination of conservation measures and entering into negotiations in giving effect to the duty of due regard. *ICJ Reports 1974*, at pp. 31–32, para. 73; *ICJ Reports 1974*, at p. 201, para. 65.

109 Fleischer, "The New Régime of Maritime Fisheries", at p. 173.

110 Article 119(2) concerning data exchange through competent international organizations also underlines the importance of institutional cooperative mechanisms.

111 J. Beer-Gabel and V. Lestang, *Les Commissions de Pêche et Leur Droit* (2003), at p. 45; Orrego Vicuña, *The Changing International Law of High Seas Fisheries*, at p. 201; M. Hayashi, "The 1995 Agreement on the Conservation and Management of Straddling and Highly Migratory Fish Stocks: Significance for the Law of the Sea Convention", 29 *Ocean & Coastal Management* (1995), at p. 58.

112 See "Strengthening RFMOs and their performances including the outcome of the 2007 Tuna RFMOs Meeting", COFI/2007/9 Rev.1, at p. 3 note 5 and its accompanying text.

113 States are obliged to enter into negotiations with a view to taking necessary measures, but the text of the LOSC does not explicitly require them to agree on collective measures or even to negotiate such measures.

North Sea Continental Shelf cases by the ICJ.[114] In the *Fisheries Jurisdiction* cases, the Court further elaborated the duty in the context of fisheries as follows:

> [t]he task before [the Parties] will be to conduct their negotiations on the basis that each must in good faith pay reasonable regard to the legal rights of the other in the waters around Iceland outside the 12-mile limit, thus bringing about an equitable apportionment of the fishing resources based on the facts of the particular situation, and having regard to the interests of other States which have established fishing rights in the area. It is not a matter of finding simply an equitable solution, but an equitable solution derived from the applicable law.[115]

The Court ruled that the parties were under mutual obligations to undertake negotiations in good faith for the equitable solution of their differences concerning their respective fishery rights, taking into account the elements specified by the Court.[116]

In view of these judgments, negotiations are considered to be sufficient as long as they are conducted in good faith and the duty does not require fishing states to reach any agreement.[117] Moreover, Article 300 of the LOSC concerning abuse of rights implies that states may not require other states to negotiate indefinitely if they have failed to agree on conservation measures. If the negotiation fails to reach an agreement on the appropriate conservation measures to be taken, it appears that because of the phrase "shall take, *or* cooperate in taking" (emphasis

114 *I.C.J. Reports 1969*, at p. 47, para. 85. The 'meaningful' or 'good faith' negotiation formula was reaffirmed in the *Case concerning the Gabcikovo-Nagymaros Project (Hungary/Slovakia)*, Judgment, 25 September 1997, *I.C.J. Reports 1997*, p. 7, at p. 78, para. 141. See also *Case concerning the Land and Maritime Boundary between Cameroon and Nigeria (Cameroon v. Nigeria: Equatorial Guinea Intervening)*, Judgment, 10 October 2002, *I.C.J. Reports 2002*, p. 303, at p. 424, para. 244 (Articles 74 and 83 of the LOSC do not require that delimitation negotiations should be successful, but, like all similar obligations to negotiate in international law, "negotiations have to be conducted in good faith"); *Arbitration Between Barbados and the Republic of Trinidad and Tobago*, Award of the Arbitral Tribunal, 11 April 2006, at para. 292 ("obliged to negotiate in good faith").

115 *I.C.J. Reports 1974*, at p. 33, para. 78; *I.C.J. Reports 1974*, at p. 202, para. 69.

116 *I.C.J. Reports 1974*, at pp. 34–35, para. 79; *I.C.J. Reports 1974*, at pp. 205–206, para. 77.

117 See, e.g., Burke, *The New International Law of Fisheries*, at p. 125; Fleischer, "The New Régime of Maritime Fisheries", at pp. 157–158; D. Freestone, "International Fisheries Law Since Rio: The Continued Rise of the Precautionary Principle", in A. Boyle and D. Freestone (eds.), *International Law and Sustainable Development: Past Achievements and Future Challenges* (1999), at p. 148; Kaye, *International Fisheries Management*, at pp. 118–119; Nandan and Rosenne (eds.), *United Nations Convention on the Law of the Sea 1982: A Commentary*, vol. III, at p. 302; O'Connell, *The International Law of the Sea*, vol. I, at p. 551. But, in the context of the 2006 Barbados/Trinidad and Tobago arbitral award, see also B. Kwiatkowska, "The 2006 *Barbados/Trinidad and Tobago* Award: A Landmark in Compulsory Jurisdiction and Equitable Maritime Boundary Delimitation", 22 *International Journal of Marine and Coastal Law* (2007), at pp. 30 and 46.

added), each fishing state is not relieved from the duty in Article 117 but is still under the duty to take necessary conservation measures for its nationals.

In relation to straddling stocks and stocks associated therewith, some states have expressed a different view. Cape Verde, Sao Tome and Principe, and Uruguay declared upon signing the LOSC that high seas fishing states had an obligation to agree with the coastal state on the necessary conservation measures.[118]

Second, fishing states are obliged, as appropriate, to cooperate to establish regional or subregional fisheries organizations under the third sentence of Article 118. It is not clear under what circumstances such organizations need to be established by virtue of the wording 'as appropriate'. Even for a fishery in which a number of states are engaged, it is still possible to conclude an arrangement for management, rather than to establish an organization (see Chapters 4 and 5).

Third, Article 119(2) demands that available scientific information, catch and fishing effort statistics, and other data relevant to the conservation of fish stocks shall be exchanged on a regular basis through competent international organizations where appropriate.[119] In view of the difficulty caused by sporadic statistics on high seas fisheries, the exchange of data is an important condition for the conservation of marine living resources.[120]

It has been argued that, given the duty to cooperate, new entrants to a fishery where conservation measures have been taken by existing fishing states, in particular through RFMOs, must apply the existing conservation measures to their nationals as far as the measures are taken in accordance with Article 119.[121]

118 As Article 309 of the LOSC prohibits reservations, declarations may not be considered as reservations unless reservations are allowed in an amending agreement; interpretive declarations can be considered as 'offers' of interpretation but lack any inherent binding character. Nelson, "Declarations, Statements and 'Disguised Reservations' with Respect to the Convention on the Law of the Sea", at pp. 775 and 784. Compare this with the following statement: "[. . .] the general obscurity and uncertainty which now are inherent in these two articles [i.e., Articles 309 and 310] will probably only be dissipated by practice and by decisions rendered in application of the dispute settlement provisions of Part XV of the Convention". Rosenne and Sohn (eds.), *United Nations Convention on the Law of the Sea 1982: A Commentary*, vol. V, at p. 227.

119 It is argued that scientific information should include biological data, the migratory habitats of the species in question, the fishing gear and methods utilized in harvesting those species, and the landing of each species, including incidental catches. See Nandan and Rosenne (eds.), *United Nations Convention on the Law of the Sea 1982: A Commentary*, vol. III, at p. 312.

120 Ibid.; Y. Tanaka, "Obligation to Co-operate in Marine Scientific Research and the Conservation of Marine Living Resources", 65 *Zeitschrift für ausländisches öffentliches Recht und Völkerrecht* (2005), at p. 944. Moreover, as Tanaka notes, because scientific work is conducted on the basis of the results of earlier investigations, the exchange of data is an important condition for the progress of scientific research in the future (p. 940).

121 See Burke, *The New International Law of Fisheries*, at p. 131; Nandan and Rosenne (eds.), *United Nations Convention on the Law of the Sea 1982: A Commentary*, vol. III, at p. 313 (arguing that, where conservation and management measures have already been established, new entrants must seek to exercise their right to fish through that mechanism and cannot ignore or flaunt

On the other hand, it has also been argued that the right to fish on the high seas in Article 116 and the requirement of non-discrimination in Article 119(3)[122] prevent existing fishing states from excluding new entrants.[123] The topic is more controversial in relation to restrictions on fishing opportunities such as the allocation of catches than in relation to the regulation of fishing gear or methods: on the one hand, new entrants should not in principle be excluded from a share in the total allowable catch (TAC), given that Article 116 allows nationals of every state to engage in fishing; on the other hand, it is possible to argue that certain states may be placed in a less favourable position in respect of the allocation because of the factors enumerated in Article 119(1).[124]

As the LOSC stipulates no criteria for the allocation of fishing opportunities,[125] it remains unclear whether certain states could be given preferential treatment in allocation by virtue of the existence of coastal state interests as referred to in Article 116. The concept of the preferential rights of coastal states outside the exclusive fishery zone under carefully circumscribed situations was recognized in the *Fisheries Jurisdiction* cases, but the concept in these cases, namely the preferential rights of Iceland up to 50 miles, was later subsumed within the 200-mile EEZ regime. The identification of such rights under customary international law has been criticized on the ground that state practice and/or *opinio juris* do not exist in this regard.[126] The LOSC does not articulate such rights with regard to high seas fisheries.[127]

2.2.2 *Post-UNCLOS III Developments*

The growing awareness of the need for cooperation through institutions has led to the elaboration of the duty of cooperation in high seas fisheries in a number

such measures simply because they have not been able to obtain an allocation); Division for Ocean Affairs and the Law of the Sea, *The Regime for High-Seas Fisheries*, at p. 34; Tahindro, "Conservation and Management of Transboundary Fish Stocks", at p. 25.

122 Article 119(3) of the LOSC requires states, in taking and implementing conservation measures, to ensure non-discrimination, in form or in fact, against the fishermen of any state.

123 See Kaye, *International Fisheries Management*, at pp. 153–154; Fleischer, "Fisheries and Biological Resources", at p. 1115; Oda, "Fisheries under the United Nations Convention on the Law of the Sea", at p. 752.

124 See also Nandan and Rosenne (eds.), *United Nations Convention on the Law of the Sea 1982: A Commentary*, vol. III, at p. 313; Division for Ocean Affairs and the Law of the Sea, *The Regime for High-Seas Fisheries*, at p. 34.

125 In addition to the literature cited in connection with coastal state interests, see E.J. Molenaar, "Regional Fisheries Management Organizations: Issues of Participation, Allocation and Unregulated Fishing", in A.G. Oude Elferink and D.R. Rothwell (eds.), *Oceans Management in the 21st Century: Institutional Frameworks and Responses* (2004), at p. 78.

126 Fleischer, "The New Régime of Maritime Fisheries", at p. 123.

127 In this regard, see Article 71 of the LOSC in the context of the EEZ.

of ways and has affected the exercise of the right of fishing on the high seas. As far as high seas fisheries are concerned, means to give effect to the duty of cooperation have centred on, if not being limited to, institutional mechanisms such as RFMO/As in recent international fisheries instruments.[128] Some authors argue that states are already bound under international law to cooperate through RFMO/As.[129]

In treaty law, the most notable example is the FSA. On the one hand, the FSA does not exclude direct cooperation between states, as confirmed, *inter alia*, by Article 8(1). On the other hand, the FSA provides that where RFMO/As have the competence to establish conservation and management measures, high seas fishing states and relevant coastal states shall give effect to their duty to cooperate by becoming members of such an RFMO or participants in such an arrangement or by agreeing to apply the conservation and management measures established by such an RFMO/A.[130] Read together, these articles imply two consequences. First, states are not relieved from the duty to cooperate simply by participating in the work of regional organizations with an advisory mandate. Second, the option of direct cooperation could be used only where no existing RFMO/A with the competence to establish conservation and management measures exists or where a given state is neither fishing for fish stocks regulated by the RFMO/A concerned nor the coastal state in the subregion or region. Article 10 enumerates the functions of RFMO/As through which states fulfil their duty to cooperate. While states shall agree on and comply with conservation and management measures to ensure the long-term sustainability of straddling fish stocks and highly migratory fish stocks,[131] states shall agree, *as appropriate*, on participatory rights such as allocations of allowable catch or levels of fishing effort within the RFMO/A.[132]

---

128 Note that while 'arrangement' is defined in Article 1(d) of the FSA, the latter does not define the term 'regional fishery management organization'. The term is considered to mean an organization having the competence to establish conservation and management measures. See, e.g., FAO Port State Measures Agreement, Article 1(i).

129 See Gjerde, "High Seas Fisheries Management under the Convention on the Law of the Sea", at p. 302; Rayfuse, "Countermeasures and High Seas Fisheries Enforcement", at pp. 56 and 59 (explicitly referring to DHSFS as well); Tahindro, "Conservation and Management of Transboundary Fish Stocks", at p. 27. See also H. Schiffman, *Marine Conservation Agreements: The Law and Policy of Reservations and Vetoes* (2008), at p. 19.

130 FSA, Article 8(3). Article 8(6) on prior consultation within a competent RFMO/A before the submission of any proposal to an intergovernmental organization reinforces this proposition. See also Tahindro, "Conservation and Management of Transboundary Fish Stocks", at p. 21. But note that the article uses 'should' rather than 'shall'. Thus, this is not a legal obligation but a desirable option.

131 FSA, Article 10(a).

132 Ibid., Article 10(b).

It is not clear, however, what implications the requirement of cooperation through institutional mechanisms under these provisions have for the implementation of the duty to cooperate. Article 8(3), as well as Article 8(4), is formulated with regard to 'regional fisheries management organization or arrangement', the term 'arrangement' in turn being defined as "a cooperative mechanism established in accordance with the [LOSC] and [the FSA] by two or more States for the purpose, *inter alia*, of establishing conservation and management measures in a subregion or region for one or more straddling fish stocks or highly migratory fish stocks" in Article 1(d). Such a broad definition may be interpreted so flexibly as to allow both treaties and non-binding agreements.[133] Then, it is not clear how this is different from direct cooperation between states. For example, Henriksen and others observe that while an arrangement can be distinguished from direct cooperation, it is not possible to identify precisely what constitutes an arrangement and makes it different from direct cooperation or an organization.[134]

The scope of the right and obligation with regard to cooperation was clarified by the introduction of the concept 'real interest in the fisheries concerned'.[135] The concept could be interpreted in two ways. The concept can be interpreted to mean that states are entitled to participate in RFMO/As on the ground that they have a real interest. Such states may become members of or participants in RFMO/As with the competence to establish conservation and management measures.[136] The terms of participation of such RFMO/As shall not preclude such states from membership or participation, nor shall they be applied in a manner which discriminates against any state or group of states having a real interest in the fisheries concerned.[137] First and foremost, high seas fishing states are qualified as 'having a real interest'. To what extent the state concerned needs to be engaged in the fishery or research and whether the intention to engage in the fishery or research or past fishing record qualify the state in this regard remain an open question. In addition to high seas fishing states, given the first sentence of Article 8(3), it is very likely that the 'relevant' coastal states would

133 See, e.g., Henriksen *et al., Law and Politics in Ocean Governance*, at p. 17.

134 Ibid.

135 FSA, Article 8(3).

136 Ibid.

137 Ibid. This does not by itself mean that RFMO constitutive instruments may not provide for approval systems. See Molenaar, "Regional Fisheries Management Organizations: Issues of Participation, Allocation and Unregulated Fishing", at p. 76; W. Edeson, "Soft and Hard Law Aspects of Fisheries Issues: Some Recent Global and Regional Approaches", in M.H. Nordquist, J.N. Moore and S. Mahmoudi (eds.), *The Stockholm Declaration and Law of the Marine Environment* (2003), at p. 178 note 25. Davies and Redgwell claim that the participation clauses of the Northwest Atlantic Fisheries Organization (NAFO) and the CBS Convention may require a reassessment in the light of Article 8(3). Davies and Redgwell, "The International Legal Regulation of Straddling Fish Stocks", at p. 264 note 365.

be considered to be entitled to take part in the regulatory system, whether or not they are engaged in the fishery concerned on the high seas. If so, this coastal state entitlement to take part in regulation, which is found in the HSFC but is not explicitly stipulated in the LOSC, constitutes an expression of coastal state interests under the LOSC.[138]

This concept might also be interpreted to imply that existing members of and participants in RFMO/As could decline a request by non-members/non-participants on the ground that they do not have a real interest.[139] This argument is perhaps based on an interpretation *a contrario* of the word 'may': as there is no provision entitling a state not having a real interest to become a member of or a participant in RFMO/As, such states 'may not' do so (at least under the FSA) unless otherwise agreed at the regional level.[140] In this regard, a question might be raised with regard to coastal states not engaged in high seas fisheries for the stock concerned and non-coastal states that have not engaged or do not plan to engage in the exploitation of the fishery resources concerned at all. Some commentators limit the scope of states having a real interest to high seas fishing states only[141] or to high seas fishing states and coastal states. Others, however, have argued that an interest in conservation, rather than in engaging in fisheries, is included within the scope of the term 'real interest'. These authors claim that Article 11 of the FSA, which provides for determining the nature and extent of participatory rights for new entrants, should not be interpreted to exclude a situation where no fishing opportunities are allocated to new members/participants at all and, as the use of 'such as' in Article 10(b) suggests, participatory rights in RFMOs are not limited to those benefiting from the utilization of fishery resources.[142]

---

138 The recognition of coastal state interests is, of course, not limited to this. Another form of recognition is the requirement of compatibility.

139 Orrego Vicuña, *The Changing International Law of High Seas Fisheries*, at p. 207. See also E.J. Molenaar, "The Concept of Real Interest and Other Aspects of Co-operation through Regional Fisheries Management Mechanisms", 15 *International Journal of Marine and Coastal Law* (2000), at pp. 497–501.

140 On the theory of treaty interpretation, including *expressio unius est exclusio alterius* and interpretation *a contrario*, see A. Aust, *Modern Treaty Law and Practice* (2000), at pp. 200–201; A.D. McNair, *The Law of Treaties* (1961), at pp. 399–410.

141 Orrego Vicuña, *The Changing International Law of High Seas Fisheries*, at p. 208.

142 Molenaar, "Regional Fisheries Management Organizations: Issues of Participation, Allocation and Unregulated Fishing", at p. 78; D.A. Balton, "Strengthening the Law of the Sea: The New Agreement on Straddling Fish Stocks and Highly Migratory Fish Stocks", 27 *Ocean Development and International Law* (1996), at p. 139 note 97. See also Gavouneli, *Functional Jurisdiction in the Law of the Sea*, at p. 110 (stating that 'third' states having a real interest are invited). Such an interpretation is not the only possible interpretation of Article 10(b), however. Conflicts may arise between a new member/particpant wishing to engage in a fishery and existing members/participants that do not want to allocate fishing opportunities to the new member/participant

The FSA as such does not provide a clear answer to the question of what is meant by a real interest.[143] The FSA does not define the term 'real interest', nor do related global or regional instruments.[144] The examination of the constitutive instruments and practice of RFMOs will shed some light on this point. First, except for the 1978 NAFO Convention, which requires members of the Fisheries Commission to be currently participating or expecting to participate in fisheries, RFMO/As allow participation in the organization or the decision-making body by coastal states regardless of engagement in exploitation or research. Second, several RFMOs do not require actual engagement in fisheries,[145] but the SEAFO Convention

---

where the stock concerned is fully-exploited. In other words, the main concern in formulating criteria for the determination of the nature and extent of participatory rights for new entrants might be how to deal with new entrants in situations of limited availability of resources, rather than the acceptance of the participation of conservation-minded states. For the drafting process of Article 11, see Orrego Vicuña, *The Changing International Law of High Seas Fisheries*, at pp. 210–211.

143 See Molenaar, "The Concept of Real Interest and Other Aspects of Co-operation through Regional Fisheries Management Mechanisms", at pp. 475–531; Davies and Redgwell, "The International Legal Regulation of Straddling Fish Stocks", at p. 270 ("[w]hether a State motivated by conservation concerns but not with imminent plans for commercial exploitation of stocks in the region would be entitled to participate in a regional [*sic*] RFO in consequence of this provision in the [FSA] is not clear and will depend upon the interpretation of 'real interest' "). Örebech and others argue that the FSA cannot be seen to protect the position of existing fishing states, while there was also a concern over undesirable consequences of open membership. P. Örebech *et al.*, "The 1995 United Nations Straddling and Highly Migratory Fish Stocks Agreement: Management, Enforcement and Dispute Settlement", 13 *International Journal of Marine and Coastal Law* (1998), at p. 122. They understand the concept of 'real interest' broadly, stating that the concept applies to the EU with regard to Norwegian Arctic cod of the high seas since the EU is very active in harvesting it within areas under national jurisdiction while not fishing for it on the high seas (pp. 122–123).

144 For example, the preamble to the SEAFO Convention provides that contracting parties recognize the relevant provisions of the FSA and desire "cooperation with the coastal States and with all other States and Organisations *having a real interest* in the fishery resources of the South East Atlantic Ocean to ensure compatible conservation and management measures" (emphasis added). SEAFO Convention, preambular paras. 3 and 9. In early stages of the negotiation, an unsuccessful attempt was made to agree on the definition of the concept of 'real interest' and to limit accession to those with a real interest. Molenaar, "The Concept of Real Interest and Other Aspects of Co-operation through Regional Fisheries Management Mechanisms", at p. 508 (citing Articles 4 and 25 of the Draft Convention discussed at the Third Session of the negotiation).

145 While the Convention on the Conservation of Antarctic Marine Living Resources may be ratified by any state interested in research or harvesting, membership of the Commission for the Conservation of Antarctic Marine Living Resources (CCAMLR) is only open to those engaged in research or harvesting. Convention on the Conservation of Antarctic Marine Living Resources (CCAMLR Convention), Canberra, 20 May 1980, Articles VII(2) and XXIX. SIOFA allows participation by those who are interested in fishing activities. Article 23(1). The Agreement for the

requires non-signatory non-coastal states to do so.[146] Where actual engagement in fisheries is not required for participation in decision-making, several RFMOs require approval by existing members.[147] A rather different category of provisions is found in the Framework Agreement for the Conservation of Living Marine Resources on the High Seas of the South Pacific (Galapagos Agreement).[148] The participants in the negotiations on the Agreement excluded distant water fishing nations from the negotiating process. The resultant Agreement does not allow distant water fishing states to participate in the Agreement before the Agreement enters into force by the ratification by all four coastal states. The coastal states have stronger positions than distant water fishing nations in the decision-making process within the Organization to be established under the Agreement.[149] In conclusion, while coastal states are entitled to participate regardless of the engagement of their vessels in the fishery concerned, other states without any intention to engage in fisheries or research are not automatically admitted to the RFMO/A concerned or its decision-making with regard to the adoption of conservation measures.

The provision concerning the institutionalization of cooperation is supplemented by the denial of access to fisheries resources without participating in RFMO/A management mechanisms. Only those states which are members/participants of such RFMO/As or which agree to apply the conservation and management measures of such RFMO/As shall have access to the fishery resources to which those measures apply.[150] The practice at the regional level shows a certain

Establishment of the General Fisheries Commission for the Mediterranean and the NEAFC Convention are silent on this issue.

146 SEAFO Convention, Articles 25–26. Molenaar points to the disadvantage of such applicants once SEAFO has adopted a conservation measure, stating that such applicants are "being confronted with measures against non-participants". Molenaar, "The Concept of Real Interest and Other Aspects of Co-operation through Regional Fisheries Management Mechanisms", at pp. 508–509.

147 For approval systems, see CCAMLR Convention, Article VII(2)(d); Agreement for the Establishment of the General Fisheries Commission for the Mediterranean (GFCM Agreement), Rome, 24 September 1949, amended lastly in November 1997, Article XI(2); Convention on Future Multilateral Co-operation in North-East Atlantic Fisheries, London, 18 November 1980, revised in 2006 (New NEAFC Convention), Article 20(4); NAFO Convention, Articles II(2), III(c), V(2) and XIII(1). An exception is SIOFA.

148 Framework Agreement for the Conservation of Living Marine Resources on the High Seas of the South Pacific, Santiago, Chile, 14 August 2000, not yet in force.

149 See, e.g., ibid., Articles 12(1)–(2), 16(2), 19(1) and 20.

150 FSA, Article 8(4). In the FSA, this provision is completed by Article 17(1), which requires a non-member/participant which does not agree to comply with the adopted conservation and management measures not to authorize its vessels to engage in fishing for the stock concerned. One author argues that the Review Conference on the Fish Stocks Agreement supported this proposition. Gavouneli, *Functional Jurisdiction in the Law of the Sea*, at p. 127.

support for this position. RFMO/As have attempted to control activities by non-contracting party vessels by giving incentives for cooperation (e.g., cooperating non-contracting party status) and/or by taking a confrontational approach (e.g., to draw the attention of non-contracting parties to the activities of their vessels in the regulatory area of RFMO/As and/or requesting full cooperation in the implementation of measures adopted by these RFMO/As).[151] Controversial measures such as port measures and trade-related measures on the basis of the blacklisting of IUU vessels have only been taken by some RFMOs. In national legislation and policies, some states prohibit fisheries in the regulatory area of RFMOs to which they are not parties.[152]

Some authors have questioned whether the provision restricting access to fishery resources and state practice in this regard are compatible with the LOSC. One commentator argues that an RFMO and its members have no absolute right to deny the fishing activities of other states.[153] If compatibility with the LOSC is recognized, then, do restrictions on access have a legal effect on non-parties to the FSA either as a rule of customary international law or as a subsequent agreement? Some commentators express concerns and/or doubts in this regard. A fear has been expressed that such restrictions become definitive definitions of the LOSC on this issue.[154] Their customary law status at this moment in time has been questioned by Birnie and others.[155]

The rather hesitant practice of many of the RFMOs other than NEAFC suggests that the provision on restricting the access of non-parties is only still applicable under treaty law. In this connection, it is also noted that the relevant provisions of the successive UNGA fisheries resolutions only relate to states engaged in fisheries for straddling or highly migratory fish stocks, and not for DHSFS, to give effect to the duty of cooperation through RFMO/As.

---

151 CCAMLR Convention, Articles X and XXII; SEAFO Convention, Articles 6(5) and (10) and 22(3)–(4); SIOFA, Article 17 (2) and (4); New NAFO Convention, Article XVI(1).

152 Australia, High Seas Fishery Temporary Order; New Zealand Plan of Action to Prevent, Deter and Eliminate Illegal, Unregulated and Unreported Fishing, at pp. 33–34. See also Iceland, Act No. 151/1996. See also Sustainable Fisheries, Including through the 1995 Agreement for the Implementation of the Provisions of the United Nations Convention on the Law of the Sea of 10 December 1982 relating to the Conservation and Management of Straddling Fish Stocks and Highly Migratory Fish Stocks, and Related Instruments, Report of the Secretary-General, A/62/260, 15 August 2007, at p. 26, para. 81.

153 M. Hayashi, "Regional Fisheries Management Organisations and Non-Members", in T.M. Ndiaye and R. Wolfrum (eds.), *Law of the Sea, Environmental Law and Settlement of Disputes: Liber Amicorum Judge Thomas A. Mensah* (2007), at p. 760. See also Davies and Redgwell, "The International Legal Regulation of Straddling Fish Stocks", at p. 265 ("[a] serious question remains").

154 See D. Freestone, "A Decade of the Law of the Sea Convention: Is It a Success?" 39 *George Washington International Law Review* (2007), at p. 528.

155 Birnie *et al.*, *International Law and the Environment*, at pp. 740–741.

Members of RFMOs may still opt out of conservation and management measures by the use of objection procedures while they have access to fishery resources. The FSA does not explicitly prohibit these procedures,[156] although it has been argued that they may be viewed as being contrary to the FSA and/or the LOSC.[157] While CCAMLR and SIOFA adopt measures by consensus and do not allow the use of objection procedures, the General Fisheries Commission for the Mediterranean (GFCM), SEAFO, NEAFC and NAFO allow for opting out of conservation measures.[158] The latter three RFMOs restrict the use of objection procedures procedurally by requiring a written explanation of the reasons and a statement of intentions following the use of objection procedures, including proposals for alternative measures. In the case of NAFO, the restriction would be more than mere procedural requirements: the reasons shall be either the inconsistency of the measure with the Convention or unjustifiable discrimination against the objecting state.[159] Under these three RFMOs, the use of objection procedures is subject to binding decisions by third party dispute settlement procedures. It is safe to say that the use of objection procedures is largely restrained by constitutive instruments or the practice of RFMOs, except for the GFCM.

Another area where cooperation between states needs to take place is the exchange of data with other high seas fishing states as well as coastal states. During the negotiation on the FSA, coastal states demanded the submission of data to them since the timely acquisition of data would greatly increase the capability of coastal states to monitor fishing occurring in an area adjacent to their EEZ, while distant water fishing states were opposed to this idea.[160] Article 7(1) of Annex I of the FSA stipulates the submission of data to other flag states and relevant coastal states, but makes it conditional on terms to be determined by RFMO/As. In the words of Orrego Vicuña, the availability of data for the coastal state is "thereby significantly limited and conditioned".[161]

---

156 For example, Article 17 is only applicable to non-members/participants.

157 See also Davies and Redgwell, "The International Legal Regulation of Straddling Fish Stocks", at p. 271; Örebech *et al.*, "The 1995 United Nations Straddling and Highly Migratory Fish Stocks Agreement: Management, Enforcement and Dispute Settlement", at pp. 125–126.

158 CCAMLR Convention, Articles IX(4) and (5) and XII(1); SIOFA, Article 8(1); GFCM Agreement, Article II(2); SEAFO Convention, Article 17(1) and 23(1); New NEAFC Convention, Articles 3(9), 5 and 12; Documents of the 23rd annual meeting of the North-East Atlantic Fisheries Commission, 8–12 November 2004, at pp. 28–29 (Annex K); New NAFO Convention, Articles XIII(2) and XIV(1) and (5). SEAFO allows opting out despite the fact that measures are taken only by consensus. SEAFO Convention, Articles 17(1) and 23 (1). On the relevant provision in a newly adopted convention for the North Pacific, see Chapter 5.

159 New NAFO Convention, Article XIV(5).

160 See Orrego Vicuña, *The Changing International Law of High Seas Fisheries*, at pp. 167–169; M. Hayashi, "The Role of the United Nations in Managing the World's Fisheries", in G.H. Blake *et al.* (eds.), *The Peaceful Management of Transboundary Resources* (1995), at p. 383.

161 Orrego Vicuña, *The Changing International Law of High Seas Fisheries*, at p. 169.

2.3 *Conservation*

The present section deals with the duty to conserve the marine living resources of the high seas. The first sub-section elaborates the provisions of the LOSC in this regard, highlighting the possibility of differing interpretations. It is divided into two parts because the LOSC stipulates regimes in two Parts (Part VII, section 2 and Part XII) without any cross-reference in spite of the fact that they are often related and the conservation of marine living resources are in fact concerned with both Parts.

The second sub-section examines international fisheries-related instruments adopted after 1982 as well as the subsequent practice of states and international organizations. The examination of these developments is particularly meaningful as far as the duty of conservation is concerned. Article 119(1) explicitly stipulates that the generally agreed international minimum standards shall be taken into account in adopting conservation and management measures. It has been argued that the generally recognized international minimum standards are set forth in various instruments including the FSA, the Convention on Biological Diversity (CBD), the FAO Code of Conduct for Responsible Fisheries (Code of Conduct), the Johannesburg Plan of Implementation (JPOI) and the FAO International Plans of Action (IPOAs).[162] The subsequent practice of states is also relevant to the interpretation of the LOSC provisions concerning conservation.

2.3.1 *LOSC Provisions*

2.3.1.1 *Conservation and Management of Marine Living Resources*

Under the LOSC, all states have the duty to take, or to co-operate with other states in taking, such measures for their respective nationals as may be necessary for the conservation of the living resources of the high seas.[163] It has been argued that this duty is part of customary international law.[164] Following a brief examination of the nature of the duty to take conservation measures, the remaining part of this sub-section elaborates on the content of conservation measures.

With respect to the nature of the duty to take conservation measures, it has been argued that this duty is linked to, and is part of, the overall flag state's duty

162 Gjerde, "High Seas Fisheries Management under the Convention on the Law of the Sea", at p. 304.

163 LOSC, Article 117. This formulation is almost identical to the HSFC. See HSFC, Article 1(2).

164 See, for example, Burke, *The New International Law of Fisheries*, at p. 100; Fleischer, "The New Régime of Maritime Fisheries", at p. 140. The ICJ already referred, in addition to the recognition of a duty to have due regard to the rights of other states, to the need for conservation for all in the *Fisheries Jurisdiction* cases. *I.C.J. Reports 1974*, at p. 31, para. 72; *I.C.J. Reports 1974*, at pp. 200–201, para. 64. For a contrary view, see Kindt and Wintheiser, "The Conservation and Protection of Marine Mammals", at pp. 356–357.

to exercise effective jurisdiction and control over its vessels on the high seas. Put differently, flag states are required to effectively exercise jurisdiction and control not only in respect of administrative, technical and social matters but also in respect of other matters regulated by the LOSC including the conservation and management of the living resources of the high seas.[165] The LOSC does not indicate how flag state responsibilities shall be discharged in respect of the exploitation of marine living resources. In other words, the method for carrying out this duty is left to the discretion of each flag state.

The term 'conservation and management measures' could include a wide range of acts. As the ICJ stated, "[a]ccording to international law, in order for a measure to be characterized as a "conservation and management measure", it is sufficient that its purpose is to conserve and manage living resources and that, to this end, it satisfies various technical requirements"; in other words, "[i]nternational law [...] characterizes "conservation and management measures" by reference to factual and scientific criteria".[166]

Neither Article 117 nor Article 118 of the LOSC stipulates what type of conservation measure shall be taken. In fact, no explicit definition of the concept of conservation or conservation measures is found in the LOSC; instead, Article 119 gives effect to these articles through the provision of criteria to determine conservation measures.[167] The *chapeau* of Article 119 ("[i]n determining the allowable catch and establishing other conservation measures") implies that the determination of the TAC is one of the conservation measures. The article does not, however, require states to determine the TAC as opposed to Article 61,

165 Fleischer, "Fisheries and Biological Resources", at p. 1114.

166 *Fisheries Jurisdiction Case (Spain v. Canada)*, Judgment (Jurisdiction), 4 December 1998, *I.C.J. Reports 1998*, p. 432, at p. 461, para. 70. The Court distinguished between a matter of definition and a question of legality, and it did not judge the legality of the specific act of Canada, but simply examined whether it had jurisdiction to entertain the dispute by interpreting the meaning attached to the expression 'conservation and management measures' in Canada's reservation to the declaration accepting the compulsory jurisdiction of the ICJ. *I.C.J. Reports 1998*, at pp. 460–461, paras. 68–69. In doing so, the Court interpreted the wording of the FSA and the FAO Compliance Agreement in this regard as meaning that "what is generally understood by "conservation and management measures" must comply with the obligations of international law that they have undertaken pursuant to these agreements, such as, compatibility with [MSY], concern for the needs of developing States, the duty to exchange scientific data, effective flag State control of its vessels, and the maintenance of detailed records of fishing vessels". In this context, the Court stated that the question of who may take such measures, and the areas to which they may relate, was neither treated as an element of their definition in international law generally nor in these agreements. *I.C.J. Reports 1998*, at pp. 461–462, para. 70.

167 Compare with Article 2 of the HSFC. Orrego Vicuña suggests that this absence of a definition was caused by the consideration that the complexity of the matter did not allow for overall definitions and it was better to identify the relevant criteria for specific determinations. Orrego Vicuña, *The Changing International Law of High Seas Fisheries*, at p. 77.

which obliges coastal states to determine the TAC in their EEZ,[168] while the explicit reference suggests that the determination of the TAC is in principle a desirable measure for the conservation of living resources in the high seas.[169] No other measure is, of course, excluded under the LOSC; the choice of measures to be taken is left to states.

The LOSC does not explicitly provide for the objectives of conservation measures. However, the procedural means for formulating such objectives are found in the LOSC.[170] Article 119 provides for factors which are relevant in determining conservation measures as follows: States shall

> (a) take measures which are designed, on the best scientific evidence available to the States concerned, to maintain or restore populations of harvested species at levels which can produce the maximum sustainable yield, as qualified by relevant environmental and economic factors, including the special requirements of developing States, and taking into account fishing patterns, the interdependence of stocks and any generally recommended international minimum standards, whether subregional, regional or global;
> (b) take into consideration the effects on species associated with or dependent upon harvested species with a view to maintaining or restoring populations of such associated or dependent species above levels at which their reproduction may become seriously threatened.

This provision is similar to that contained in Article 61(3) and (4) concerning conservation measures for marine living resources in the EEZ. In fact, these provisions were developed in parallel at UNCLOS III. It appears to have been the intention of the drafters of the LOSC that the same concept of conservation is applied both in EEZ fisheries and in high seas fisheries.[171]

During the negotiations at UNCLOS III, as the concept of EEZ was accepted, the main outstanding issue was the responsibility of coastal states in the conservation of fisheries resources in the EEZ. African and Latin American states generally referred to the authority of coastal states, while Western European and Socialist states favoured the involvement of international fisheries organizations.[172] A revised proposal by the United States, while recognizing the regulatory authority of coastal states, incorporated criteria and factors such as 'on the best evidence

---

168 LOSC, Article 61(1).

169 See also Orrego Vicuña, *The Changing International Law of High Seas Fisheries*, at p. 46. But see Chapters 3–5 in the context of deep-sea fisheries.

170 The distinction between 'content' and 'procedural' is borrowed from Burke, who uses this distinction in analyzing the compatibility of US legislation with the EEZ fisheries regime in the (draft) LOSC. W.T. Burke, "U.S. Fishery Management and the New Law of the Sea", 76 *American Journal of International Law* (1982), at p. 36.

171 See Second Committee, 44th meeting (1974) in *Official Records*, vol. II, at p. 298, para. 21 (US).

172 Nandan and Rosenne (eds.), *United Nations Convention on the Law of the Sea 1982: A Commentary*, vol. II, at pp. 600–601.

available to the coastal State', 'levels which can produce the maximum sustainable yield', 'taking into account relevant environmental and economic factors, and any generally agreed global and regional minimum standards', and 'take into account effects on species associated with or dependent upon harvested species'.[173] Many of the criteria and factors in this proposal, with a number of later modifications, eventually became part of Article 61(3) and (4) and, in the context of the high seas, Article 119(1). In view of the drafting history, it is reasonable to assume that the factors and criteria in Article 119(1) were particularly influenced by the concerns over the exploitation of transboundary stocks.

The conservation measures shall be aimed at, in the first place, maintaining or restoring populations of harvested species. The primary reference point to be employed in the conservation measures is levels at which the MSY can be maintained. The term 'optimum sustainable yield' – a term similar to 'MSY' – found in the HSFC did not specify how to arrive at the 'optimum' position. To the contrary, the provision in the LOSC specifies which factors qualify the MSY to achieve the goal of conservation and management measures: relevant environmental and economic factors as well as fishing patterns, the interdependence of stocks and any generally recommended international minimum standards.[174]

The way in which these factors are applied in determining the TAC is not specified in the LOSC. In other words, even if the MSY is determined on the basis of scientific data,[175] the determination of the degree of departure from the MSY inevitably calls for agreement between the states concerned and there is no criterion to govern the degree of departure, nor is a provision made concerning balancing between the factors.[176] For example, in theory, the consideration of economic factors could lead to deviations in opposite directions.[177] One might predict more abundant populations of the harvested stocks by pursuing the maximum economic

173 Article 12(2)(a) and (b) of the revised US proposal, reproduced in *Official Records*, vol. III, at p. 223. It is noted that the words 'to the coastal State' are found after the words 'on the best evidence available'.

174 In addition, Article 61 makes an explicit reference to the "economic needs of coastal fishing communities" as a factor to be taken into account in designing conservation measures in the EEZ. While there is no equivalent provision in Article 119, it might be possible to argue that this factor is also relevant in high seas fisheries as a part of relevant 'economic factors' in Article 119.

175 According to Tahindro, "most [fisheries bodies] are hampered by issues such as disagreement over scientific advice concerning management". Tahindro, "Conservation and Management of Transboundary Fish Stocks", at p. 27.

176 It is not clear to what extent states can derogate from the MSY by taking action in favour of the conservation of fisheries resources. There is no requirement of optimum utilization with regard to high seas fisheries, but as the conservation status of a given fish stock could affect other stocks in the same marine ecosystem, stricter measures might result in putting the interests of other states in high seas fisheries in jeopardy under some circumstances.

177 For the meaning of relevant environmental factors, see Section 2.3.1.2 below.

yield.[178] On the other hand, one might also depart from the MSY to increase short-term socio-economic benefits sacrificing long-term sustainability. Commentators differ in their views on this point: for example, Burke argues that both directions could be pursued,[179] while Kaye argues that "no catch limit beyond the MSY would be consistent with the [LOSC]".[180]

States shall also take into consideration the effects on associated or dependent species. Since the effects only need to be taken into consideration in determining conservation measures for harvested stocks, Article 119 as such does not explicitly require states either to determine the TAC or to take other conservation measures for these species unless they are also targeted.[181] In taking into consideration the effects, the desirable goal is to maintain or restore populations of those species above levels at which their reproduction may become seriously threatened.[182] This provision effectively broadens the duty to take conservation measures by encompassing the conservation of these species,[183] and there is, arguably, a duty to ensure the sustainability of those species.[184] In addition, the need to consider the effects on these species presupposes the duty to seek the relevant information in advance.[185]

The measures shall be designed on the best scientific evidence available to the states concerned. On the one hand, the wording 'designed, on the best scientific evidence available' in Article 119 appears to leave less room for discretion than the wording 'taking into account' (which seems to allow a deviation) in Article 61(2).[186] Furthermore, the phrases 'taking into account the best scientific

---

178 See Koers, *International Regulation of Marine Fisheries*, at pp. 54–63; Kaye, *International Fisheries Management*, at p. 100. However, while the overall cost of the fishery concerned will decrease, this is not necessarily the case for each and every fishing state. Burke, *The New International Law of Fisheries*, at p. 112.

179 Burke, "U.S. Fishery Management and the New Law of the Sea", at p. 31. See also T.A. Clingan, Jr, "An Overview of Second Committee Negotiations in the Law of the Sea Conference", 63 *Oregon Law Review* (1984), at p. 57; Fleischer, "Fisheries and Biological Resources", at p. 992.

180 Kaye, *International Fisheries Management*, at p. 100. See also Tahindro, "Conservation and Management of Transboundary Fish Stocks", at p. 9.

181 Despite this, it could be argued that other articles, such as Articles 192 and 194(5), oblige states to take measures in any case. For these articles, see further below.

182 It seems that the criteria are much more lenient than those for harvested stocks. See also Kindt and Wintheiser, "The Conservation and Protection of Marine Mammals", at p. 361 ("[t]his requirement could be interpreted to justify a stock size below even MSY levels"). See also Orrego Vicuña, *The Changing International Law of High Seas Fisheries*, at p. 47.

183 Nandan and Rosenne (eds.), *United Nations Convention on the Law of the Sea 1982: A Commentary*, vol. III, at p. 311. See also Freestone, "International Fisheries Law Since Rio", at p. 147.

184 Division for Ocean Affairs and the Law of the Sea, *The Regime for High-Seas Fisheries*, at p. 9.

185 Burke, *The New International Law of Fisheries*, at p. 113; Nandan and Rosenne (eds.), *United Nations Convention on the Law of the Sea 1982: A Commentary*, vol. III, at p. 311.

186 Burke, *The New International Law of Fisheries*, at p. 115 note 87. Burke argues that even the phrase 'taking into account' in Article 61 is not intended to leave the coastal state free to refer

evidence available' and 'designed' are separated in two paragraphs in Article 61.[187] On the contrary, the phrase 'designed, on the best scientific evidence available' in Article 119 appears to link the designation of measures and the scientific basis more closely than Article 61. On the other hand, it has been argued that measures need not necessarily be based solely on scientific evidence: other factors may be taken into consideration.[188] With regard to the word 'designed', although it is not defined in the LOSC, all the provisions using the word 'designed' relate to situations where the action concerned is aimed at a particular vague objective and where it is difficult to evaluate whether the objective has been achieved by that action.[189] The wording in Article 119 certainly indicates the importance of a scientific basis in the determination of conservation measures for high seas fisheries. The phrase used in Article 119 requires the states concerned, at the very least, to use, in good faith, the best scientific evidence available in designing conservation measures.[190]

The states concerned are not required to seek the best scientific evidence at a given time, but are required to use the best scientific evidence *available to them*. In the negotiation at UNCLOS III, a suggestion was made by the EEC delegation to include "available to it [i.e., the coastal state]" after the reference to "best scientific evidence" in Article 61(2).[191] By virtue of these qualifying words, it became clear that coastal states would not be under an obligation to actively conduct scientific investigations in order to obtain the best scientific evidence, but could take conservation measures on the basis of scientific information to which they

---

to evidence of lesser quality and the phrase appears to be a diplomatic euphemism for 'using'. Burke, "U.S. Fishery Management and the New Law of the Sea", at p. 34. For the importance of slight differences in modifiers in the LOSC in general, see P. Allott, "Power Sharing in the Law of the Sea", 77 *American Journal of International Law* (1983), at p. 11.

187 Paragraph 3 reads in part: "Such measures shall *also* be designed to maintain or restore populations of harvested species at levels which can produce [MSY]" (emphasis added).

188 See, e.g., Kaye, *International Fisheries Management*, at pp. 103 and 150; Nandan and Rosenne (eds.), *United Nations Convention on the Law of the Sea 1982: A Commentary*, vol. III, at p. 310; Freestone, "International Fisheries Law Since Rio", at p. 147. See also Section 3.2 below on 'technical' information.

189 The word 'designed' is used on seven occasions in the LOSC, including twice for living resources as described above. See Articles 60(5), 155(2), 194(3), 207(5) and 211(1).

190 Some authors infer rather stronger implications from the 'best scientific evidence' standard. For example, Schiffman argues that "[s]ince the great majority of conservation and management measures are informed by scientific input, vetoes that block them, and specific reservations that allow individual states to depart from them, would appear to undermine the 'best scientific evidence' requirement". Schiffman, *Marine Conservation Agreements: The Law and Policy of Reservations and Vetoes*, at p. 189.

191 A.H.A. Soons, "Regulation of Marine Scientific Research by the European Community and Its Member States", 23 *Ocean Development and International Law* (1992), at pp. 259–260.

have access at a given moment.[192] Moreover, this qualification leaves room for the varying capabilities of the states concerned.[193] It seems that this analysis could also be applied to high seas fisheries.

The need for a scientific basis for conservation measures poses a dilemma in the conservation of marine living resources. On the one hand, it is difficult to require 'full' understanding, and, moreover, such a requirement risks delaying any conservation measures to be taken. Article 119 does not suggest that no measures should be taken until the adequate information is accessible, while it is safe to say that states cannot take measures while disregarding the existing scientific evidence. On the other hand, if the wording 'evidence' is distinguished from 'information' by the criterion of credibility, one might argue that the phrase may imply that no action should be taken, unless scientific information is reliable enough to be 'evidence', because action based on poor data can also have harmful consequences.[194]

The provisions of Part VII, section 2 fall short of explicitly demanding the conducting of scientific research or the collection of scientific information.[195] Nevertheless, the requirement of the best scientific evidence available appears to assume that states have made some efforts to obtain scientific information in the first place since the availability of data and other scientific information is essential for accurate stock assessment and for the design of sound conservation measures.[196]

2.3.1.2 *Protection and Preservation of the Marine Environment*

The importance of the protection of the marine environment in the LOSC has been stressed by a number of authors in the context of fisheries.[197] Environmental

---

192 This should not be interpreted to imply that states are not required to collect scientific information at all. See the subsequent paragraphs of this section.

193 Soons, "Regulation of Marine Scientific Research by the European Community and Its Member States", at p. 260.

194 Burke, "U.S. Fishery Management and the New Law of the Sea", at p. 34. Burke himself is not convinced that the slight difference between 'information' and 'evidence' can be given such significance that this alone warrants a refusal to take action (p. 35).

195 See also Kaye, *International Fisheries Management*, at p. 150. But see Burke, *The New International Law of Fisheries*, at pp. 92 and 125–127. As opposed to Article 119, Article 206 explicitly requires states to conduct impact assessments when "States have reasonable grounds for believing that planned activities under their jurisdiction or control may cause substantial pollution of or significant and harmful changes to the marine environment". See also Buck, CRS Report for Congress (January 7, 2004), at p. 8.

196 See Burke, *The New International Law of Fisheries*, at pp. 92, and 125–127. For the importance of data, see S.M. Garcia and J. Majkowski, "State of High Seas Resources", in T. Kuribayashi and E.L. Miles (eds.), *The Law of the Sea in the 1990s: A Framework for Further International Cooperation* (1992), at p. 218.

197 See, e.g., Freestone, "International Fisheries Law Since Rio", at p. 149; E. Hey, "The Provisions of the United Nations Law of the Sea Convention on Fisheries Resources and Current International

protection has become integrated into the general principles of fisheries management. One commentator aptly notes that "a balance must be sought between the objectives of the provisions on the protection and preservation of the marine environment and the objectives of the fisheries provisions of [the LOSC], i.e. the sustainable exploitation of the stocks must be ensured".[198] The following paragraphs examine the relevance of environmental considerations to the conservation of the marine living resources of the high seas under the LOSC in the light of the United Nations Conference on the Human Environment (the Stockholm Conference), which was held just before UNLOS III started.[199]

The development of international environmental law led to the inclusion of various categories of activities under the term 'environment'. The conservation of marine living resources was perceived as part of environmental problems in the Stockholm Declaration. The second sentence of Principle 4 of the Stockholm Declaration posits this by stating that "[n]ature conservation, including wildlife, must [...] receive importance in planning for economic development". In addition, Principle 3 stipulates that "[t]he capacity of the earth to produce vital renewable resources must be maintained and, wherever practicable, restored or improved".

The LOSC confirms this wide notion of environmental problems. The term 'environment' is not defined in the LOSC: during the negotiations at UNCLOS III, proposals were made to define the term 'environment', but none of them was adopted.[200] Nevertheless, agreement existed at UNCLOS III that the term appears to include 'marine life' as stated by the Chairman of the Third Committee.[201] Besides, Article 194(5) explicitly refers to measures which are necessary to protect

Fisheries Management Needs", *The Regulation of Driftnet Fishing on the High Seas: Legal Issues (FAO Legislative Study No. 47)* (1991), at p. 9; Orrego Vicuña, *The Changing International Law of High Seas Fisheries*, at p. 25 (the full meaning of the LOSC comes to light when the provisions relating to the protection of the marine environment are also taken into account).

198 Hey, "The Provisions of the United Nations Law of the Sea Convention on Fisheries Resources and Current International Fisheries Management Needs", at p. 9.

199 Held in Stockholm, 5–16 June 1972. For a comprehensive analysis of the Conference and the Stockholm Declaration on the Human Environment, see L.B. Sohn, "The Stockholm Declaration on the Human Environment", 14 *Harvard International Law Journal* (1973), at pp. 423–515.

200 S. Rosenne and A. Yankov (eds.), *United Nations Convention on the Law of the Sea 1982: A Commentary*, vol. IV (1991), at p. 42. The term 'marine environment' is defined in the Regulations on Prospecting and Exploration for Polymetallic Nodules in the Area, adopted at the sixth session of the International Seabed Authority (20–31 March and 3–14 July 2000), Regulation 1(3)(c). The Regulations define the term in a non-exhaustive way as "the physical, chemical, geological and biological components, conditions and factors which interact and determine the productivity, state, condition and quality of the marine ecosystem, the waters of the seas and oceans and the airspace above those waters, as well as the seabed and ocean floor and subsoil thereof".

201 *Official Records*, vol. X, at p. 97.

and preserve certain categories of marine life. Therefore, the general duty of marine environmental protection under Article 192 extends to the conservation of marine living resources from fisheries as a part of environmental protection. The duty in Article 192 should not be considered as an absolute prohibition of any impact on the environment.[202] In the context of fisheries, therefore, measures taken to this end are not necessarily a total ban on the fishing activities concerned.

Subsequent articles give effect to the general duty stipulated in Article 192. Article 194 provides for the duty to take measures to prevent, reduce and control pollution of the marine environment. Some commentators are sceptical about the applicability of Article 194(5) to fisheries as such,[203] while others appear to see no problem in this regard.[204] The LOSC defines pollution as follows:

> "pollution of the marine environment" means the introduction by man, directly or indirectly, of substances or energy into the marine environment, including estuaries, which results or is likely to result in such deleterious effects as harm to living resources and marine life, hazards to human health, hindrance to marine activities, including fishing and other legitimate uses of the sea, impairment of quality for use of sea water and reduction of amenities.[205]

It appears to be incompatible with the ordinary meaning of the term to assume that the removal of fish as such inherently falls under the term 'pollution' as defined in this way. However, energy could be considered to include 'physical disturbance such as anchoring or grounding'.[206] Defined in this way, fishing could be considered to generate pollution not only in situations such as an oil spill, lost or abandoned gear and discards but also through the impacts of

202 See, e.g., Rosenne and Yankov (eds.), *United Nations Convention on the Law of the Sea 1982: A Commentary*, vol. IV, at p. 36.

203 See J. Morishita, "What is the Ecosystem Approach for Fisheries Management", 32 *Marine Policy* (2008), at p. 20; Mossop, "Protecting Marine Biodiversity on the Continental Shelf Beyond 200 Nautical Miles", at p. 289; G.S. Stone *et al.*, "Seamount Biodiversity, Exploitation and Conservation", in L.K. Glover and S.A. Earle (eds.), *Defying Ocean's End: An Agenda for Action* (2004), at p. 65; L. Glowka, "Putting Marine Scientific Research on a Sustainable Footing at Hydrothermal Vents", 27 *Marine Policy* (2003), at p. 305, note 5.

204 See E. Hey, "Global Fisheries Regulations in the First Half of the 1990s", 11 *International Journal of Marine and Coastal Law* (1996), at pp. 484–485. See also Owen, "The Application of the Wild Birds Directive Beyond the Territorial Sea", at p. 61; Owen, *The Powers of the OSPAR Commission and Coastal State Parties to the OSPAR Convention to Manage Marine Protected Areas on the Seabed Beyond 200 nm From the Baseline: A Report for WWF Germany*, at p. 13.

205 LOSC, Article 1(1)(4).

206 See E.J. Molenaar, *Coastal State Jurisdiction over Vessel-Source Pollution* (1998), at p. 17. But, see, *contra*, K.M. Gjerde and D. Ong, "Protection of Particularly Sensitive Sea Areas under International Marine Environmental Law: Report of the International Meeting of Legal Experts on Particularly Sensitive Sea Areas University of Hull, 20–21 July 1992", 26 *Marine Pollution Bulletin* (1993), at p. 11.

bottom trawling on the benthic environment. On that account, states are obliged to take measures to prevent such pollution from occurring during fishing activities under this provision.

International environmental law has refined the concept of the conservation of marine living resources. Principle 2 of the Stockholm Declaration states:

> [t]he natural resources of the earth, including the air, water, land, flora and fauna and especially representative samples of natural ecosystems, must be safeguarded for the benefit of present and future generations through careful planning or management, as appropriate.

Two points of importance follow from this pronouncement. First, the purpose of management from this perspective is to ensure the long-term sustainability of resources as such.[207] While the phrase 'sustainability of resources' does not appear in the LOSC, it is possible to argue that the relevant provisions of the LOSC shall be dictated by the need to ensure the sustainability of marine living resources as well as the sustainability of their yields.[208] Second, the Principle stipulates that the natural resources listed in that Principle are to be safeguarded for the benefit of mankind.[209]

Besides, international environmental law broadened the scope of factors to be considered in the concept 'conservation'. The explicit reference in Principle 2 of the Stockholm Declaration to 'representative samples of natural ecosystems'[210] implies a new kind of consideration in fisheries management. Principle 4 states, in part, that "[m]an has a special responsibility to safeguard and wisely manage the heritage of wildlife *and its habitat,* which are now gravely imperilled by a combination of adverse factors" (emphasis added). The Conference, *inter alia,* referred to the need for information on living marine resources and their environment and fisheries activities as well as the evaluation and monitoring of

---

207 The concept has been weakened in that the original "idea that states hold their resources 'in trust' for present and future generations has been replaced by the vaguer notion of an unspecified somebody safeguarding the resources for the 'benefit' of these generations" (footnote omitted). Sohn, "The Stockholm Declaration on the Human Environment", at p. 457.

208 See also G.J. Schram and A. Tahindro, "Developments in Principles for the Adoption of Fisheries Conservation and Management Measures", in E. Hey (ed.), *Developments in International Fisheries Law* (1999), at p. 265.

209 Sohn, "The Stockholm Declaration on the Human Environment", at p. 457. Interestingly, Sohn considers that the Conference followed UNGA Resolution 2749 (XXV) on the Declaration of Principles Governing the Sea-bed and the Ocean Floor, and the Subsoil Thereof, Beyond the Limits of National Jurisdiction in stating that the natural resources listed in the Stockholm Declaration are to be safeguarded for the benefit of mankind.

210 Being not satisfied with the word 'representative' inserted upon a suggestion by Brazil, the United States made an interpretative statement that "the phrase means retention of a complete system with all of the complex interrelationships intact, not a portion thereof". Ibid., at p. 457 note 110.

environmental conditions.[211] The protection of habitats and taking into account other environmental conditions therefore emerged as part of the principles of fisheries management.

It is submitted that relevant environmental factors in Article 119(1) of the LOSC should be construed so as to include the protection of ecosystems and habitats, at least vulnerable ones.[212] This interpretation is supported by Article 194(5) which, in the context of the protection and preservation of the marine environment, specifically mentions marine living resources, their habitats and ecosystems as a whole ("rare or fragile ecosystems as well as the habitat of depleted, threatened or endangered species and other forms of marine life").[213] While the explicit introduction of such concepts as ecosystems and habitats into the LOSC is significant, the phrases such as 'rare or fragile' ecosystems and 'depleted, threatened or endangered' species and other forms of marine life[214] appear to set a rather high threshold for protection.[215]

2.3.2 *Post-UNCLOS III Developments*

The duty to conserve the living resources of the high seas has been developed through the refinement of the concept of 'conservation' in accordance with the parallel developments in international environmental law.[216] In this

---

211 Action Plan for the Human Environment, adopted at the United Nations Conference on Human Environment, Recommendations for action at the international level, Recommendation 46(b)–(c). See also Recommendations 32 and 46(e) on the protection of species inhabiting international waters or those which migrate from one country to another and on management of stocks of fish and other aquatic animals, respectively.

212 See Birnie *et al.*, *International Law and the Environment*, at p. 6.

213 The US proposal adding paragraph 5 to the then Article 195 (now Article 194(5)) appears to be aimed at extending the *object* of protection rather than activities subject to regulation. Oxman writes that "these amendments [i.e., the addition of paragraph 5 and the proposal to add in 'Use of terms' a sentence reading "'Marine environment' includes marine life"] expand and clarify the rights and obligations to take measures to minimize ship accidents and damage resulting from such accidents". See B.H. Oxman, "The Third United Nations Conference on the Law of the Sea: The Seventh Session (1978)", 73 *American Journal of International Law* (1979), at pp. 24–25.

214 The words 'marine life' are qualified by 'depleted, threatened or endangered'. This point is clear in the French text reading "l'habitat des espèces et autres organismes marins en régression, menacés ou en voie d'extinction".

215 Hey, "Global Fisheries Regulations in the First Half of the 1990s", at p. 485.

216 Another set of development is the strengthening of flag state responsibilities, notably through the FAO Compliance Agreement. Agreement to Promote Compliance with International Conservation and Management Measures by Fishing Vessels on the High Seas (FAO Compliance Agreement), Rome, approved by the FAO Conference on 24 November 1993. On this Agreement, see G. Moore, "The Food and Agriculture Organisation Compliance Agreement", 10 *International Journal of Marine and Coastal Law* (1995), at pp. 412–425; D.A. Balton, "The Compliance

respect, four interrelated elements are worth noting: sustainable fisheries, ecosystem approaches, the conservation of marine biodiversity and the precautionary approach.[217] It could be argued that the elaboration of the concept of conservation modified standards which were applicable to giving effect to the duty to take necessary conservation measures. In the following paragraphs these topics are considered in turn.

2.3.2.1 *Sustainable Fisheries*

The concept of sustainable development has acquired a prevalent position in the field of environmental protection and development. This sub-section first articulates the concept of sustainable development in international law and then examines its application to high seas fisheries.

On the international stage, the concept of sustainable development was first introduced in the seminal report by the World Commission on Environment and Development (the 'Brundtland Commission' or 'WCED') entitled "Our Common Future".[218] Drawing upon the report, the Rio Declaration provides that "environmental protection shall constitute an integral part of the development process and cannot be considered in isolation from it",[219] and Agenda 21 stipulates various ways to implement this concept. The concept became the main topic of the World Summit on Sustainable Development (WSSD) held in Johannesburg in 2002.

Various aspects of the concept have been analyzed in detail in a number of international instruments and in the literature, including the ILA New Delhi Declaration of Principles of International Law relating to Sustainable Development.[220]

Agreement", in E. Hey (ed.), *Developments in International Fisheries Law* (1999), at pp. 31–53; G. Moore, "The FAO Compliance Agreement", in M.H. Nordquist and J.N. Moore (eds.), *Current Fisheries Issues and the Food and Agriculture Organization of the United Nations* (2000), at pp. 77–91.

217 As far as the term is concerned, 'precautionary principle' and 'precautionary approach' are both in use. This study uses 'approach' when it refers to the concept in the context of fisheries. On the implications of distinguishing between these two terms, see Birnie *et al.*, *International Law and the Environment*, at p. 155. Note that some authors insist that the precautionary approach employed in fisheries is different from the precautionary principle in other areas in its normative content. See, e.g., Orrego Vicuña, *The Changing International Law of High Seas Fisheries*, at p. 157 *et seq.*

218 World Commission on Environment and Development, *Our Common Future* (1987).

219 Rio Declaration, Principle 4.

220 ILA New Delhi Declaration of Principles of International Law relating to Sustainable Development, New Delhi, 6 April 2002. For the literature, see, e.g., articles in A. Boyle and D. Freestone (eds.), *International Law and Sustainable Development: Past Achievements and Future Challenges* (1999).

Although commentators vary in their understanding of the concept,[221] the following principles that were articulated in the above-mentioned ILA Declaration constitute a useful list of principles which are relevant to high seas fisheries: the integration of social, economic and environmental objectives, the duty of states to ensure the sustainable use of natural resources, equity (intergenerational equity and intragenerational equity), common but differentiated responsibilities and the precautionary principle.[222] Most of these principles are analyzed below while related concepts such as the precautionary approach and ecosystem approaches are examined in detail in a separate sub-section.

The objective of sustainable development involves a comprehensive and integrated approach to economic, social and political processes.[223] In the words of the ILA, states "should strive to resolve apparent conflicts between competing economic, financial, social and environmental considerations".[224]

The duty of states to ensure the sustainable use of natural resources derives from this overall objective. States are under a duty to manage natural resources "in a rational, sustainable and safe way so as to contribute to the development of their peoples [...] and to the conservation and sustainable use of natural resources and the protection of the environment, including ecosystems".[225]

Intergenerational equity requires each generation to use and develop its natural and cultural heritage in such a manner that it can be passed on to future generations in no worse condition than it was received.[226] The definition of the concept of sustainable development by the Brundtland Commission emphasizes

---

221 Compare, e.g., Birnie *et al., International Law and the Environment*, at pp. 116–123; P. Sands, *Principles of International Environmental Law*, 2nd edition (2003), at pp. 253–266.

222 Note also that the Institut de Droit International referred to sustainable development in parallel with the concepts of intergenerational equity, the precautionary approach, environmental security and the principle of shared but differentiated responsibility. Resolution on Responsibility and Liability under International Law for Environmental Damage, adopted at the session of Strasbourg in 1997, preambular para. 8.

223 ILA New Delhi Declaration of Principles of International Law relating to Sustainable Development, preambular para. 13. This process includes the reconciliation between the economic development and environmental protection, as confirmed by the ICJ in the *Case Concerning Gabcikovo-Nagymaros Project* and by the arbitral tribunal in the *Iron Rhine Railway* case. See *I.C.J. Reports 1997*, at p. 78, para. 140; *Arbitration Regarding the Iron Rhine ("Ijzeren Rijn") Railway between the Kingdom of Belgium and the Kingdom of the Netherlands*, Award of the Arbitral Tribunal, 24 May 2005, at pp. 28–29, para. 59.

224 ILA New Delhi Declaration of Principles of International Law relating to Sustainable Development, para. 7.2.

225 Ibid., para. 1.2.

226 Birnie *et al., International Law and the Environment*, at p. 119. See also A. Kiss, "The Rights and Interests of Future Generations and the Precautionary Principle", in D. Freestone and E. Hey (eds.), *The Precautionary Principle and International Law: The Challenge of Implementation* (1996), at pp. 19–28.

the centrality of the concept of intergenerational equity.[227] The main implication of the concept is to prevent irreversible harm such as the depletion of a species of natural resources (e.g., fish and the components of their ecosystem) so as to leave an option for future generations. The concept may not be interpreted to require a blanket ban on any activity unless it is considered necessary to preserve the option concerned.

The concept of intragenerational equity implies "the right of all peoples within the current generation of fair access to the current generation's entitlement to the Earth's natural resources".[228] Furthermore, it should be noted that the concept 'food security' became a central concept of the work of the FAO.[229] Marine-captured fish provide an important source of nutrition for human beings and may contribute to the improvement of the nutritional situation of people in developing countries. Fisheries are now perceived as a tool to achieve 'food security'.

In addition to the principle of equity, the principle of common but differentiated responsibilities works in two different ways. First, the differentiation of responsibilities must take into account the economic and developmental situation of developing states, and the special needs and interests of developing countries should be recognized.[230] Second, developed countries bear a special burden of responsibility in reducing and eliminating unsustainable patterns of production and consumption and in contributing to capacity-building in developing countries, including financial assistance and the transfer of environmentally sound technology.[231] Developed states are thus required both to strive towards achieving sustainable development in relation to their nationals and to assist developing states in achieving the same objective.

The concept of sustainable development shall be applied to high seas fisheries.[232] Fisheries are one of the areas which may contribute to achieving the overall objective of sustainable development through, among other things, food security.[233] In this sense, fisheries are integral to the concept

227 *Our Common Future*, at pp. 8–9 and 43–66.

228 ILA New Delhi Declaration of Principles of International Law relating to Sustainable Development, para. 2.1.

229 See, e.g., Kyoto Declaration, adopted at the International Conference on the Sustainable Contribution of Fisheries to Food Security, 4–9 December 1995, Kyoto, Japan, para. 1.

230 ILA New Delhi Declaration of Principles of International Law relating to Sustainable Development, paras. 3.2–3.3.

231 Ibid., para. 3.4.

232 Orrego Vicuña, *The Changing International Law of High Seas Fisheries*, at pp. 145–153. See also Hey, "The Provisions of the United Nations Law of the Sea Convention on Fisheries Resources and Current International Fisheries Management Needs", at p. 9.

233 See A/60/63, at pp. 47–62.

of sustainable development as one of the development activities in order to achieve sustainable development.

The concept of sustainable development has two implications for fisheries. First, fisheries as such need to be 'sustainable'. In other words, fisheries may not proceed without incorporating environmental concerns. Thus, the elaboration of the meaning of the sustainable development of living marine resources as well as the formulation of the means to achieve that objective is required. The FAO, at the 94th session of the Council in 1988, adopted a definition of sustainable development for aquatic and terrestrial systems as follows:

> [s]ustainable development is the management and conservation of the natural resource base, and the orientation of technological and institutional change in such a manner as to ensure the attainment and continued satisfaction of human needs for present and future generations. Such development conserves land, water, plant and genetic resources, is environmentally non-degrading, technologically appropriate, economically viable and socially acceptable.[234]

It is clear that the definition by the FAO emphasizes the importance of satisfying human needs. The definition also takes into consideration the prevention of environmental degradation as an element of sustainable development. Second, sustainability relates not only to the resources and the environment but also to the fishery that exploits living marine resources.[235] Fisheries are sensitive to the state of the resources to be exploited;[236] adverse impacts on marine ecosystems caused by a fishery may render the fishery itself more difficult or less profitable. Thus, the sustainable development of living marine resources shall aim at two interrelated objectives of the sustainability of resources and that of fisheries themselves, the former being the necessary requisite for the latter. The concept of sustainable fisheries embracing both objectives came to be used in the UN Secretary-General's reports on fisheries and FAO documents as a goal to be achieved.[237] In these documents, the concept of sustainable fisheries is framed as the integration of the need for exploitation and for conservation.

---

234 Cited in J.F. Caddy and R.C. Griffiths, Living Marine Resources and Their Sustainable Development: Some Environmental and Institutional Perspectives, Fisheries Technical Paper No. 353.

235 Ibid.

236 S.M. Garcia, The Precautionary Approach to Fisheries and its Implications for Fishery Research, Technology and Management: An Updated Review", in Precautionary Approach to Fisheries, FAO Fisheries Technical Paper No. 350–2.

237 See, e.g., Sustainable Fisheries, Including through the Agreement for the Implementation of the Provisions of the United Nations Convention on the Law of the Sea of 10 December 1982 relating to the Conservation and Management of Straddling Fish Stocks and Highly Migratory Fish Stocks, and Related Instruments, Report of the Secretary-General, A/59/298, 26 August 2004, *passim*.

A number of post-UNCLOS III global fisheries instruments aim to achieve 'sustainable use' and 'conservation'.[238] Furthermore, in view of the increasing introduction of ecosystem considerations in fisheries management, as seen below, one might discern that the objective of obtaining a maximum yield is paralleled by the goal of conducting fisheries without causing harm to ecosystems. It has even been argued that the optimum utilization concept is not obligatory in high seas fisheries and the LOSC itself simply requires states engaged in high seas fishing to ensure that stocks are maintained or restored to levels which have the *potential* to produce the MSY.[239] Despite the advancement of the conservation concerns, however, the utilization aspect is still part of the fisheries management objective as found in the FSA, the Code of Conduct, the JPOI and UNGA fisheries resolutions.[240]

Reflecting changes in fisheries management objectives, the criteria for establishing reference points for target species are also changing. With the understanding of the uncertainty in fisheries research, fisheries management has adopted new reference points for target species, which are precautionary in nature, for the purpose of avoiding a failure in fisheries management. The FSA obliges states to apply the precautionary approach, and stipulates that in implementing the precautionary approach, states shall, *inter alia*, determine the stock-specific reference points.[241] The Code of Conduct also provides that states and RFMOs and arrangements should determine stock-specific target reference points and limit reference points and the actions to be taken if these reference points are exceeded.[242] Target reference points are the desirable state of the stock. Limit reference points set safe biological limits for harvesting beyond which the stock

---

[238] See F. Orrego Vicuña, "The Law Governing High Seas Fisheries: In Search of New Principles", 18 *Ocean Yearbook* (2004), at p. 392; E. Hey, "Reviewing Implementation of the LOS Convention and Emerging International *Public* Law", in A.G. Oude Elferink (ed.), *Stability and Change in the Law of the Sea: The Role of the LOS Convention* (2005), at p. 79 note 24; Gavouneli, *Functional Jurisdiction in the Law of the Sea*, at p. 121. It could even be argued that it may be inferred from the provisions of Articles 63(2), 64 and 117 that the obligation to conserve high seas living marine resources is to be dictated by the need to ensure the long-term sustainability of those resources. Tahindro, "Conservation and Management of Transboundary Fish Stocks", at p. 9.

[239] Freestone, "International Fisheries Law Since Rio", at p. 147.

[240] FSA, Articles 2 and 5(a)–(b); Code of Conduct for Responsible Fisheries Rome, adopted by the FAO Conference on 31 October 1995, Articles 7.1.1 and 7.2.1; Johannesburg Plan of Implementation, para. 31(a), UNGA Resolution 66/68, para. 1.

[241] FSA, Article 6(3)(b). See also FSA, Annex II. However, note that the FSA still maintains the MSY as one of its reference points. FSA, Article 5(b).

[242] Code of Conduct, Article 7.5.3. Threshold reference points may also be used to provide an early warning to reduce the risk of reference points being exceeded. The threshold reference point indicates that the state of a fishery and/or a resource is approaching a target or limit reference point. Garcia, The Precautionary Approach to Fisheries and its Implications for Fishery Research, Technology and Management, at section 2.

cannot produce MSY.[243] When the target reference point is exceeded and the limit reference point is approached, measures should be taken to ensure that it will not be exceeded.[244] Therefore, it is observed that precautionary reference points, particularly target reference points, are increasingly used as the principal reference point of fisheries management.

2.3.2.2 *Ecosystem Approaches*

Fishing entails a risk of causing several kinds of impacts on marine ecosystems.[245] For example, the UN Secretary-General identifies four kinds of ecosystem impacts that may be caused by fisheries: a reduction of target biomass in the ecosystem, impacts on non-target fisheries and bycatch taken in fisheries operations, impacts on the habitat as a result of the use of destructive fishing gear, and indirect impacts on other species through food-chain effects.[246] It is widely recognized that "the conventional approaches to fisheries management, focusing more or less exclusively on the target species and the objective of sustainable yields, has been inadequate for conservation and sustainable use of ecosystems as a whole".[247]

A number of ecosystem considerations have already been incorporated in treaties and other international instruments. Both the LOSC and the HSFC require states to take into account impacts on associated or dependent species. A number of post-UNCLOS III instruments significantly extended factors reflecting ecosystem considerations and emphasized the need to incorporate

243 See Garcia, The Precautionary Approach to Fisheries and its Implications for Fishery Research, Technology and Management, at section 2. See also FSA, Annex II, para. 2.

244 See Code of Conduct, Article 7.5.3(b).

245 The term 'ecosystem' refers to a functional complex of living and non-living resources. For example, the CBD defines 'ecosystem' as "a dynamic complex of plant, animal and micro-organism communities and their non-living environment interacting as a functional unit". Convention on Biological Diversity (CBD), Rio de Janeiro, 5 June 1992, Article 2. See also the definition of 'the Antarctic marine ecosystem' in the CCAMLR Convention. CCAMLR Convention, Article I(3), which defines it as "the complex of relationships of Antarctic marine living resources with each other and with their physical environment".

246 A/59/298, at pp. 22–26. Similarly, another report by the Secretary-General describes three types of impacts on VMEs: impacts on target species, impacts on non-target and associated species and impacts on benthic ecosystems. Impacts of Fishing on Vulnerable Marine Ecosystems: Actions Taken by States and Regional Fisheries Management Organizations and Arrangements to Give Effect to Paragraphs 66 to 69 of General Assembly Resolution 59/25 on Sustainable Fisheries, Regarding the Impacts of Fishing on Vulnerable Marine Ecosystems, Report of the Secretary-General, A/61/154, 14 July 2006, at pp. 13–17.

247 Implementing the Ecosystem Approach to Fisheries, Including Deep-sea Fisheries, Biodiversity Conservation, Marine Debris and Lost or Abandoned Fishing Gear, COFI/2007/8, December 2006, at p. 8.

such ecosystem considerations in fisheries management.[248] Notably, the Reykjavik Declaration on Responsible Fisheries in the Marine Ecosystem (the Reykjavik Declaration) brought ecosystem considerations to the forefront of fisheries management.[249] In the Declaration, states declare that they "will individually and collectively work on incorporating ecosystem considerations into that management".[250] Moreover, it emphasizes the importance of incorporating ecosystem considerations in the work of RFMOs and improving cooperation between RFMOs and regional environmental organizations.[251]

Recently, terms such as 'ecosystem approach to fisheries' (EAF) or 'ecosystem-based approach to fisheries management' have been introduced. Decision V/6 of the Conference of the Parties (COP) to the CBD called upon governments and international organizations to apply the ecosystem approach.[252] More importantly, the JPOI confirms the necessity of EAF, referring to actions at all levels to

> [e]ncourage the application by 2010 of the ecosystem approach, noting the Reykjavik Declaration on Responsible Fisheries in the Marine Ecosystem and decision V/6 of the Conference of Parties to the Convention on Biological Diversity.[253]

Subsequent documents such as the recommendations of the United Nations Open-ended Informal Consultative Process on Oceans and the Law of the Sea (ICP)[254] and successive UNGA fisheries resolutions reiterated the need to implement this commitment in the context of fisheries. At the United Nations Conference on Sustainable Development, or the Rio+20, the participants committed themselves to effectively apply an ecosystem approach in the management of activities having an impact on the marine environment.[255] In the context of fisheries, among other actions, the participants committed themselves to enhancing action to manage bycatch, discards and other adverse ecosystem impacts from fisheries, including by eliminating destructive fishing practices as well as actions to protect vulnerable marine ecosystems from significant adverse impacts (SAIs), including through the effective use of impact assessments.[256]

---

248 Agenda 21, para. 17.46(f); FSA, preambular paras. 2 and 7; Code of Conduct, Articles 6.1, 6.5, 6.6 and 6.8.

249 Reykjavik Declaration on Responsible Fisheries in the Marine Ecosystem, Reykjavik, Iceland, 4 October 2001.

250 Report of the Reykjavik Conference on Responsible Fisheries in the Marine Ecosystem, Reykjavik, Iceland, 1–4 October 2001, Fisheries Report 658, FIID/R658, at p. 106.

251 Reykjavik Declaration, para. 3.

252 Decision V/6, para. 2.

253 Johannesburg Plan of Implementation, para. 30(d) (footnotes omitted). See also para. 32(c).

254 Report on the Work of the United Nations Open-Ended Informal Consultative Process on Oceans and the Law of the Sea at its Seventh Meeting, A/61/156, 17 July 2006, at para. 5(a).

255 The Future We Want, Rio de Janeiro, Brazil, 22 June 2012, para. 158.

256 Ibid., para. 168. See also para. 176.

No uniform definition of the term 'ecosystem approach' has been agreed upon.[257] Its definition was a source of disagreement when the Seventh meeting of the ICP attempted to address the problem. Paragraph 6 of the report of the ICP recognized that "there is no universally agreed definition of an ecosystem approach".[258] Some documents use 'approaches' or 'an approach' instead of 'the approach'.[259] It has even been suggested that the application of an ecosystem approach "does not need to follow a single blueprint but be consistent with local context, means and culture".[260]

Nevertheless, some global instruments are useful to identify the implications of this approach, offering a useful list of considerations to be taken into account in fisheries management.[261] First of all, the Reykjavik Declaration clarified the concept of an 'ecosystem approach to fisheries' in the following terms:

> *[a]ffirming* that incorporation of ecosystem considerations implies more effective conservation of the ecosystem and sustainable use and an increased attention to interactions, such as predator-prey relationships, among different stocks and species of living marine resources; furthermore that it entails an understanding of the impact of human activities on the ecosystem, including the possible structural distortions they can cause in the ecosystem.[262]

The Declaration further specified actions to be taken in fisheries management. Second, the Review Conference on the FSA recommended states to enhance the understanding of ecosystem *approaches* and referred to the incorporation of ecosystem considerations in fisheries management, singling out actions to conserve associated and dependent species and to protect habitats of special concern.[263] Third, the FAO Technical Guidelines for Responsible Fisheries commented on the purpose of an ecosystem approach to fisheries and adopted a working definition of an ecosystem approach.[264]

---

257 COFI/2007/8, at p. 2.

258 A/61/156, at p. 2. The first meeting of the network of the secretariats of regional fisheries organizations (RSN-1), held on 12–13 March 2007, recognized that different approaches to the ecosystem approach to fisheries would be taken in different types of fisheries, and noted that as a consequence of the perceived difficulty in defining an 'ecosystem approach', concerns have been raised about expressly elaborating ecosystem approach principles in regional fishery body Conventions or Agreements. Report of the First Meeting of Regional Fishery Body Secretariats Network, Rome, 12–13 March 2007, FAO Fisheries Report No. 837, FIEL/R837 (En), at p. 15, paras. 81 and 83.

259 See Morishita, "What is the Ecosystem Approach for Fisheries Management", at p. 19.

260 COFI/2007/8, at p. 2.

261 For the definition of 'ecosystem approach to fisheries management' at the regional level, see, e.g., SPRFMO Convention, Article 3(2)(b). See also Chapters 4–5 below.

262 Reykjavik Declaration, preambular para. 16.

263 Outcome of the Review Conference, para. 18(d).

264 FAO Technical Guidelines for Responsible Fisheries 4, Suppl. 2 (2003), at p. 14. See also the Report of the Bergen Conference on Implementing the Ecosystem Approach to Fisheries,

Other international instruments also provide definitions of an ecosystem approach that is of relevance to high seas fisheries. First, COP Decision V/6 of the CBD describes that the ecosystem approach is "a strategy for the integrated management of land, water and living resources that promotes conservation and sustainable use in an equitable way", helping to reach a balance between the three objectives of the CBD.[265] It also identifies principles of the ecosystem approach: consideration of the actual or potential effects of activities on adjacent and other ecosystems (Principle 3); the conservation of ecosystem structure and functioning, in order to maintain ecosystem services, as a priority target (Principle 5); objectives for ecosystem management to be set for the long term (Principle 8); recognizing that change is inevitable (Principle 9); seeking the appropriate balance between, and the integration of, conservation and use of biological diversity (Principle 10). Second, the ICP stated that ecosystem *approaches* to oceans management "should be focused on managing human activities in order to maintain and, where needed, restore ecosystem health to sustain goods and environmental services, provide social and economic benefits for food security, sustain livelihoods [...], and conserve marine biodiversity", and listed a number of actions to be included in *an* ecosystem approach.[266]

Components of ecosystems include marine living resources such as fish, marine mammals, seabirds and plants, as well as seamounts and hydrothermal vents. The following paragraphs analyze representative ecosystem considerations to be taken into account in fisheries management such as the interaction of species (fish and non-fish) and the protection of habitats.[267]

A number of post-UNCLOS III documents deal with interactions among species.[268] Interactions to be taken into account comprise two elements: impacts on associated or dependent species, and bycatch of non-target species. The FSA and the Code of Conduct provide for conservation and management measures for associated or dependent species in a number of provisions.[269] Furthermore, the Reykjavik Declaration refers to the undertaking to "identify and describe the structure, components and functioning of [...] diet composition and food webs, species interactions and predator-prey relationships".[270] These provisions contrast

26–28 September 2006, Bergen, Norway, available at <http://cieaf.imr.no/_data/page/6218/CIEAF_Conference_Report_230207.pdf>.

265 Section A, para. 1. For the objectives of the CBD, see Section 2.3.2.3 below.

266 A/61/156, at paras. 4 and 6.

267 In addition, a form of ecosystem approach includes addressing the effects of environmental factors such as climate change. Morishita, "What is the Ecosystem Approach for Fisheries Management", at p. 24.

268 Agenda 21, para. 17.46(b)–(c); FSA, Article 5(f).

269 See FSA, Article 5(d)–(f); Code of Conduct, Article 6.2.

270 Reykjavik Declaration, para. 5(b).

with those of Part VII of the LOSC, which do not explicitly lay down the obligation to conserve associated or dependent species.[271]

As regards bycatch, it is required in a number of instruments that the bycatch of non-target species needs to be minimized and waste shall be avoided. The FSA provides for the minimization of waste, discards and catch of non-target species.[272] The Code of Conduct also refers to the minimization of waste, discards, and the catch of non-target species.[273] The Reykjavik Declaration refers to the undertaking to improve the monitoring of bycatch and discards, and to support research and technology developments in fishing gear and practices to improve gear selectivity.[274] UNGA resolutions have urged states, RFMOs and arrangements, and other relevant international organizations to take action to reduce or eliminate bycatch and fish discards, including, as appropriate, technical measures, and encouraged participation in regional mechanisms with mandates to conserve non-target species taken incidentally in fishing operations.[275]

Specific regimes have been developed at the global and/or regional levels. First, seabirds, cetaceans, sharks and sea turtles have been identified as being seriously affected by fishing activities, and are subject to special protection given in several international instruments, including treaties and memoranda of understanding.[276] Second, there have been initiatives to regulate the use of certain types of fishing gear or techniques. Fishing practices that have been identified as destructive or inadequate include gears such as large-scale driftnet fishing and bottom-trawling and techniques such as the use of poison and explosives.[277] Some legislation addresses destructive fishing practices by means of the requirement of permits: directly, some of the most destructive fishing practices are prohibited or regulated outright in basic fishery laws and regulations; indirectly, destructive fishing practices are taken into account in considering applications for permits. The former

271 But the obligation is implicit in Article 119 under certain circumstances as well as Article 192.

272 FSA, Article 5(f).

273 Code of Conduct, Articles 6.6, 7.2.2(g), 7.6.9 and 8.5.1.

274 Reykjavik Declaration, para. 5(d)–(e).

275 E.g., UNGA Resolution 63/112, paras. 75–78.

276 Although these instruments were developed to address by-catch in pelagic fisheries, these marine living resources are taken as by-catch also in deep-sea fisheries. For example, the Secretary-General in his report identifies by-catch species affected by deep-water fisheries including not only benthic invertebrates and fish species, but also migrating cetaceans and sea turtles, seabirds and deep-sea sharks. A/61/154, at pp. 15–16, paras. 46–49.

277 See e.g., information available at FAO, "Destructive fishing practices", <http://www.fao.org/fi/website/FIRetrieveAction.do?dom=topic&fid=12353>. The term 'destructive fishing practice' could be understood to include both impacts on biomass of target species caused by indiscriminate fishing methods such as driftnet fishing and those on the marine ecosystem, including associated or dependent species, habitats and biodiversity.

way of regulation includes fishing employing explosives, poisons or dynamite,[278] driftnet fishing[279] and bottom trawling.[280] In considering applications for permits, several states take into account, among other things, involvement in IUU fishing and/or impacts on benthic ecosystems. Third, another attempt to protect ecosystems is the action currently undertaken against fishing by lost or abandoned gear, so-called 'ghost fishing'. The UNGA has been working on this issue in cooperation with other organizations such as the FAO.[281]

The attention of states has been increasingly focused on the protection of habitats. The CBD, for example, defines the word 'habitat' as "the place or type of site where an organism or population naturally occurs".[282] The Code of Conduct refers to wetlands, mangroves, reefs, lagoons, nursery and spawning areas as examples of critical habitats in marine and fresh water ecosystems to be protected and rehabilitated.[283] This issue is extensively dealt with in the Reykjavik Declaration. Among other things, states will undertake to identify and describe the role of habitat and the biological, physical and oceanographic factors affecting ecosystem stability and resilience, and to support research and technology developments concerning fishing gear and practices to reduce adverse impacts of fishing practices on habitat.[284] The issue has drawn more attention recently in relation to the impacts on VMEs of bottom fisheries. Ecosystem features of concern include seamounts, hydrothermal vents and cold water corals. Chapters 3–5 extensively deal with this issue in the sections on deep-sea fisheries.

---

278 South Korea, Fisheries Act, Article 73; Namibia, Marine Resources Act, Article 47(1); South Africa, Marine Living Resources Act, Section 44(1)(a).

279 Australia, Fisheries Management Act, Section 13(2)–(4); Namibia, Marine Resources Act, Article 47(2); South Africa, Marine Living Resources Act, Section 47. See also 16 U.S. Code 1826 *et seq.*

280 Namibia, Regulations relating to the Exploitation of Marine Resources, Regulation 14(1). Interestingly, in July 2012, the European Commission proposed a Regulation for the management of deep-sea fisheries in the North-east Atlantic, which intends to introduce a gradual phase-out of bottom trawls and bottom-set gillnets for European Union vessels in international waters in the Eastern Central Atlantic and the Regulatory Area of NEAFC. Proposal for a Regulation of the European Parliament and of the Council establishing specific conditions to fishing for deep-sea stocks in the North-East Atlantic and provisions for fishing in international waters of the North-East Atlantic and repealing Regulation (EC) No 2347/2002, 19 July 2012, COM(2012) 371 final, 2012/0179 (COD), Article 9.

281 See UNGA Resolution 58/14, Sustainable Fisheries, Including through the 1995 Agreement for the Implementation of the Provisions of the United Nations Convention on the Law of the Sea of 10 December 1982 relating to the Conservation and Management of Straddling Fish Stocks and Highly Migratory Fish Stocks, and Related Instruments, 24 November 2003, para. 44 and the relevant provisions of subsequent fisheries resolutions.

282 CBD, Article 2.

283 Code of Conduct, Article 6.8. The FSA refers only to the protection of habitats of special concern as a part of the application of the precautionary approach. FSA, Article 6(3)(d).

284 See Reykjavik Declaration, para. 5(a)–(b) and (e).

Another ecosystem consideration relates to a spatial aspect (the compatibility of conservation and management measures within the same ecosystem). In other words, this consideration requires fisheries to be managed in a compatible manner within the same marine ecosystem.[285] This requirement may play a role in adopting conservation and management measures for high seas fisheries in general, going beyond the management of fisheries for straddling and highly migratory fish stocks.[286] As the convention area of RFMOs is not necessarily established in accordance with the location of ecosystems, discrete stocks may straddle regulatory areas of two or more RFMOs.[287] As in the case for straddling stocks and highly migratory species, without compatible conservation and management measures, fisheries management may fail to achieve its purpose on either or both sides of the border between the RFMOs. Thus, cooperation between relevant RFMOs is necessary in this case. Direct cooperation between RFMOs (e.g., NAFO and NEAFC) and the coordination of measures between such RFMOs have already taken place. FAO-hosted meetings of secretariats of RFMOs as well as future consultations of members of RFMOs, as proposed by UNGA resolution 61/105,[288] may contribute to improved cooperation and the coordination of measures among the RFMOs concerned.

2.3.2.3 *Marine Biodiversity*

Biological diversity has been defined as "the variability among living organisms from all sources including [...] marine and other aquatic ecosystems and the ecological complexes of which they are part", and includes diversity "within species, between species and of ecosystems".[289] Biodiversity is linked to fisheries in

---

285 For example, see Orrego Vicuña, *The Changing International Law of High Seas Fisheries*, at p. 171 *et seq.*; Davies and Redgwell, "The International Legal Regulation of Straddling Fish Stocks", at p. 263.

286 In the context of straddling fish stocks and highly migratory fish stocks, states have articulated the requirement of compatibility between measures adopted for the EEZ and those for the high seas, in giving effect to the rights, duties and interests of coastal states. See, e.g., FSA, Article 7(2); Agenda 21, paras. 17.1 and 17.49.

287 This would be the case of, *inter alia*, NAFO and NEAFC.

288 UNGA Resolution 61/105, Sustainable Fisheries, Including through the 1995 Agreement for the Implementation of the Provisions of the United Nations Convention on the Law of the Sea of 10 December 1982 relating to the Conservation and Management of Straddling Fish Stocks and Highly Migratory Fish Stocks, and Related Instruments, 8 December 2006, para. 71. See also Report of the Twenty-Seventh Session of the Committee on Fisheries, Rome, 5–9 March 2007, at para. 85.

289 CBD, Article 2. The definitions of terms such as 'biological diversity', 'ecosystem' and 'habitat' in the CBD are incorporated into or reiterated almost verbatim in other international instruments. E.g., see Annex V on the Protection and Conservation of the Ecosystems and Biological Diversity of the Maritime Area to the Convention for the Protection of the Marine Environment of the North-East Atlantic, Sintra, 22–23 July 1998, Article 1.

two senses: destructive fishing practices could pose a risk of decreasing biodiversity; the loss of biodiversity causes the decreased productivity of fish.[290]

Treaties regulating high seas fisheries have traditionally dealt with problems concerning resources in the oceans in a piecemeal fashion. For example, despite the statement that states parties to the LOSC are "*[c]onscious* that the problems of ocean space are closely interrelated and need to be considered as a whole",[291] the LOSC deals with the conservation and management of the marine living resources of the high seas and the protection of the marine environment in different Sections and there is no cross-reference between them.

The Expert Group on Environmental Law of the WCED was the first to articulate legal principles requiring states to maintain 'maximum biodiversity' by ensuring the survival and promoting the conservation of all species of flora and fauna in their natural habitat. Their proposals led to the drafting of and, eventually, the adoption of the CBD in 1992.[292] The CBD now enjoys almost universal participation.[293]

The objectives of the CBD are "the conservation of biological diversity, the sustainable use of its components and the fair and equitable sharing of the benefits arising out of the utilization of genetic resources".[294] The CBD reaffirms that the conservation of biological diversity is "a common concern of humankind".[295]

The CBD stipulates that its provisions apply differently in areas within the limits of national jurisdiction and in areas beyond national jurisdiction. In the case of 'biological components', the provisions apply only in relation to areas within the limits of national jurisdiction. On the other hand, in the case of 'processes and

290 The impacts of biodiversity loss on fisheries are documented in B. Worm *et al.*, "Impacts of Biodiversity Loss on Ocean Ecosystem Services", 314 *Science* (2006), at pp. 787–790 (arguing for positive relationships between diversity and ecosystem functions and services). The term 'biodiversity loss' was defined in CBD COP Decision VII/30, for the purpose of assessing progress towards the target to achieve a significant reduction in the current rate of biodiversity loss by 2010, as "the long-term or permanent qualitative or quantitative reduction in components of biodiversity and their potential to provide goods and services, to be measured at global, regional and national levels". COP Decision VII/30, para. 2.

291 LOSC, preambular para. 3.

292 M. Bowman and C. Redgwell, "Introduction", in M. Bowman and C. Redgwell (eds.), *International Law and the Conservation of Biological Diversity* (1996), at pp. 1–4.

293 There are 193 parties to the CBD as of 27 July 2012. Information available at the website of the CBD, "List of Parties" <http://www.cbd.int/convention/parties/list>.

294 CBD, Article 1. The CBD also confirms that states have the responsibility to ensure that activities within their jurisdiction or control do not cause damage to the environment of areas beyond the limits of national jurisdiction. CBD, Article 3.

295 Ibid., preambular para. 3. On the content of and differences between the CHM concept and the concept of Common Concern of Humankind, see S. Mahmoudi, "Common Heritage of Mankind, Common Concern of Humanity", in J.-P. Beurier, A. Kiss and S. Mahmoudi (eds.), *New Technologies and Law of the Marine Environment* (2000), at pp. 215–223.

activities', the provisions apply to those carried out under the jurisdiction or control of the states parties in areas under *and* beyond national jurisdiction, regardless of where their effects occur.[296] It is not clear from the text of the Convention as such which provisions are applicable to the conservation and sustainable use of marine biodiversity beyond the limits of national jurisdiction. In the context of genetic resources, the Secretary-General in his report stated that "[t]he provisions of the CBD therefore do not apply to genetic resources beyond national jurisdiction".[297] It has also been pointed out that while "bioprospecting activities in the deep seabed carried out under the control of a Party to the CBD are in principle subject to the CBD provisions", "the latter provisions are essentially based on a provider/user countries relationship which seems ill-suited for the deep seabed".[298]

In relation to the marine environment, moreover, the application of the CBD is circumscribed. Parties to the CBD "shall implement [the CBD] with respect to the marine environment consistently with the rights and obligations of States under the law of the sea".[299] In fact, the United States expressed its disappointment in this regard in its declaration, stating "we are disappointed with the development of issues related to [...] the legal relationship between this Convention and other international agreements, and the scope of obligations with respect to the marine environment".[300] But, it has also been suggested that the CBD provisions exist in parallel with and supplement the LOSC with respect to the marine environment: because Article 22(2) of the CBD is qualified by the words 'rights and obligations', the provisions of the CBD could be applied to the extent that it does not infringe upon the rights or obligations of states under the LOSC.[301]

Notwithstanding the potential obstacles associated with the conservation of marine biodiversity beyond national jurisdiction in the text of the Convention, parties to the CBD identified marine and coastal management as a priority area. COP2 issued Decision II/10, known as the 'Jakarta Mandate', and assigned a number of tasks in this regard. As noted earlier, COP Decision V/6 endorsed the 'ecosystem approach' as a useful tool for the conservation of biodiversity, and elaborated the concept in this context. Protected areas as a tool for biodiversity conservation also became a focal point in the work of the CBD regime, and several decisions and studies were issued, including COP Decision VIII/24.

296 CBD, Article 4.

297 A/62/66, at para. 201.

298 European Commission, Background Paper No. 3, at p. 8.

299 CBD, Article 22(2).

300 Declaration (Upon adoption), para. 5, reproduced in Secretariat of the Convention on Biological Diversity, *Handbook of the Convention on Biological Diversity Including its Cartagena Protocol on Biosafety* (3rd edition, 2005), at p. 394.

301 Wolfrum and Matz, "The Interplay of the United Nations Convention on the Law of the Sea and the Convention on Biological Diversity", at p. 476.

The Ad Hoc Open-Ended Working Group on Protected Areas held its first meeting in June 2005, considering, among other things, 'Options for cooperation for the establishment of marine protected areas in marine areas beyond the limits of national jurisdiction'. Subsequent developments within and outside the CBD regime will be dealt with in Chapter 3.

A number of recent international instruments of relevance to fisheries management also refer to marine biodiversity, in association with ecosystem approaches. The JPOI makes it necessary to "[m]aintain the productivity and biodiversity of important and vulnerable marine and coastal areas, including in areas within and beyond national jurisdiction",[302] and "[d]evelop and facilitate the use of diverse approaches and tools, including the ecosystem approach, the elimination of destructive fishing practices", and the establishment of MPAs consistent with international law and based on scientific information.[303] Among others, at the 2005 World Summit, this commitment was confirmed by heads of state and government.[304] The Secretary-General recommended the incorporation of the commitment to significantly reduce the rate of loss of biodiversity by 2010 into a new target under Millennium Development Goal 7 for the follow-up on the Millennium Declaration.[305] The ICP underlined the importance of the conservation of biodiversity to an ecosystem approach, inviting states to consider that an ecosystem approach should "[s]eek to minimize adverse impacts of human activities on marine ecosystems and biodiversity".[306] The UNGA has established an open-ended working group to study issues relating to the conservation and sustainable use of biodiversity beyond areas of national jurisdiction, which held its first meeting in 2006 and four more meetings in 2008, 2010, 2011 and 2012, respectively.[307] Last but not least, the Rio+20 renewed the commitment to maintain the biodiversity of oceans and marine ecosystems.[308] In particular, the participants committed "to address, on an urgent basis, the issue of the conservation and sustainable use of marine biological diversity of areas beyond national jurisdiction including by taking a decision on the development of an international

302 Johannesburg Plan of Implementation, para. 32(a).
303 Ibid., para. 32(c).
304 UNGA Resolution 60/1, 2005 World Summit Outcome, 16 September 2005, para. 56(c). In the same paragraph, it also referred to the negotiation and elaboration of an international regime to promote and safeguard the fair and equitable sharing of benefits arising out of the utilization of genetic resources.
305 Report of the Secretary-General on the Work of the Organization, A/61/1(SUPP), 16 August 2006, at para. 24.
306 A/61/156, para. 6(n).
307 See Chapter 3 below.
308 The Future We Want, para. 158.

instrument under the [LOSC]", "[b]uilding on the work of the ad hoc working group and before the end of the sixty-ninth session of the General Assembly".[309]

Various international fisheries instruments also stipulate the conservation of biodiversity. The FSA refers to the protection of biodiversity in the marine environment as one of the general principles.[310] The Code of Conduct also explicitly prescribes the conservation of biological diversity.[311] The Reykjavik Declaration, in the context of the need to advance scientific knowledge to promote the incorporation of ecosystem considerations, referred to the objective to reduce the adverse impacts of fishing practices on biodiversity.[312] The UNGA in its Resolutions on sustainable fisheries also recognized 'the immense importance and value' of biodiversity deep-sea ecosystems contain.[313] Furthermore, it started calling upon states, including through RFMOs and arrangements, to play an active role in global efforts to conserve and sustainably use marine living resources so as to contribute to marine biological diversity.[314] The protection of biodiversity in the context of fisheries will be further investigated in Chapters 3–5 with regard to deep-sea fisheries and the protection of VMEs.

2.3.2.4 *Precautionary Approach*

In the field of environmental protection, state responsibility for injury *ex post facto* is not a desirable solution to the problem. In some cases, environmental damage is irreversible or it costs much more to restore the environment to its prior condition. In other words, precautionary action is desirable or necessary to prevent the deterioration of the environment. Besides, the application of the precautionary principle is dictated by the very concept of sustainable development,[315] since the latter concept requires states to take into account the needs of future generations – a cautious attitude is a logical way of behaviour.

Although the definition of the precautionary principle is different from one document to another, the basic character of the concept is expressed as "erring on the side of caution".[316] Features common to the concept of precaution relate to (1) a risk of harm to the environment; (2) uncertainty in the causal

309 Ibid., para. 162.

310 FSA, Article 5(g).

311 Code of Conduct, Articles 6.6, 7.2.2(d), 8.4.8 and 12.10.

312 Reykjavik Declaration, para. 5(e).

313 See, e.g., UNGA Resolution 66/68, para. 121. See also para. 105.

314 Ibid., para. 128.

315 ILA New Delhi Declaration of Principles of International Law relating to Sustainable Development, para. 4.

316 See A. Trouwborst, *Precautionary Rights and Duties of States* (2006), at p. 29 (text accompanying note 49 and the literature cited therein).

relationship between the activity and the potential harm; and (3) action *ex ante* to prevent or reduce the risk.[317]

Applying the precautionary principle does not in itself decide what action is needed for the purpose of risk abatement.[318] The precautionary principle plays a role in addressing a risk of potential harm, rather than stipulating a particular type of action. As elaborated below, action required to be taken under the precautionary principle is therefore determined by considering factors such as the type of activity taking place and the risk involved.

Although the precautionary principle originates in a field other than fisheries, it can be applied to fisheries management. Initially, especially until the adoption of the FSA, views on the compatibility between the precautionary approach and the international law of high seas fisheries varied considerably and some argued that the precautionary principle would not be applicable to fisheries management. For example, the EC maintained that the application of the precautionary principle was limited to marine pollution.[319] However, most, if not all, recent international fisheries instruments oblige or recommend states and RFMOs to apply the precautionary approach to fisheries and indicate ways to implement it.[320]

In applying the precautionary approach to fisheries management, common features of the precautionary principle need to be adjusted to the context of fisheries. The following three elements will be examined below: harm to the environment, the role of scientific information, and the types of precautionary action.[321]

---

317 Ibid., at pp. 30–33. Some others also refer to the reversal of the burden of proof as one of the most characteristic attributes. See, e.g., Garcia, The Precautionary Approach to Fisheries and its Implications for Fishery Research, Technology and Management, at section 3.1. On this concept, see subsequent paragraphs.

318 Trouwborst, *Precautionary Rights and Duties of States*, at p. 165 *et seq.*; Birnie *et al.*, *International Law and the Environment*, at pp. 161–164.

319 A. Kanehara, "A New or Pathological Tendency in the International Regulation of Sovereign States: From the Perspective of the International Regulation of Whaling", 47 *Japanese Annual of International Law* (2004), at p. 49 n. 54; G.J. Hewison, "The Precautionary Approach to Fisheries Management: An Environmental Perspective", 11 *International Journal of Marine and Coastal Law* (1996), at p. 310.

320 For example, FSA, Articles 5–6; Code of Conduct, Articles 6.5 and 7.5; UNGA Resolution 66/68, para. 7. In addition, the CBD, while not explicitly referring to the precautionary approach, notes in its preamble that where there is a threat of significant reduction or loss of biological diversity, lack of full scientific certainty should not be used as a reason for postponing measures to avoid or minimize such a threat. CBD, preambular para. 9.

321 Note that, in this regard, terminologies are different from one commentator to another: e.g., different measures, strict application and strong version. See, generally, A. Trouwborst, *Evolution and Status of the Precautionary Principle in International Law* (2002), at p. 3 *et seq.*

First, in the practice of states, thresholds of harm which trigger the application of the precautionary principle are two-fold: serious or irreversible harm.[322] The term 'irreversible' could include 'virtually irreversible'.[323] The term 'serious' is more context-specific.[324] What amounts to serious or irreversible harm depends on the concrete situation.

Impacts of fisheries on the marine environment could be divided into two types of impacts requiring different actions, depending on the causal links between fisheries and impacts, and the likelihood and the potential magnitude of the harm.[325]

On the one hand, fisheries always result in the reduction of target biomass, at least in the short-term before the recruitment of juveniles restores the biomass to the original point. The Secretary-General in his report observes as follows: "it should be noted that even a sustainable fisheries harvest has an impact on the target fisheries biomass in the ecosystem [...]. One element of effective ecosystem management is therefore setting and enforcing sustainable catch limits".[326] Therefore, the reduction of biomass as such should not be regarded to constitute unacceptable harm to the environment.[327]

If a fishery results in the over-exploitation, depletion or extinction of the stock concerned, the impact of the fishery may be qualified as 'serious or irreversible harm' to the environment. For example, ITLOS in the *Southern Bluefin Tuna*

---

322 Trouwborst, *Precautionary Rights and Duties of States*, at pp. 53–56. But see N. Oral, "Protection of Vulnerable Marine Ecosystems in Areas Beyond National Jurisdiction: Can International Law Meet the Challenge?" in A. Strati, M. Gavouneli and N. Skourtos (eds.), *Unresolved Issues and New Challenges to the Law of the Sea: Time Before and Time After* (2006), at p. 98, note 72. She seems to suggest that the precautionary principle itself does not require the threshold of 'serious harm'.

323 Trouwborst, *Precautionary Rights and Duties of States*, at pp. 61–62. He includes in this category damage that "cannot be undone in the course of several human generations", and points to harm to ecosystems with a slow recovery rate such as coral reefs. A. Trouwborst, "The Precautionary Principle in General International Law: Combating the Babylonian Confusion", 16 *Review of European Community & International Environmental Law* (2007), at p. 189. In this regard, see Section 3.2.2.1 below.

324 The term 'serious harm to the marine environment' is defined in the Regulations on Prospecting and Exploration for Polymetallic Nodules in the Area as "any effect from activities in the Area on the marine environment which represents a significant adverse change in the marine environment determined according to the rules, regulations and procedures adopted by the Authority on the basis of internationally recognized standards and practices". Regulation 1(3)(f).

325 See also W. Howarth, "The Interpretation of 'Precaution' in the European Community Common Fisheries Policy", 20 *Journal of Environmental Law* (2008), at pp. 219–220 (arguing for the need for contrasts between conservation and natural resource management in analyzing the application of the precautionary principle to fisheries).

326 A/59/298, at p. 22, para. 75.

327 See also Howarth, "The Interpretation of 'Precaution' in the European Community Common Fisheries Policy", at p. 216.

cases considered "that, in the view of the Tribunal, the parties should in the circumstances act with prudence and caution to ensure that effective conservation measures are taken to prevent serious harm to the stock of southern bluefin tuna".[328] As far as this type of impact is concerned, fishing activities have a direct causal relationship with the possible harm; uncertainty is limited to the magnitude and likelihood of the harm. To address this type of impact, recent fisheries-related instruments use precautionary reference points to indicate thresholds of acceptable impacts on target species. For example, the FSA articulates two types of reference points and the target reference points are designed to have a buffer zone taking into account uncertainty in stock assessment and recruitment rates.[329]

On the other hand, there could be indirect outcomes engendered by fishing activities such as risks of harm to non-target species and habitats. Fisheries entail a varying degree of bycatch; habitats are also susceptible to damage from certain types of fishing gear and/or techniques.[330] In this case, the link between the fishing activity and the possible harm is somewhat indirect and ambiguous. Moreover, the amount of risks causing the harm is open to challenge.

Second, one feature of the precautionary approach is to take action *despite* uncertainty: in other words, action needs to be taken to protect the environment whether or not uncertainty exists.[331] This is not to say that no scientific information is required for the precautionary approach to be applied and all risks are unacceptable: *some* information is needed at all times in order to identify the risk of harm and to develop a solution to the problem. In fact, the importance of scientific information in fisheries management has been emphasized in recent international fishery-related instruments.[332] The formulation of the relationship between scientific information and the application of the precautionary

328 *Southern Bluefin Tuna Cases (New Zealand v. Japan; Australia v. Japan)*, 27 August 1999, Cases Nos. 3 and 4, *Order (Request for provisional measures)*, at para. 77.

329 FSA, Annex II, para. 2.

330 Four fishing practices which may have impacts on deep-sea habitats have been identified: bottom-trawling and dredging, bottom-set longlines, bottom-set gillnets, and pots and traps. A/61/154, at pp. 9–10, paras. 20–23.

331 See Trouwborst, *Precautionary Rights and Duties of States*, at pp. 91–96. On the absence of the requirement of scientific uncertainty under the FSA, see also Orrego Vicuña, *The Changing International Law of High Seas Fisheries*, at p. 162; Schiffman, *Marine Conservation Agreements: The Law and Policy of Reservations and Vetoes*, at p. 185. On the concept of uncertainty in the context of fisheries, see Howarth, "The Interpretation of 'Precaution' in the European Community Common Fisheries Policy", at pp. 220–221.

332 UNGA Resolution 44/225, Large-Scale Pelagic Driftnet Fishing and Its Impact on the Living Marine Resources of the World's Oceans and Seas, 22 December 1989, preambular para. 6; FSA, Article 5(b); Code of Conduct, Article 7.2.1; Reykjavik Declaration, para. 5.

approach in fisheries-related instruments is not uniform.[333] But, in any case, it is safe to say that the availability of at least some information is assumed.[334]

Third, there is no single way of implementing the precautionary approach in a given context, although the concept 'precautionary principle' has often been associated with the reversal of the burden of proof and a ban on the activity concerned.[335] The concept 'burden of proof', however, is of limited significance in this context.[336] In legal proceedings, if the burden of proof has not been discharged, the assertion of the party is dismissed by impartial third parties (e.g., judges and arbitrators) who decide the case.[337] As the ICJ stated, "[u]ltimately, [...] it is the litigant seeking to establish a fact who bears the burden of proving it; and in cases where evidence may not be forthcoming, a submission may in the judgment be rejected as unproved [...]".[338] However, there is not always an impartial third party in decision-making processes in a fisheries context.[339] In RFMOs, states collectively decide what conservation measures should be taken in accordance with their policy and evaluation of the situation, while

---

333 Compare, e.g., UNGA Resolution 59/25, para. 66 and UNGA Resolution 61/105, para. 70.

334 See Birnie *et al.*, *International Law and the Environment*, at p. 156. In the words of Trouwborst, the precautionary principle does not take effect as long as there are no 'reasonable grounds for concern'. Trouwborst, *Precautionary Rights and Duties of States*, at pp. 115–117.

335 For example, see UNGA Resolution 44/225 concerning large-scale driftnet fishing. This is considered to be a representative example of the reversal of the burden of proof as applied to fisheries management in association with the application of the precautionary approach. For example, see Hey, "Global Fisheries Regulations in the First Half of the 1990s", at p. 466; Burke, *The New International Law of Fisheries*, at p. 121.

336 On the definition of the 'burden of proof' as a legal term, see B.A. Garner (ed.), *Black's Law Dictionary* (Second pocket edition, 2003), at p. 80. If one seeks to invoke the concept in a non-legal context, the inherent uncertainty over impacts of fisheries on marine ecosystems, which is the norm in fisheries management, makes it impossible to provide conclusive proof in the first place.

337 See also S. Rosenne, *The Law and Practice of the International Court, 1920–2005*, Fourth edition (2006), at p. 1040. But also note that he points out that "[i]n many cases the absence of a strict plaintiff/defendant relationship between the parties makes it difficult to determine with any degree of finality which party is the *actor*" (Italics in the original).

338 *Military and Paramilitary Activities in and against Nicaragua (Nicaragua v. United States of America)*, Judgment (Jurisdiction of the Court and Admissibility of the Application), 26 November 1984, *I.C.J. Reports 1984*, p. 392, at p. 437, para. 101. This general principle of law has been confirmed by the jurisprudence of the ICJ. For recent confirmation, see *Sovereignty over Pedra Branca/Pulau Batu Puteh, Middle Rocks and South Ledge (Malaysia/Singapore)*, Judgment, 23 May 2008, at para. 45. But also see *Case concerning the Frontier Dispute (Burkina Faso/Republic of Mali)*, Judgment, 22 December 1986, *I.C.J. Reports 1986*, p. 554, at pp. 587–588, para. 65 ("[...] the rejection of any particular argument on the ground that the factual allegations on which it is based have not been proved is not sufficient to warrant upholding the contrary argument").

339 Dispute settlement procedures involving impartial third parties may be provided by either the LOSC or RFMO constitutive instruments or practice.

they themselves assert the desirability of certain measures. Therefore, the burden of proof has no utility in this case.

Notwithstanding the above considerations, the reversal of the burden of proof has implications in procedural terms where those who benefit from the activity need to convince others that the activity is harmless or the harm is acceptable. In other words, the activity might be assumed to be harmful to the resource unless proven otherwise, and the responsibility to argue that the activity is harmless or the impacts are acceptable might lie on those who intend to derive benefits from the ecosystem.[340] The reversal of the burden of proof in this sense has implications particularly for new or exploratory fisheries.

As discussed above, the concept of (the reversal of) the burden of proof as a ban on fishing activities could be applied only in limited circumstances.[341] As Orrego Vicuña suggested, since scientific uncertainty is always the rule in fisheries activities, activities cannot be paralysed until full scientific certainty is achieved that no environmental damage will ensue.[342] Other types of precautionary action include: setting precautionary reference points for target species, determining catch limits for non-target species, controlling new or exploratory fisheries, implementing interim arrangements such as interim closure, emergency measures and collecting data and other scientific information.[343]

It is submitted that where the causal relationship between the fishery activity and harm to the environment appears to be more uncertain, less strict measures are likely to be adopted; where the potential harm may be more serious, more stringent measures are likely to be taken. In fact, the requirement of

---

340 See Garcia, The Precautionary Approach to Fisheries and its Implications for Fishery Research, Technology and Management, at section 5.3.

341 Commenting on the provisions of the FSA on the precautionary approach, Davies and Redgwell observe that "[t]he application of the precautionary approach in Article 6 is particularly welcome, though it does not go so far as to prevent fishing even where stocks are threatened". Davies and Redgwell, "The International Legal Regulation of Straddling Fish Stocks", at p. 261.

342 Orrego Vicuña, "The Law Governing High Seas Fisheries: In Search of New Principles", at p. 387.

343 FSA, Article 6 and Annex II; Code of Conduct, Article 7.5. Measures applying the precautionary approach may also contribute to the protection of ecosystems. In addition, an ecosystem approach to fisheries influences precautionary measures by requiring adaptive management: as Howarth points out, "precaution looks less like a one-off response to an uncertain ecological threat and more like an iterative process involving a sequence of actions each of which is dependent on those preceding it and the monitoring and evaluation of their ecological success or failure". Howarth, "The Interpretation of 'Precaution' in the European Community Common Fisheries Policy", at p. 242. The consequence of this proposition would be that more stringent measures, including a moratorium, will be required if precautionary measures taken in the past have not proved to be effective in the conservation of fishery resources and ecosystems.

proportionality demands that actions correspond to the size of the risk involved, so as to avoid the adoption of excessively strict measures.[344]

2.4 *Concluding Remarks*

At the beginning of this chapter, a question was posed whether or not general principles of high seas fisheries have changed fundamentally in the LOSC. The analysis of this chapter indicates that the general principles have neither remained unchanged nor been fundamentally altered by the provisions of the LOSC. The principles identified in Chapter 1 constitute the general principles of high seas fisheries also under the LOSC: circumscribed freedom of fishing, cooperation between states and the conservation of marine living resources.

The freedom of fishing on the high seas continues to be restrained by the due regard clause as well as other restrictions of a general nature such as the prohibition of an abuse of rights. A major addition is the requirement of due regard for interests with regard to the activities of the Area. Under the LOSC, high seas fisheries and mining activities in the Area need to be conducted in a mutually reconcilable way, paying reasonable regard to each other. In reality, this balancing cannot be always easy to achieve. Bottom fisheries have more likelihood of contact with the seafloor or areas around it than pelagic fisheries. In order to comply with the spirit of the Convention, limitations need to be imposed, taking into account various factors. There is a clear need to accommodate both interests through the enactment of regulations by competent international organizations.

Fishing under Articles 87 and 116 should not be considered in the narrow sense of the term. A growing trend is discernible where states and international organizations have been concerned with the regulation of support activities. Even if a fishery-related activity is not regarded as fishing in the sense of Articles 87 and 116, the LOSC provides a basis for regulating such an activity by virtue of the provisions of Part VII.

The right to fish on the high seas is subject to, among other things, the rights, duties and interests of coastal states. The phrase 'subject to' is open to various interpretations, including the view that coastal states may take unilateral conservation measures for fisheries on the adjacent high seas by vessels of third states. However, under the LOSC, as opposed to the HSFC, exceptions to exclusive flag state jurisdiction are not expressly stipulated with regard to high seas fisheries. In addition, there is a clear difference in the formulations of the special interests

344 Trouwborst, *Precautionary Rights and Duties of States*, at pp. 149–153. See also Howarth, "The Interpretation of 'Precaution' in the European Community Common Fisheries Policy", at pp. 224–225.

of coastal states between the HSFC and the LOSC: the former concerns the 'area' adjacent to their territorial sea while the latter, *inter alia*, covers 'stocks' occurring in the adjacent high seas area. On the other hand, the right of high seas fishing is subject to the interests of a coastal state where the fishery takes place over the continental shelf of the coastal state. In particular, it is arguable that high seas fisheries could be restricted by coastal states on the following grounds: (1) by virtue of sovereign rights for the purpose of exploring and exploiting natural resources of the continental shelf; (2) jurisdiction of coastal states over MSR on the continental shelf if it is a requisite for engaging in such fisheries; and (3) the protection of the marine environment of the continental shelf from high seas fisheries in the area superjacent to it. Such measures are restricted, among others, by Article 78(2) in order not to infringe on or result in any unjustifiable interference with high seas fishing. Since pelagic fisheries are not likely to have contact with the seafloor in the normal course of operations, there has been little practice on this matter. But, it is not surprising that high seas deep-sea fisheries give rise to conflicts between distant water fishing nations and coastal states in this regard.

As noted in Chapter 1, the regime of high seas fisheries in Section 2 of Part VII of the LOSC was negotiated with fishery resources linked to the EEZ in mind. Nevertheless, the stock and species covered by Article 116 are not limited to these resources. The wording of the LOSC and its drafting history indicate that stocks discrete to the high seas are not excluded from the scope of the high seas fisheries regime under the LOSC. To what extent the general principles of high seas fisheries under the LOSC impact on the regulation of fisheries for such stocks would need a detailed analysis of the practice of states and international organizations, however.

As in the HSFC, freedom of fishing is elaborated under the LOSC as the right of states for their nationals to engage in fishing. The term 'nationals' under the LOSC is not clear, but the drafting history suggests that it means vessels flying the flag of the state concerned. Nevertheless, it should be noted that, in recent practice, the term 'nationals' has been increasingly used in the sense of natural and juridical persons, rather than fishing vessels.

Cooperation between states is the essential element of the legal regime of high seas fisheries. The range of states which may take part in the regulatory system of high seas fisheries is not unequivocal in the LOSC. The FSA and the practice of RFMO/As indicate that coastal states are entitled to do so, while this is not clear for non-fishing non-coastal states which have concerns for the conservation of the fishery resources concerned. The type of cooperation is not specified under the LOSC. Cooperation may be pursued directly or through regional or subregional fisheries organizations. The FSA provision obliging high seas fishing states to participate in RFMO/A regimes or agree to comply with their measures is a significant step in this regard. While the institutionalization of cooperation is achieved through this provision, the loose definition of 'arrangement' could be interpreted to make it no different from direct cooperation. In any case, although an emerging general trend exists pointing to development in the direction contemplated by the FSA, one would still hesitate to conclude its general acceptance by the international community. As regards the duty of negotiation, it is widely accepted that what is required by this duty is to negotiate in good faith, while some states have argued that high seas fishing states shall agree to coastal state conservation and management measures for straddling stocks of the high seas.

The innovative aspect of the legal regime of high seas fisheries under the LOSC is that environmental concerns are integrated into the concept of conservation, and this advancement has been elaborated in subsequent instruments. Under the LOSC the reference point of the MSY is qualified by environmental factors and conservation measures need to take into account the impacts on associated or dependent species. While the LOSC does not offer criteria governing how to depart from the MSY using the factors mentioned and two diametrically opposed interpretations are possible, the subsequent international instruments virtually oblige stricter conservation measures by introducing precautionary reference points. The rather moderate provision on the conservation of associated or dependent species in the LOSC has been amplified in subsequent practice. In addition, the concept of the conservation of marine ecosystems is introduced through the provisions of Part XII, in particular Article 194(5). The LOSC stipulates that a scientific basis is needed for the determination of conservation and management measures, while subsequent instruments have introduced the concept of the precautionary approach in parallel with the best scientific information available: these two considerations are not conflicting but complementary. Contrary to the perceptions of some commentators, the precautionary approach does not by itself determine the action to be taken for the conservation of marine living resources.

The duty of conservation has been significantly elaborated by the adoption of new concepts and approaches in recent international fisheries-related instruments, including the objective of sustainable fisheries, ecosystem approaches, the protection of marine biodiversity and the precautionary approach. The objective of fisheries management has moved to the sustainable use and conservation of marine living resources and, arguably, conservation measures are being gradually taken for the purpose of conservation as such, rather than conservation with a view to utilization. In other words, the underlying value of conservation is in the process of change. This is reflected in the adoption of the precautionary approach in general and precautionary reference points in particular. Enhanced ecosystem considerations and biodiversity conservation are other aspects of the contemporary change in the concept of conservation. All in all, the traditional concept of conservation is being replaced by a new, elaborated and broadened concept.

As a matter of fact, the new concept was already partly heralded by some provisions of the LOSC, that is, the regime for marine mammals under Articles 65 and 120. The latter article allows for stricter conservation measures to be taken for marine mammals in the high seas. The regime for straddling and highly migratory fish stocks does not yet go as far as the new concept of conservation, but it is not unlikely that the international community will establish a new, separate regime for the harvesting of resources other than marine mammals of the high seas (or a part of them) on the basis of the provisions of the LOSC incorporating new concepts and approaches.

The next three chapters will examine the practice of states and international organizations with a view to investigating the interplay between the general principles of high seas fisheries and international regulation, focusing on three issues: DHSFS, deep-sea fisheries on the high seas and the protection of the marine ecosystem in areas beyond national jurisdiction.

Chapter Three

Practice at the Global Level

Chapter 2 has concluded that the general principles of high seas fisheries as reflected in the LOSC provide a legal framework for all high seas fisheries, including fisheries for DHSFS. At the same time, it has been pointed out that there is a potential that the general principles of high seas fisheries under the LOSC are implemented differently for fisheries for DHSFS and those for straddling fish stocks. Chapters 3–5 investigate how general principles are being implemented in the face of new challenges in high seas fisheries. Practice at the global and regional levels is examined in these chapters.

This chapter considers new challenges in high seas fisheries under three headings. The first section analyzes initiatives that call for the application of some principles of the FSA to discrete high seas stocks. Many of the new trends in fisheries management described in Chapter 2 have primarily been developed in the context of fisheries for straddling and highly migratory fish stocks or the protection of the environment in general. It is not certain to what extent new approaches under the FSA are considered to be applicable to high seas fisheries in general. The section therefore examines whether and, if so, which principles of the FSA are applicable to DHSFS. The second section seeks to examine a new category of fisheries, i.e., deep-sea fisheries, and to consider the possible implications of such a new classification for the legal regime of high seas fisheries under the LOSC. In other words, the main questions are whether the special characteristics of deep-sea fisheries on the high seas require states and RFMOs and arrangements to implement the general principles of high seas fisheries differently; whether such differences have led to the creation of a new sub-set of principles for deep-sea fisheries. The third section deals with MPAs and other area-based management tools in the high seas and their potential impacts on the legal regime for fisheries on the high seas. In other words, it considers whether the need for coordination in establishing multi-purpose MPAs in the high seas has modified the jurisdictional framework set out in the LOSC and the FSA.

3.1 *Initiatives to Address Fisheries for Discrete High Seas Fish Stocks by Applying Principles of the FSA*

The issue of DHSFS surfaced at the global level in 2003. At the 25th session of the FAO Committee on Fisheries (COFI) in February 2003, several members drew attention to DHSFS in discussing the need for an improved management of deep-sea fisheries and noted the need to further develop international law in this regard.[1]

The 2004 Reports of the UN Secretary-General analyzed the issue briefly. The Reports pointed out that "there is no legally binding global agreement containing conservation and management measures for discrete high seas stocks"[2] and some of these stocks "generally remain outside existing regulatory frameworks".[3]

As the Review Conference on the FSA in May 2006 was considered to be a potential venue for considering the issue of DHSFS,[4] preparatory meetings for the Review Conference, notably a conference held in St. John's, Newfoundland and Labrador, Canada to discuss high seas fisheries issues in preparation for the Review Conference (the St. John's Conference), discussed this issue in detail.[5]

---

1 Report of the Twenty-Fifth Session of the Committee on Fisheries, Rome, 24–28 February 2003, Fisheries Report No. 702, FIPL/R702, at para. 26. Similarly, during the discussion of the Informal Consultations of States Parties to the FSA (ICSP) in 2003, a number of delegations identified the absence of a regulatory regime for demersal species as a major gap in present high seas fisheries management. An extension of the FSA's regulatory regime was considered a possible solution to addressing the gap, while a state party pointed out that other fora had the mandate to oversee the work of RFMOs. Report of the Second Informal Consultations of States Parties to the Agreement for the Implementation of the Provisions of the United Nations Convention on the Law of the Sea of 10 December 1982 relating to the Conservation and Management of Straddling Fish Stocks and Highly Migratory Fish Stocks (New York, 23–25 July 2003), ICSP2/UNFSA/REP/INF.1, August 2003, at p. 13, para. 61. See also L.A. Kimball, "Deep-Sea Fisheries of the High Seas: The Management Impasse", 19 *International Journal of Marine and Coastal Law* (2004), at p. 267.

2 Oceans and the Law of the Sea, Report of the Secretary-General, A/59/62/Add.1, 18 August 2004, at p. 77.

3 A/59/298, at p. 22. UNGA Resolution 59/25 of that year did not expressly deal with the issue of DHSFS. However, some of its parts have implications for the conservation and management of these stocks. For example, participants in negotiations to establish RFMOs were urged to "apply provisions of the Convention *and the Agreement* to their work" (emphasis added). UNGA Resolution 59/25, para. 54. Among ongoing negotiations at that time, those for the Southern Indian Ocean primarily concerned DHSFS. Note that UNGA Resolutions on sustainable fisheries reiterate the previous year's recommendations unless circumstances change. Unless necessary, this chapter refrains from referring to provisions in subsequent years' resolutions that have the substantially same content.

4 See A/59/298, at p. 10; Report of the Twenty-Sixth Session of the Committee on Fisheries, Rome, 7–11 March 2005, Fisheries Report No. 780, FIPL/R780, at p. 15, para. 93.

5 While the Consultations did not discuss the substantive issues related to DHSFS, the Chairman's background paper on possible initiatives for strengthening the substance and methods

The participating ministers at the St. John's Conference issued a Declaration setting out their commitment to specific actions to improve the governance of high seas fisheries, which identified the issue of the sustainable management of discrete high seas fisheries as one of the possible gaps and agreed to work to address it.[6] Furthermore, participants in the workshop shared the view that "States should apply the fundamental management principles" of the FSA to DHSFS. According to the Conference Report, such application can be "confirmed formally" at the Review Conference and, based on the outcome of the Conference, states should consider developing a legal instrument based on this commitment.[7]

At the same time, discussions on the issue of DHSFS also took place in the UNGA process. The sixth meeting of the ICP in June 2005, welcoming the outcome of the St. John's Conference, proposed to the General Assembly to encourage "States, as appropriate, to recognize that the general principles of the [FSA] should also apply to discrete fish stocks in the high seas".[8] Following this recommendation, the UNGA adopted Resolution 60/31, paragraph 12 of which "*[e]ncourages* States, as appropriate, to recognize that the general principles of the [FSA] should also apply to discrete fish stocks in the high seas".

The issue of DHSFS was also discussed within the framework of the CBD. Concerns for the conservation and management of DHSFS were also raised in the context of the protection of biological diversity. In preparation for the first meeting of the Working Group on Protected Areas, the Secretariat of the CBD provided a document entitled "Options for Cooperation for the Establishment of

---

of the implementation of the provisions of the Agreement included the issue of DHSFS in the list, including initiatives to "[a]pply basic [FSA] provisions to high seas discrete stocks and/or deep sea fisheries" and to "address gaps in the regulation of high seas fisheries, broaden competence of RFMOs and/or establish new RFMOs". Report of the Fourth Informal Consultations of States Parties to the Agreement for the Implementation of the Provisions of the United Nations Convention on the Law of the Sea of 10 December 1982 relating to the Conservation and Management of Straddling Fish Stocks and Highly Migratory Fish Stocks (New York, 31 May–3 June 2005), ICSP4/UNFSA/REP/INF.1, 18 July 2005, at Annex V, pp. 20–21. One observer (Greenpeace International) expressed concerns about threats to these stocks and asked for the FSA to be extended to cover all high seas stocks (p. 8).

6 Ministerial Declaration at the Conference on the Governance of High Seas Fisheries and the UN Fish Agreement: Moving from Words to Action, St. John's, Newfoundland and Labrador, 2 May 2005, para. 13(A).

7 St. John's Conference Report, at p. 19. However, it has been pointed out that, while this view was agreed by consensus at the workshop, subsequent discussion at the plenary about the status and purpose of the summary of the workshops shows hesitation on the part of states to take action in line with the view expressed at the workshop. See E.J. Molenaar, "Addressing Regulatory Gaps in High Seas Fisheries", 20 *International Journal of Marine and Coastal Law* (2005), at p. 555.

8 Report on the Work of the United Nations Open-Ended Informal Consultative Process on Oceans and the Law of the Sea at its Sixth Meeting, A/60/99, 7 July 2005, at p. 4, para. 6(f). Note that Chile opposed specific references to FSA articles, noting that these stocks are already addressed in the LOSC. *ENB*, vol. 25, No. 18, at p. 6.

Marine Protected Areas in Marine Areas beyond the Limits of National Jurisdiction", which stated, on the basis of a legal study undertaken in collaboration with the IUCN:[9] "there is growing awareness of discrete high-seas fish stocks associated, for example, with seamounts, which was not the case when the [LOSC] and the [FSA] were adopted".[10] The document identified a number of options for cooperation regarding the establishment of MPAs beyond the limits of national jurisdiction, including the further use and improvement of existing instruments. In particular, it was stated that the scope of the FSA could be expanded to include all high seas fish stocks, an option which requires that "precautionary and ecosystem approaches are applied in conservation and management measures for discrete stocks".[11] However, the first meeting of the Working Group did not reach consensus on this point, and the proposed sentence remained for consideration at the second meeting of the Working Group.[12]

At the Review Conference on the FSA, Canada proposed to request the FAO to produce guidelines for discrete high seas stocks.[13] In the end, an agreement was not reached on this proposal. The Conference, in line with UNGA Resolution 60/31, merely observed as follows:

> [RFMOs] with competence to regulate straddling fish stocks have the necessary competence to conserve and manage discrete high seas stocks. There is no obstacle for [them] to adopt management measures in respect of discrete high seas stocks in accordance with the general principles set forth in the [FSA].[14]

The subsequent UNGA fisheries Resolutions clarified in more detail what should be done with respect to DHSFS. First, the UNGA called upon states "to adopt necessary measures to ensure the long-term conservation, management and sustainable use of [DHSFS] in accordance with the [LOSC] and consistent

9 The International Legal Regime of the High Seas and the Seabed Beyond the Limits of National Jurisdiction and Options for Cooperation for the Establishment of Marine Protected Areas (MPAs) in Marine Areas Beyond the Limits of National Jurisdiction, UNEP/CBD/WG-PA/1/INF/2, 28 April 2005.

10 Options for Cooperation for the Establishment of Marine Protected Areas in Marine Areas beyond the Limits of National Jurisdiction, UNEP/CBD/WG-PA/1/2, 20 April 2005, at p. 9.

11 Ibid., at p. 11.

12 See Report of the First Meeting of the ad hoc Open-ended Working Group on Protected Areas, UNEP/CBD/WG-PA/1/6, 20 June 2005, at pp. 14–16 and 28 (Annex I, Recommendations Adopted by the Ad Hoc Open-Ended Working Group on Protected Areas at its First Meeting, 1/1., para. 4(s)(vii)). The second meeting of the Working Group took place in February 2008. But, as the issue of options for cooperation was dropped from the agenda, the issue of DHSFS did not fall under the scope of the meeting. See Section 3.3 below.

13 *ENB*, vol. 7 No. 57 (23 May 2006), at p. 2.

14 Outcome of the Review Conference, para. 16.

with the general principles set forth in the [FSA]".[15] Second, the UNGA called on all states, directly or through RFMO/As, to apply widely, in accordance with international law and the Code of Conduct, the precautionary approach and an ecosystem approach to the conservation, management and exploitation of fish stocks, including DHSFS.[16] Furthermore, the UNGA in its Resolution 63/112 called upon all states to apply stock-specific precautionary reference points to ensure that populations of harvested stocks, in particular *inter alia* DHSFS, and, where necessary, associated or dependent species are maintained at, or restored to, sustainable levels and to use these reference points for triggering conservation and management action.[17] Third, the UNGA called for the collection and reporting of catch and effort data and fishery-related information in a complete, accurate and timely way, including for DHSFS.[18] The Resolutions also requested the FAO to revise its global fisheries statistics database to provide information, among other things, on DHSFS on the basis of where the catch was taken.[19]

3.2 *Deep-Sea Fisheries on the High Seas*

There is no universally agreed definition of deep-sea fisheries. For example, instead of defining deep-sea fisheries, the FAO Expert Consultation in 2006

15 UNGA Resolution 61/105, para. 19. A reference to the Code of Conduct was added in UNGA Resolution 62/177, Sustainable Fisheries, Including through the 1995 Agreement for the Implementation of the Provisions of the United Nations Convention on the Law of the Sea of 10 December 1982 relating to the Conservation and Management of Straddling Fish Stocks and Highly Migratory Fish Stocks, and Related Instruments, 18 December 2007, para. 23. The 2006 Expert Consultation on deep-sea fisheries in the high seas also recognized the need to apply the principles and relevant provisions of the FSA to the management of discrete high-seas fish stocks. Report and Documentation of the Expert Consultation on Deep-sea Fisheries in the High Seas, Bangkok, Thailand, 21–23 November 2006, at para. 56.

16 UNGA Resolution 61/105, para. 5. References to specific fish stocks, including DHSFS, in this context are not found in UNGA Resolution 64/72 and subsequent resolutions, but nothing implies any intention to change the substance of the provision. UNGA Resolution 64/72, Sustainable Fisheries, Including through the 1995 Agreement for the Implementation of the Provisions of the United Nations Convention on the Law of the Sea of 10 December 1982 relating to the Conservation and Management of Straddling Fish Stocks and Highly Migratory Fish Stocks, and Related Instruments, 4 December 2009, para. 6.

17 UNGA Resolution 63/112, para. 8. Subsequent resolutions do not explicitly refer to DHSFS in this context.

18 UNGA Resolution 61/105, para. 8. See also UNGA Resolution 60/31, Sustainable Fisheries, Including through the 1995 Agreement for the Implementation of the Provisions of the United Nations Convention on the Law of the Sea of 10 December 1982 relating to the Conservation and Management of Straddling Fish Stocks and Highly Migratory Fish Stocks, and Related Instruments, 29 November 2005, para. 65.

19 UNGA Resolution 61/105, para. 28.

considered deep-sea fisheries to be "those fisheries that are centred at depths below 200 metres". It recognized the need for further classifying deep-sea fisheries on the basis of their biological characteristics.[20] In fact, as noted in Section 3.2.2 below, the determination of fisheries falling under the scope of the FAO International Guidelines on deep-sea fisheries was one of the controversial issues in the FAO Technical Consultation in February 2008.

The characteristics of many, but not all, deep-sea species include geographically restricted populations, high longevity and low fecundity.[21] These characteristics lead to the vulnerability of target species. In addition, as noted in the Introduction, deep-sea fisheries, especially bottom trawling, are likely to cause adverse impacts on vulnerable benthic ecosystems and deep-sea biodiversity.

Due to these special characteristics and impacts of deep-sea fisheries, one may wonder whether they can be managed through the traditional regulatory framework intended for pelagic (straddling) fish stocks.[22] In this regard, discussions at the FAO Expert Consultation held in 2006 are insightful. On the one hand, the Expert Consultation recognized that "many of the issues associated with the effective management of deep-sea fisheries differ in degree rather than substance from those associated with management of other fisheries" and, therefore, "recommendations for management that have been applied to fisheries generally are also applicable to deep-sea fisheries". On the other hand, the Expert Consultation referred to the special characteristics of deep-sea fisheries and observed additional conservation challenges: (1) the frequent lack of information needed to apply many of the usual tools for the assessment of stocks and management of fisheries; (2) lesser knowledge of the structural and functional characteristics of deepwater ecosystems; (3) because of their lower productivity and life histories capable of sustaining only low exploitation rates, the consequences of perturbations of deepwater ecosystem components pose a higher risk of serious or irreversible harm.[23] On this basis, the Expert Consultation suggested the need

---

20 Report and Documentation of the Expert Consultation on Deep-sea Fisheries in the High Seas, Bangkok, Thailand, 21–23 November 2006, at para. 47. See also an article by ICES, "Is time running out for deepsea fish?", available at <http://www.ices.dk/marineworld/deepseafish.asp>, which states "[f]isheries carried out in waters deeper than about 400 m are generally considered to be deep-sea fisheries". Cautious remarks were made against the generalization of deep-sea species. See, e.g., J.D.M. Gordon, "Environmental and Biological Aspects of Deepwater Demersal Fishes", in R. Shotton (ed.), *Deep Sea 2003: Conference on the Governance and Management of Deep-sea Fisheries: Part 1: Conference Reports. Queenstown, New Zealand, 1–5 December 2003* (2005), at pp. 81–82.

21 See, e.g., FAO Fisheries Department, SOFIA 2004, at p. 93.

22 As noted in the Introduction, deep-sea species are the bulk of DHSFS currently targeted in high seas fisheries.

23 Report and Documentation of the Expert Consultation on Deep-sea Fisheries in the High Seas, Bangkok, Thailand, 21–23 November 2006, at paras. 67–70.

for an approach to the management of such fisheries which is different from the management of other types of fisheries:

> management actions should be more precautionary than those implemented for shelf fisheries [...] Decisions on these individual tools [for management] should support and be consistent with a strict application of the precautionary approach and an ecosystem approach to fisheries, because of the characteristics of deep-sea fisheries.[24]

States have been addressing deep-sea fisheries at the global level mainly through the UNGA and the FAO.[25] The ICP started discussions on deep-sea fisheries on the high seas in the context of protecting benthic marine ecosystems in 2002.[26] On the basis of the recommendations by the ICP, the UNGA adopted some recommendations in Resolution 59/25 and since then has been addressing the regulation of bottom fisheries on the high seas in its annual fisheries resolutions. In parallel with the UNGA process, COFI of the FAO has called for attention to deep-sea fisheries on the high seas since 2003.[27] COFI recommended further actions at subsequent sessions. So far, while policy discussions have mainly been held at the UNGA, the FAO has taken on the technical aspects in the implementation of the recommendations by the UNGA. As elaborated in Chapters 4 and 5, these actions were interwoven with those at the regional and national levels.

The remainder of this section describes the actions taken by these two main fora in addressing deep-sea fisheries on the high seas. The first sub-section analyzes the recommendations contained in the successive UNGA Resolutions on sustainable fisheries, focusing on actions to be taken by states and RFMO/As. The second sub-section deals with the development of the FAO International Guidelines on deep-sea fisheries on the high seas, initiated by COFI of the FAO upon the request by the UNGA. Although the Guidelines are not intended to be

---

24 Ibid., at paras. 72 and 75.

25 Other initiatives include, *inter alia*, actions by CBD COP and those at the Rio+20 meeting. While the CBD COP once urged the UNGA, other relevant international organizations and states parties to take urgent actions to protect VMEs from destructive practices, its subsequent involvement remained modest. Compare CBD COP Decision VII/5, UNEP/CBD/COP/DEC/VII/5 (2004), paras. 30 and 60–62 as well as suggested activities concerning operational objective 2.4; Decision VIII/24, UNEP/CBD/COP/DEC/VIII/24 (2006), paras. 35–37; Decision X/29, UNEP/CBD/COP/DEC/X/29 (2010), para. 54. The outcome document of the Rio+20 does not explicitly refer to bottom fisheries or deep-sea fisheries, but its commitment to enhance actions to protect VMEs from SAIs, including through the effective use of impact assessments clearly encompasses actions addressing bottom fisheries on the high seas. The Future We Want, para. 168.

26 Report on the Work of the United Nations Open-ended Informal Consultative Process on Oceans and the Law of the Sea at Its Third Meeting, A/57/80, 2 July 2002, at p. 7, paras. 19–25.

27 Report of the Twenty-Fifth Session of the Committee on Fisheries, Rome, 24–28 February 2003, at paras. 26 and 106.

legally binding, they will likely be consulted in the process of implementing the UNGA Resolutions at the regional and national levels.

3.2.1 *UNGA Resolutions*

Since 2003, the UNGA has been addressing various ecosystem impacts of fisheries in the section entitled "Responsible fisheries in the marine ecosystem" in its annual resolutions on sustainable fisheries. Among others, the General Assembly has repeatedly called upon states to take action, individually and/or through RFMO/As, to address the impacts of fisheries on VMEs. Although the principal activity that is being targeted is bottom trawling on the high seas, the scope of these recommendations is not limited to bottom trawling, but extends to destructive fishing practices in general, in areas both under and beyond national jurisdiction. First, Resolution 59/25 called upon states, either by themselves or through RFMO/As, to take action urgently.[28] Resolution 61/105 called upon states to take immediate action, individually *and* through RFMO/As, to sustainably manage fish stocks and to protect VMEs from destructive fishing practices.[29] Following the adoption of the FAO International Guidelines on Deep-sea Fisheries in 2008 as described below, the recommended action has been more focused: the UNGA has been calling upon states to take action immediately "to implement the [FAO] International Guidelines [...]".[30]

With regard to specific action dealing with bottom fisheries, including bottom trawling, urgent actions as well as long-term solutions have been articulated in the successive UNGA resolutions. Despite pressures from environmental NGOs and some governments suggesting a moratorium on all bottom trawling fisheries on the high seas, states did not agree on a moratorium either in Resolution 59/25 of 2004 or in Resolution 61/105 of 2006.

---

28 UNGA Resolution 59/25, para. 66. The following year, Resolution 60/31 called upon states to urgently accelerate their cooperation in establishing interim targeted protection mechanisms for VMEs in regions. UNGA Resolution 60/31, para. 72. Also note that in 2004 when Resolution 59/25 was adopted, the resolution on oceans and the law of the sea also mentioned the protection of biodiversity in deep-sea areas. UNGA Resolution 59/24, Oceans and the Law of the Sea, 17 November 2004, para. 68.

29 UNGA Resolution 61/105, para. 80. See also UNGA Resolution 62/177, para. 97; UNGA Resolution 63/112, para. 102.

30 UNGA Resolution 64/72, para. 113; UNGA Resolution 65/38, Sustainable Fisheries, Including through the 1995 Agreement for the Implementation of the Provisions of the United Nations Convention on the Law of the Sea of 10 December 1982 relating to the Conservation and Management of Straddling Fish Stocks and Highly Migratory Fish Stocks, and Related Instruments, 7 December 2010, para. 117. Note that paragraph 117 of UNGA Resolution 64/72 and paragraph 120 of UNGA Resolution 65/38 also refer to the need for consistency with the Guidelines in implementing Resolutions 61/105 and 64/72. See also UNGA Resolution 66/68, para. 121 ("to continue implementing the 2008 International Guidelines").

Resolution 59/25 called upon states, either by themselves or through RFMO/As, to consider the interim prohibition of destructive fishing practices until appropriate conservation and management measures are adopted in accordance with international law.[31] This recommendation is markedly different from the recommendation of a moratorium in UNGA Resolution 44/225 on large-scale pelagic driftnet fishing in one important aspect: in Resolution 59/25, considering a moratorium on a case-by-case basis was recommended, while Resolution 44/225 recommended a moratorium *per se*.[32] The phrase 'on a case-by-case basis' implies that a blanket ban was not considered to be desirable and the socio-economic effects of a prohibition should be weighed against the merits to the ecosystems concerned, presumably taking into account the difficulties associated with monitoring and enforcing the prohibition in relation to deep-sea fisheries.[33]

The phrases 'on a scientific basis' and 'including the application of the precautionary approach' in Resolution 59/25 indicate that science-based fisheries management and the application of the precautionary approach are not mutually exclusive in the context of high seas deep-sea fisheries. The precautionary approach was thus to be applied in such a way as to accommodate and utilize scientific data and not to exclude the relevance of scientific information in decision-making. The possible prohibition was to last for a limited duration, at least in theory, since the *interim* prohibition will be lifted when appropriate conservation and management measures will have been adopted in accordance with international law.

While the Resolution did not exclude destructive fishing practices other than bottom trawling, it obviously focused on bottom trawling on the high seas *that has adverse impacts on VMEs*. Resolution 59/25 referred to ecosystem features such as seamounts, hydrothermal vents and cold water corals located beyond national jurisdiction as examples of VMEs. It is clear from the discussion at the ICP as well as Decision VII/5 of the CBD Conference of Parties that measures were, first of all, directed at the impacts on VMEs beyond national jurisdiction.[34]

---

31 UNGA Resolution 59/25, para. 66.

32 In fact, at the ICP in 2004 most states were not in a position to support proposed recommendations to consider an interim prohibition of bottom trawling in areas beyond national jurisdiction, and the proposal to recommend the General Assembly to urgently consider an interim prohibition on bottom trawling was deleted. See *ENB*, vol. 25, No. 12, at p. 5. The agreed elements of the ICP called for the consideration of an interim prohibition of destructive practices, without direct reference to bottom trawling. Report on the Work of the United Nations Open-Ended Informal Consultative Process on Oceans and the Law of the Sea at Its Fifth Meeting, A/59/122, 1 July 2004, at para. 6(a).

33 On such difficulties, see M.W. Lodge, "Improving International Governance in the Deep Sea", 19 *International Journal of Marine and Coastal Law* (2004), at p. 300, note 7 and its accompanying text.

34 See *ENB*, vol. 25, No. 12, at p. 5. See also CBD COP Decision VII/5, paras. 60–61.

One question is whether bottom trawling occurring in the water column above the continental shelf of coastal states is included in the scope of the recommendation in this regard. As noted in Chapter 2 (as well as in Section 3.2.2 below), the extent to which coastal states may exercise jurisdiction to regulate bottom fisheries on the high seas superjacent to their continental shelf is controversial. Since the UNGA Resolutions refer to seamounts, hydrothermal vents and cold water corals located beyond national jurisdiction only for illustrative purposes, fishing activities on the high seas having adverse impacts on ecosystem features located within national jurisdiction are not excluded from the scope of the recommendation.

The Resolutions recognized the crucial importance of RFMO/As in the regulation of bottom fisheries in the mid to long term. First of all, RFMO/As having competence to regulate bottom fisheries were called upon to urgently adopt appropriate conservation and management measures in accordance with international law (and to ensure compliance therewith).[35] Bottom trawling to be addressed by such urgent conservation and management measures of RFMOs was, as in the case of the interim prohibition in paragraph 66 of Resolution 59/25, limited to that having "adverse impacts on vulnerable marine ecosystems". Resolution 60/31 clarified the measures to be taken to protect VMEs, indicating spatial and temporal measures.[36] After a review in 2006, again failing to impose an immediate moratorium despite continued calls to do so,[37] the General Assembly specified a series of concrete measures to be implemented no later than 31 December 2008:

> (a) To assess, on the basis of the best available scientific information, whether individual bottom fishing activities would have significant adverse impacts on vulnerable marine ecosystems, and to ensure that if it is assessed that these activities would have significant adverse impacts, they are managed to prevent such impacts, or not authorized to proceed;
> (b) To identify vulnerable marine ecosystems and determine whether bottom fishing activities would cause significant adverse impacts to such ecosystems and the long-term sustainability of deep sea fish stocks, inter alia, by improving scientific research and data collection and sharing, and through new and exploratory fisheries;

35 UNGA Resolution 59/25, para. 67. See also A/59/122, at para. 6(b); *ENB*, vol. 25, No. 12, at p. 5.

36 UNGA Resolution 60/31, para. 70.

37 See, e.g., The World Conservation Congress REC 3.099 The protection of seamounts, deep-sea corals and other vulnerable deep-sea habitats from destructive fishing practices, including bottom trawling, on the high seas, Bangkok, Thailand, 17–25 November 2004, para. 5; UN Millennium Project, Environment and Human Well-being: A Practical Strategy, Report of the Task Force on Environmental Sustainability, at p. 87; T. Murombo, "The Role of International Environmental Diplomacy in the Sustainable Use of Marine Biodiversity in Areas Beyond National Jurisdiction: Ending Deep Sea Trawling", 40 *Comparative and International Law Journal of Southern Africa* (2007), at pp. 172–192.

(c) In respect of areas where vulnerable marine ecosystems, including seamounts, hydrothermal vents and cold water corals, are known to occur or are likely to occur based on the best available scientific information, to close such areas to bottom fishing and ensure that such activities do not proceed unless conservation and management measures have been established to prevent significant adverse impacts on vulnerable marine ecosystems;

(d) To require members of the regional fisheries management organizations or arrangements to require vessels flying their flag to cease bottom fishing activities in areas where, in the course of fishing operations, vulnerable marine ecosystems are encountered, and to report the encounter so that appropriate measures can be adopted in respect of the relevant site [...][38]

The key terms such as VMEs and SAIs are not defined in the Resolution. The standards and criteria for identifying them are critical for the proper implementation of these provisions. The UNGA recommended that the FAO should develop such standards and criteria as needed to implement them.[39]

In addition, Resolution 59/25 called upon members of existing RFMO/As without the competence to regulate bottom fisheries or the impacts of fishing on VMEs to expand their competence and it also called upon states urgently to cooperate in the establishment of new RFMO/As with the competence to regulate bottom fisheries and the impacts of fishing on VMEs where necessary and appropriate, if no relevant RFMO/A existed.[40] In 2006, the General Assembly further called upon participants in such negotiations to expedite such negotiations and, by no later than 31 December 2007, to adopt and implement interim measures consistent with paragraph 83 of Resolution 61/105.[41] These resolutions were ambiguous as to which states are entitled to participate in such negotiations (e.g., high seas fishing states, coastal states not engaged in the high seas fishery concerned and/or other states).

As described above, the successive UNGA Resolutions clearly pointed out that RFMO/As are assigned a primary role in addressing the impacts of high seas bottom fisheries on VMEs. In other words, circumstances in which flag states have the discretion to determine what action is necessary are limited to those where there is no RFMO/A or where the existing RFMO/A has not adopted interim measures. In fact, the recommendations in Resolution 61/105 support this proposition. Following the recommendations concerning RFMO/A-related action, the Resolution called upon flag states either to adopt and implement measures in accordance with paragraph 83, *mutatis mutandis*, or to cease authorizing fishing

---

38 UNGA Resolution 61/105, para. 83.

39 Ibid., para. 89. This issue will be examined in the next sub-section.

40 UNGA Resolution 59/25, paras. 68–69.

41 UNGA Resolution 61/105, para. 85. See also UNGA Resolution 64/72, para. 124.

vessels flying their flag to conduct bottom fisheries pending the establishment of a competent RFMO/A or interim measures.[42]

As pointed out in Chapter 2, coastal states may exercise rights or jurisdiction with regard to the continental shelf subjacent to the high seas and, in certain circumstances, high seas fishing states shall meet the requirements concerning the conservation of sedentary species and the regulation of MSR activities. In this regard, it is remarkable that recent Resolutions include a proviso concerning the implementation of UNGA Resolution 61/105. For example, paragraph 104 of Resolution 63/112 stated as follows: the General Assembly

> [recalls] that nothing in paragraphs 83 to 86 of its resolution 61/105 is to prejudice the sovereign rights of coastal States over their continental shelf or the exercise of the jurisdiction of coastal States with regard to that shelf under international law as reflected in the [LOSC].[43]

As examined in Chapter 2, the rights and jurisdiction of coastal states with regard to the outer continental shelf are not clearly understood in connection with activities taking place on the high seas. In discussions on the UNGA Resolutions in the future, a clarification of legal ambiguities is inevitable, perhaps from a broad perspective of the law of the sea in general rather than in the context of deep-sea fisheries, with a view to preventing disputes from occurring on this issue.

In 2009, the General Assembly reviewed the implementation of Resolution 61/105 and found that further actions were needed.[44] First of all, it called upon RFMO/As with the competence to regulate bottom fisheries, states participating in negotiations to establish such organizations or arrangements, and flag states to take urgent actions in areas beyond national jurisdiction. It reiterated and in many respects advanced actions stipulated in paragraph 83 and other paragraphs of UNGA Resolution 61/105 concerning impact assessments, MSR, VME encounter protocols and the adoption of conservation and management measures including monitoring, control and surveillance measures.[45] This is complemented by the provision not to authorize bottom fishing activities until such measures (as well as measures stipulated in Resolution 61/105) have been adopted and implemented.[46] Second, the Resolution referred to the special circumstances and requirements of developing states and in this regard it stated

42 UNGA Resolution 61/105, para. 86. The resolution also called upon states to make publicly available a list of those vessels authorized to conduct bottom fisheries in areas beyond national jurisdiction and the measures adopted pursuant to paragraph 86 (para. 87).

43 Subsequent resolutions, reiterating this non-prejudice clause, refer to Article 77 of the LOSC. See UNGA Resolution 64/72, para. 115.

44 Ibid., paras. 118–124.

45 Ibid., para. 119.

46 Ibid., para. 120.

that implementation by such states should proceed in a manner that gives full consideration to section 6 of the FAO Guidelines.[47] Third, the Resolution called upon states and RFMO/As to enhance efforts to cooperate in the collection and exchange of scientific and technical data and information related to the implementation of the measures called for in Resolutions 61/105 and 64/72.[48] In particular, paragraph (a) called for exchanging best practices and developing, where appropriate, regional standards for use by States engaged in bottom fisheries in areas beyond national jurisdiction and regional fisheries management organizations or arrangements.

In 2011, the UNGA conducted a further review.[49] While noting the progress made in the implementation of the relevant paragraphs of these Resolutions, it found that the urgent actions called for in Resolutions 61/105 and 64/72 had not been fully implemented and further actions in accordance with the precautionary approach, ecosystem approaches and international law and consistent with the Guidelines were needed to strengthen the continued implementation.[50] In this regard, it called upon states, through RFMO/As with the competence to regulate bottom fisheries, states participating in negotiations to establish such organizations or arrangements, and flag states to take urgent actions regarding bottom fishing in areas beyond national jurisdiction as follows:

(a) Strengthen procedures both for carrying out assessments to take into account individual, collective and cumulative impacts, and for making the assessments publicly available, recognizing that doing so can support transparency and capacity-building globally;
(b) Establish and improve procedures to ensure that assessments are updated when new conditions or information so require;
(c) Establish and improve procedures for evaluating, reviewing and revising, on a regular basis, assessments based on best available science and management measures;

47 Ibid., para. 121. This was repeated in UNGA Resolution 66/68, para. 134.

48 UNGA Resolution 64/72, para. 122. The development and strengthening of data collection standards, procedures and protocols and research programmes for the identification of VMEs and the assessment of impacts are further provided in paragraph 123 of the Resolution.

49 UNGA Resolution 66/68, paras. 121–137. Note that, in parallel with negotiations for the UNGA Resolution, a workshop was held to discuss the implementation of the relevant paragraphs of Resolutions 61/105 and 64/72 on sustainable fisheries addressing the impacts of bottom fishing on VMEs and the long-term sustainability of deep-sea fish stocks. See 'Summary by the Moderator of the discussions held at the Workshop to discuss implementation of paragraphs 80 and 83 to 87 of resolution 61/105 and paragraphs 117 and 119 to 127 of resolution 64/72 on sustainable fisheries, addressing the impacts of bottom fishing on vulnerable marine ecosystems and the long-term sustainability of deep-sea fish stocks', reproduced in Letter dated 27 October 2011 from the Moderator of the Workshop to the President of the General Assembly, A/65/566, 18 November 2011.

50 UNGA Resolution 66/68, para. 129.

(d) Establish mechanisms to promote and enhance compliance with applicable measures related to the protection of vulnerable marine ecosystems, adopted in accordance with international law[.][51]

The UNGA also dealt with a number of issues which had so far not been sufficiently touched upon. First, the UNGA strengthened the link between the management of bottom fisheries and research activities: it referred to the importance of MSR such as seabed mapping and recommended further research, including MSR.[52] Second, the UNGA referred to the role of coastal states in the management of bottom fisheries beyond their EEZ. In addition to the repetition of the non-prejudice provision in paragraph 123, it noted "the adoption by coastal States of conservation measures regarding their continental shelf to address the impacts of bottom fishing on vulnerable marine ecosystems, as well as their efforts to ensure compliance with those measures".[53] Third, the UNGA not only welcomed the ongoing work of the FAO related to the management of deep-sea fisheries in the high seas, but also invited the FAO to consider undertaking a number of activities in facilitating the implementation of the Guidelines by states and RFMOs and arrangements.[54]

The UNGA decided to conduct a further review in 2015 of the actions taken by states and RFMO/As in response to the relevant paragraphs of UNGA Resolutions 64/72 and 66/68, with a view to ensuring the effective implementation of the measures and to make further recommendations, where necessary.[55]

3.2.2 *FAO International Guidelines on Deep-sea Fisheries on the High Seas*

The Technical Consultation on the International Guidelines for the Management of Deep-sea Fisheries in the High Seas was held in Rome in February and August 2008. It was convened in order to assist states and RFMOs and arrangements in sustainably managing deep-sea fisheries and implementing the relevant provisions of UNGA Resolution 61/105. In particular, COFI instructed that the Guidelines "should include standards and criteria for identifying vulnerable marine ecosystems beyond areas under national jurisdiction and the impacts of fishing activities on such ecosystems".[56] The Technical Consultation adopted the International Guidelines in August 2008.[57] The FAO started to implement a

51 Ibid.

52 Ibid., paras. 131–133.

53 Ibid., para. 124.

54 Ibid., para. 135. See also para. 136.

55 Ibid., para. 137.

56 Report of the Twenty-Seventh Session of the Committee on Fisheries, Rome, 5–9 March 2007, at para. 77.

57 There was no suggestion to develop follow-up instruments such as an international plan of action or a legally-binding instrument, although the use of the Guidelines in the upcoming

programme on the implementation of the Guidelines, including specific activities to increase collaboration and communication between the different stakeholders on deep-sea fisheries on the high seas.[58] The following paragraphs briefly discuss the Guidelines, focusing on the aspects which are relevant to the implementation of UNGA Resolution 61/105.

At the outset, it is useful to provide an overview of the scope of the Guidelines in terms of both the area of application and the type of fisheries. Among the recurring topics during the meeting were jurisdictional issues concerning deep-sea fisheries on the high seas over the continental shelf. As the title of the Guidelines suggests, the meeting was primarily concerned with fisheries on the high seas. It was extensively discussed whether the Guidelines should stipulate two different regimes for fisheries on the high seas superjacent to the continental shelf and those superjacent to the international seabed area (the Area). Some delegations argued for the recognition of different regimes between them on the basis of the sovereign rights and jurisdiction of coastal states over the continental shelf.[59] Instead of confining the application of the Guidelines to areas beyond the EEZ *and the continental shelf*, a non-prejudice clause was inserted regarding the sovereign rights and jurisdiction of coastal states with regard to the continental shelf as paragraph 25.

This issue is particularly relevant where no RFMO/A exists, especially where negotiations to establish an RFMO/A have not yet started such as the South West Atlantic. In that case, the coastal state might attempt to control high seas fisheries, claiming sovereign rights with regard to sedentary species (e.g., the protection of corals) or requiring its prior consent with regard to conducting MSR on its continental shelf.[60]

Paragraph 13 defines the type of fishing activities to which the Guidelines are applicable. It allows the inclusion of certain fisheries of concern while excluding pelagic fisheries on the high seas. First, the Guidelines apply to fisheries with the following two characteristics: (1) the total catch includes species that can only sustain low exploitation rates; and (2) the fishing gear is likely to contact

---

meetings to establish new RFMOs in the South Pacific and the North-Western Pacific was mentioned by some delegations.

58 See COFI/2011/7, FAO's Role for Improved Integration of Fisheries and Aquaculture Development and Management, Biodiversity Conservation and Environmental Protection, at para. 39. This includes a workshop held in Busan, Republic of Korea, on 10–12 May 2010 to examine issues in the implementation of the Guidelines and ways forward.

59 In fact, explicit reference was hardly made to the issue of 'straddling' deep-sea fish stocks.

60 The Guidelines stipulate that states should gather scientific information, such as "baseline information on the ecosystems, habitats and communities in the fishing area, against which future changes are to be compared" in paragraph 47(ii), in order to establish whether deep-sea fisheries are likely to cause SAIs. In view of the lack of the definition of MSR, a legal regime applicable to the collection of such information is subject to different interpretations.

the seafloor during the normal course of fishing operations. Second, the Guidelines are also applied as appropriate to similar fisheries, including those targeting medium productivity species.

3.2.2.1 *Common Understandings of Key Concepts*

As stated at the beginning of this sub-section, COFI explicitly instructed the Technical Consultation to include standards and criteria for identifying VMEs and the impacts of fishing activities on such ecosystems in the Guidelines. Thus, establishing definitions or common understandings of these key concepts (i.e., VMEs and SAIs) was a crucial issue for the meeting.

Paragraphs 14–16 of the Guidelines describe VMEs. The Guidelines recognized the need to take into account not only the likely effect of various threats but also the probability of those threats as one of the factors that determine the risks to a marine ecosystem.[61]

A list of characteristics to be used as criteria in the identification of VMEs was introduced in paragraph 42, and the list of examples of potentially vulnerable species groups, communities and habitats as well as features that potentially support them are contained in Annex I. The latter list includes those which were not explicitly mentioned in UNGA Resolutions 59/25 and 61/105 such as sponges as well as cold seeps.

In designating an ecosystem as 'vulnerable', the decision should evaluate habitats and ecosystems against the criteria in paragraph 42, "individually or in combination using the best available scientific and technical information".[62]

In paragraph 17 of the Guidelines, SAIs are described as:

> those that compromise ecosystem integrity (i.e. ecosystem structure or function) in a manner that: (i) impairs the ability of affected populations to replace themselves; (ii) degrades the long-term natural productivity of habitats; or (iii) causes, on more than a temporary basis, significant loss of species richness, habitat or community types.[63]

Impacts should be evaluated 'individually, in combination and cumulatively'. Among the six factors stipulated for consideration when determining the scale and significance of an impact is "the spatial extent of the impact relative to the availability of the habitat type affected".[64] Temporary impacts are those that are limited in duration and that allow the particular ecosystem to recover over

---

61 Para. 16.

62 Para. 46.

63 Para. 17.

64 Para. 18(ii). One might argue on this basis that the removal of a coral reef on a seamount during a particular fishery operation does not constitute an SAI if the same habitat type is available in the vicinity of the affected feature in sufficient abundance.

an acceptable timeframe, which are decided on a case-by-case basis and should be in the order of 5–20 years.[65]

Where the impact assessment concludes that the area does not contain VMEs or that SAIs are not likely, additional assessments are still required in limited circumstances. When there have been significant changes to the fishery or other activities in the area, or when natural processes are thought to have undergone significant changes, impact assessments should be repeated.[66]

3.2.2.2 *Conservation and Management Principles*

The Guidelines stipulate that the main objectives of the management of deep-sea fisheries are to promote responsible fisheries that provide economic opportunities while ensuring the conservation of marine living resources and the protection of marine biodiversity, by ensuring the long-term conservation and sustainable use of marine living resources in the deep seas and preventing SAIs on VMEs.[67] The objectives of the management of deep-sea fisheries as agreed for the Guidelines clearly indicate an emphasis on environmental protection, in particular the protection of benthic ecosystems.

The meeting agreed that states and RFMO/As should adopt and implement measures in accordance with the precautionary approach and an EAF.[68] As explored below, these principles find expression in many paragraphs of the agreed text. In addition, the meeting discussed the possibility to take more precautionary conservation and management measures for deep-sea fisheries than for pelagic fisheries. For instance, paragraph 23 states that "[i]n recognition of the potential vulnerability of deep-sea resources and their ecosystems, conservation and management measures for [deep-sea fisheries] should ensure that, while knowledge is low, harvest rates are kept low enough to minimise risk to sustainability and harvests only increase as knowledge, management capacity, and [monitoring, control and surveillance] increases".

Paragraph 21 stipulates that states and RFMO/As should recognize the need, in managing deep-sea fisheries, to do so in a manner which is consistent with the Code of Conduct and general principles set forth in the FSA. In this respect, the Guidelines enumerated actions to be taken in a non-exhaustive list in the same paragraph, including some notable additions to those found in the Code of Conduct and the FSA. First, subparagraph (ii) requires states and RFMO/As to identify areas or features where VMEs are known or are likely to occur, and the location of fisheries in relation to these areas and features. Second, management should be based on the best scientific and *technical* information available, as opposed to

---

65 Para. 19.

66 Para. 53.

67 Para. 11.

68 Para. 12.

the best scientific information (or evidence) available in the FSA and the Code of Conduct, and the fisher's knowledge is to be taken into account, where appropriate (subparagraph (iv)).

The action to be taken with regard to fishing gear and techniques under the Guidelines could not be interpreted to prohibit less selective gear such as bottom trawling since the term 'cost-effective' was inserted as in Article 5(f) of the FSA and Article 7.2.2(g) of the Code of Conduct.[69] Another point of note in this subparagraph is the reference to "recognizing the difficulties of managing fisheries with mixed species or high bycatch".

3.2.2.3 *Management in the Absence of Any Competent RFMO/A*

Paragraph 28 refers to the cooperation of states in the establishment of new RFMOs and arrangements. It calls on states to cooperate to develop interim measures prior to their establishment. It stipulates that

> [p]rior to the establishment of such a new RFMO/A States participating in the negotiations should cooperate to adopt and implement interim conservation and management measures to ensure sustainable management of [deep-sea fisheries] in the long-term and to prevent significant adverse impacts to VMEs, taking full account of these Guidelines.

The potential role of the FAO is of particular relevance to areas where no RFMO/A exists. The Guidelines stipulate the submission of impact assessments and conservation and management measures to the FAO as well as the role of the FAO as an information clearing-house (i.e., making it publicly available).[70]

3.2.2.4 *Prior Impact Assessments*

As indicated in paragraph 83 of UNGA Resolution 61/105, an impact assessment is an important element in managing deep-sea fisheries. The meeting agreed that "[f]lag States and RFMO/As should conduct assessments to establish if deep-sea fishing activities are likely to produce significant adverse impacts in a given area" and enumerated examples of information to be assessed.[71] As the type of information in the list indicates, the meeting recognized the need for impact assessments for proposed fisheries.[72]

---

69 Para. 23(v).

70 See, e.g., paras. 34, 52 and 57.

71 Para. 47.

72 The first meeting of the RFB Secretariats Network (RSN-1) agreed that there is an associated need to collect relevant data to better understand the potential impact of bottom trawling; as a precautionary measure, RSN-1 also noted that stringent data and information requirements should be put in place before opening any deepwater trawl fishery in the future. Report of the First Meeting of Regional Fishery Body Secretariats Network, Rome, 12–13 March 2007, at p. 5, para. 17.

Risk assessments referred to in paragraph 47(vi) do not need to be conducted in a uniform manner. They "should take into account, as appropriate, differing conditions prevailing in areas where [deep-sea fisheries] are well established and in areas where [deep-sea fisheries] have not taken place or only occur occasionally".[73] This provision appears to be intended to avoid burdening established fisheries too much. However, it has been pointed out that areas that have been subject to deep-sea fisheries for a long time might still contain unknown VMEs so that the differentiation should not be interpreted to imply that lower standards are applied to risk assessments in established fisheries. Put differently, fisheries in existing fishing areas might continue unchecked while risks to VMEs are not mitigated.

The Guidelines recognize that "there may be circumstances in which States may have to rely on information and data obtained only from vessels flying their flags or their own research activities when assessing [deep-sea fisheries] that take place in areas where no competent RFMO/A is in place".[74]

3.2.2.5 *Authorization of Fishing*

In line with paragraph 83 of UNGA Resolution 61/105, paragraph 73 of the Guidelines stipulates that if deep-sea fisheries activities would have SAIs, states and RFMOs and arrangements should ensure that these activities are managed to prevent such impacts or not authorized to proceed. Furthermore, the Guidelines formulated paragraph 74 on situations of substantial uncertainty. If the presence of VMEs or the likelihood that individual deep-sea fisheries activities would cause SAIs on VMEs cannot be adequately determined, states should only authorize them to proceed in accordance with: (i) precautionary conservation and management measures to prevent SAIs as described in paragraph 65; (ii) a protocol for encounters with VMEs consistent with paragraphs 67–69;[75] and (iii) measures to reduce the uncertainty. In other words, although the provision was not formulated in a negative way (e.g., "should not be authorized to proceed"), deep-sea fisheries are prohibited when substantial uncertainty remains unless and until conservation and management measures to prevent SAIs have been taken.

---

[73] Para. 48.

[74] Para. 49.

[75] Paras. 67–69 stipulate actions to be taken in order to implement paragraph 83(d) of UNGA Resolution 61/105.

3.3 *High Seas MPAs and Other Area-Based Management Tools*

Area-based management tools have been widely employed to protect and preserve the marine environment, including living resources, through the prohibition or regulation of certain activities in a defined area. The name of area-based management tools varies in accordance with the regulated activities and the purposes of the regulation. Among others, the term 'MPA' has been used on many occasions.[76] So far, MPAs have been mainly established within areas under national jurisdiction for various purposes, while existing high seas MPAs have been limited in terms of their number and regulatory purposes.[77] There is no single global high seas MPA treaty or global network of MPAs encompassing all regulatory purposes. In a number of recent international fora, the establishment of MPAs on the high seas and the regulation of activities in such areas have been advocated as a new management tool.[78] First and foremost, the JPOI called for the use of diverse approaches and tools and referred to the establishment of MPAs consistent with international law and based on scientific information, including representative networks by 2012 and time/area closures for the protection of nursery grounds and periods.[79] The CBD stipulates that "[e]ach Contracting Party shall, as far as possible and as appropriate [...] [e]stablish a system of protected areas or areas where special measures need to be taken to conserve biological diversity".[80] One commentator argues that "[r]ead in conjunction with each other, articles 4, 5 and 8 [of the CBD] not only supported the establishment

[76] There is no generally accepted definition of the term 'MPA'. For example, the IUCN defines it as "[a]ny area of intertidal or subtidal terrain, together with its overlying water and associated flora, fauna, historical and cultural features, which has been reserved by law or other effective means to protect part or all of the enclosed environment". Resolution 17.38, General Assembly of the IUCN (1988), reconfirmed in Resolution 19.46 (1994). For present purposes, this term is used in an all-encompassing way, without excluding any particular purpose of regulation.

[77] In fact, all global treaties except the CBD and most regional treaties purport to regulate only particular types of activities. A notable exception among the regional mechanisms is the Mediterranean Network. See Protocol Concerning Specially Protected Areas and Biological Diversity in the Mediterranean, Barcelona, 10 June 1995, which is applied to all the maritime waters of the Mediterranean, whether internal waters, historical waters, territorial seas, EEZ, fishing zones, ecological zones or high seas. See also T. Scovazzi, "Marine Protected Areas on the High Seas: Some Legal and Policy Considerations", 19 *International Journal of Marine and Coastal Law* (2004), at pp. 11–15. For the regional mechanism for the protection of the marine environment in the North-East Atlantic, see Section 4.4.2.3 below.

[78] Nevertheless, some states challenged the compatibility of MPAs on the high seas with the law of the sea, particularly the LOSC. *ENB*, Vo. 25 No. 6 (9 June 2003), at pp. 4–5.

[79] Johannesburg Plan of Implementation, para. 32(c). This commitment has been generally followed in other global fora. See, e.g., UNGA Resolution 66/231, Oceans and the Law of the Sea, 24 December 2011, para. 176; CBD COP Decision X/29, paras. 4, 8(d), 13(a), 29, 33 and 45.

[80] CBD, Article 8(a).

of high seas MPAs but rather under certain circumstances actually required their establishment".[81]

In a fisheries context, RFMO/As have established area closures for certain fisheries in many areas. As these tools may play a role in preserving not only spawning biomass but also critical habitats and sensitive life stages of species,[82] they have been adopted by RFMO/As with a view to conserving target species and/or protecting marine ecosystems. The implications and benefits of MPAs for fisheries are less clear than for biodiversity conservation, though. With regard to the establishment of closed areas for fisheries by RFMO/As, the main legal issue is whether the RFMO/A concerned has the competence to do so under its constitutive instrument.

As for integrated or multi-purpose MPAs, additional legal and policy issues arise. In fact, at the meeting of the biodiversity working group established by the UNGA, while a number of delegations stated that multi-purpose MPAs would be 'a key tool' to manage biodiversity beyond areas of national jurisdiction,[83] other delegations took a cautious approach to the utility of high seas MPAs and their compatibility with the law of the sea. In their view, "there should be a strong causal link between the impacts being addressed and the management measures proposed, consistent with customary international law, as reflected in the [LOSC]".[84] At the second meeting of the Working Group, states were more inclined to recognize the need for MPAs but there was still some hesitation: the joint statement of the Co-Chairpersons states that, while "[a]rea-based management tools are essential and effective tools in the conservation and sustainable use of marine biological diversity", it was emphasized that "marine protected areas were only one of the several available tools and needed to be consistent with international law".[85] In subsequent meetings, although developments such as the adoption of the criteria for the identification of ecologically or biologically significant marine areas, the criteria for identifying VMEs developed by the FAO and the establishment of MPAs in areas beyond national jurisdiction in the North-East Atlantic helped to accelerate focused discussions on MPAs in areas beyond national jurisdiction, differences in the views of participating states remain.[86]

---

81 Oral, "Protection of Vulnerable Marine Ecosystems in Areas Beyond National Jurisdiction: Can International Law Meet the Challenge?" at p. 97.

82 Fisheries management, Technical Guidelines for Responsible Fisheries 4, at p. 47.

83 A/61/65, at p. 16, para. 61.

84 Ibid., at pp. 16–17, para. 62.

85 A/63/79, at p. 7, para. 26.

86 See Letter dated 16 March 2010 from the Co-Chairpersons of the Ad Hoc Open-ended Informal Working Group to the President of the General Assembly, A/65/68, 17 March 2010, at pp. 4 and 12–14, paras. 17–18 and 58–69; Letter dated 30 June 2011 from the Co-Chairs of the Ad Hoc Open-ended Informal Working Group to the President of the General Assembly, A/66/119,

International legal issues arising in relation to the freedom of the high seas include the compatibility between the establishment of such areas on the high seas and the freedom of the high seas as well as the invalidity of claims of sovereignty over the high seas and in relation to the Area.[87] Proponents have argued that the freedom of the high seas is not absolute and the exercise of freedoms is subject to obligations under the law of the sea such as the conservation of the marine living resources of the high seas and the protection of the marine environment.[88] Accordingly, as Articles 87 and 92(1) of the LOSC must be understood in this context, multi-purpose MPAs are not incompatible with the law of the sea. In addition, as noted earlier, the CBD stipulates the establishment of MPAs in Article 8.[89] All in all, it can be argued that the establishment of multi-purpose high seas MPAs is not in itself incompatible with the law of the sea, but they are only binding on states that have consented.[90]

Another controversial issue for multi-purpose MPAs may be potential conflicts over which entity (or entities) is entitled to designate such MPAs, to determine the applicable regulatory measures and to enforce such measures therein: e.g., states (an individual state or a group of states); regional bodies (RFMO/As or regional seas programmes under the United Nations Environment Programme (UNEP)); global bodies (the International Maritime Organization (IMO), the FAO, the CBD, the UNGA or any other entity to be established for that purpose). In fact, various international institutions have been working on the issue of area-based management tools simultaneously, including COPs of the CBD, the FAO, the IMO and the Ad Hoc Open-ended Informal Working Group to Study Issues Relating to the Conservation and Sustainable Use of Marine Biological Diversity beyond Areas of National Jurisdiction established by the UNGA. Questions

30 June 2011, at pp. 6–7, paras. 23–29; Letter dated 8 June 2012 from the Co-Chairs of the Ad Hoc Open-ended Informal Working Group to the President of the General Assembly, A/67/95, 13 June 2012, at p. 6, paras. 20–23.

87 LOSC, Articles 87, 89 and 137(1).

88 As regards measures necessary for protecting and preserving VMEs and species, Article 194(5) has been invoked as a legal basis for the designation of certain categories of areas, which may require higher standards of environmental protection. European Commission Background Paper No. 3, at p. 16. Scovazzi points out that although a possible convention on high seas MPAs could not be applicable to third party vessels, every state is already under the obligations arising from customary international law and from the LOSC to protect and preserve rare or fragile ecosystems and to cooperate for this purpose. T. Scovazzi, "New International Instruments for Marine Protected Areas in the Mediterranean Sea", in A. Strati, M. Gavouneli and N. Skourtos (eds.), *Unresolved Issues and New Challenges to the Law of the Sea: Time Before and Time After* (2006), at p. 120.

89 For the relationship between the provisions of the LOSC and the CBD, see Section 2.3.2.3 above.

90 For a similar view, see, e.g., Wolfrum and Matz, "The Interplay of the United Nations Convention on the Law of the Sea and the Convention on Biological Diversity", at p. 468.

include: which institution should be allowed to designate multi-purpose MPAs and, if two or more institutions have the competence to employ area-based management tools to regulate different activities falling under their respective competence in the same area, which institution, if any, should play a primary role and how potential conflicts should be avoided, including the cooperation and coordination in developing criteria for the designation of the area as well as in respect of the determination of measures to regulate activities therein.

With regard to the cooperation and coordination among relevant international organizations, policy questions have not yet been settled at the global level. The Task Force on Biodiversity in Marine Areas beyond National Jurisdictions has been established within UN-Oceans in order to coordinate information input to the General Assembly, the CBD, and other international processes dealing with biodiversity in marine areas beyond national jurisdiction.[91] A new Task Force on Marine Protected Areas and Other Area-based Management Tools was established at the 5th Meeting of UN-Oceans.[92] The composition of the lead agencies confirms the need for cooperation and coordination on this issue: while the Biodiversity Task Force is led by the CBD Secretariat and DOALOS, the Task Force on MPAs and other area-based management tools is led by four co-lead agencies (i.e., the CBD Secretariat, UNESCO, the FAO and UNEP). It remains to be seen how the Task Forces will contribute to the development of these issues in the future.

Under the 1995 Jakarta Mandate, parties to the CBD have been working on this issue, and designated the topic 'marine and coastal protected areas' as one of the thematic areas.[93] The Ad Hoc Open-Ended Working Group on Protected Areas in its first meeting in June 2005 considered, among others, 'Options for cooperation for the establishment of marine protected areas in marine areas beyond the limits of national jurisdiction'. In 2006, the eighth COP decided to defer to the UNGA on this issue, recognizing the UNGA's "central role in addressing issues relating to the conservation and sustainable use of biodiversity in marine areas beyond national jurisdiction".[94] The COP did not include the issue of MPAs in

91 Report of the First Inter-Agency Meeting of UN-OCEANS, 25–26 January 2005, UNESCO-IOC, Paris, at p. 6. Reports and other information concerning the UN-Oceans are available at the website of the UN-Oceans <http://www.unoceans.org/Index.htm>.

92 Report of the Fifth Session of UN-OCEANS, UNESCO Headquarters, Paris 21–22 May 2007, at pp. 8–9.

93 'Marine and coastal protected areas' was designated as programme element 3. See Decision IV/5 and its Annex.

94 Decision VIII/24, first recital after the heading '*Options for cooperation for the establishment of marine protected areas in marine areas beyond the limits of national jurisdiction*'. See also European Commission Background Paper No. 3, at p. 19. The subsequent COP also supported this position by focusing on scientific and technical aspects and called for cooperation in the context of the UNGA processes. See, e.g., Decision IX/20, paras. 8 and 18–19. It should not,

areas beyond national jurisdiction in the provisional agenda for the second meeting of the ad hoc Working Group, but it agreed to consider, at its ninth meeting, progress and further supporting actions as appropriate concerning biodiversity beyond national jurisdiction, including MPAs. Thus, although the CBD has deferred to the UNGA with respect to policy issues, it might become active again in policy discussions on this issue in the future.[95] It should be noted that the CBD adopted the scientific criteria for identifying ecologically or biologically significant marine areas in need of protection and the scientific guidance for designating representative networks of MPAs at its ninth COP.[96] The identification of ecologically or biologically significant marine areas is ongoing in various areas of the world's oceans and seas.

Upon a recommendation by COFI in 2005, the FAO started developing technical guidelines on the design, implementation and testing of MPAs, building on the best available knowledge on fisheries science and management and the role and requirements of MPAs, with particular emphasis on their potential contribution to an ecosystem approach to fisheries.[97] The UNGA has welcomed the proposed work of the FAO on this issue.[98] Nevertheless, it did not give the coordinating or leading role to the FAO with regard to MPAs for fisheries conservation purposes.[99] Therefore, while the work of the FAO may set scientific criteria and operational standards in respect of fisheries, the FAO will not be able to designate such areas in regions where there exist competent RFMO/As. Nor is it competent to develop scientific or technical criteria for MPAs for non-fisheries purposes or to play a leading role in policy discussions.[100]

---

however, be forgotten that the COP emphasized the critical role of the CBD by observing as follows: "*Recognizes* that the Convention on Biological Diversity has a key role in supporting the work of the General Assembly with regard to marine protected areas beyond national jurisdiction, by focusing on provision of scientific and, as appropriate, technical information and advice relating to marine biological diversity, the application of the ecosystem approach and the precautionary approach, and in delivering the 2010 target". Decision VIII/24, para. 42.

95 See also E. Morgera, "Competence or Confidence? The Appropriate Forum to Address Multi-Purpose High Seas Protected Areas", 16 *Review of European Community & International Environmental Law* (2007), at p. 11.

96 CBD COP Decision IX/20, Marine and coastal biodiversity, UNEP/CBD/COP/DEC/IX/20 (2008), para. 14.

97 COFI/2007/8, at p. 5. The Guidelines were finally published in 2011. FAO, "Marine Protected Areas and Fisheries", FAO Technical Guidelines for Responsible Fisheries No. 4, Suppl. 4 (2011).

98 See, e.g., UNGA Resolution 63/112, para. 109.

99 Ibid. (the UNGA "urges coordination and cooperation among all relevant international organizations and bodies").

100 The FAO, in the process of developing the technical guidelines, has confined itself to technical issues involved in MPAs for fisheries management and emphasized the need for cooperation between other interest groups, including the CBD. See, e.g., paras. 22, 31 and 38 of the

In the meeting of the biodiversity working group established by the UNGA, on the one hand, it was "reaffirmed that the [UNGA] has a central role in addressing issues relating to the conservation and sustainable use of marine biological diversity beyond areas of national jurisdiction [and other] organizations, processes and agreements also have an essential complementary role within areas of their respective competence".[101] On the other hand, while "[i]t was proposed that the General Assembly could assume a leading role in the identification of criteria for the establishment of marine protected areas, and that such underutilized bodies as the Meeting of States Parties of the [LOSC] could be considered for this purpose", there appears to have been no consensus on assigning such a role to the UNGA or the meeting of the parties to the LOSC.[102] In fact, the above proposal was made in the following context:

> it was noted that further cooperation was necessary to further develop criteria for the identification of ecologically and biologically significant areas, the development of systems of marine protected areas and biogeographic classification systems. Further consideration should also be given to ways of establishing, managing and enforcing such designations, recognizing in this context the existing role and mandate of such bodies as FAO, IMO and the Convention on Biological Diversity and of regional seas conventions.[103]

Read together, the above statements imply that delegations were reluctant to recognize such a coordinating or leading role for the UNGA. The same view was expressed at the second meeting. At that meeting, some delegations stressed that it was also important to recognize regional differences and the need to develop area-based management tools on a case-by-case basis, based on the best available scientific information, taking into account the purpose of the area-based management tools and the specific ecological and biological features of the area.[104] All in all, legal and policy frameworks for coordination and cooperation in establishing multi-purpose MPAs are ambiguous at the moment.

It is in this context that an initiative to adopt an implementation agreement to the LOSC on the protection of marine biodiversity in areas beyond national

---

conclusions and recommendations by 2006 Workshop on MPAs and Fisheries Management (Fisheries Report No. 825). Among other things, it stated that the FAO could contribute to the improved management of fisheries in areas beyond national jurisdiction by providing technical guidance on effective means of management and that greater collaboration between the FAO and the CBD would help to integrate initiatives to achieve the conservation of biodiversity with efforts to ensure sustainable use of fishery resources (para. 22).

101 A/61/65, at p. 21 (Annex I, para. 2).

102 No such statement is recorded in the Co-Chairpersons' summary of trends. See also ibid., at p. 16, para. 60.

103 Ibid.

104 A/63/79, at p. 8, para. 31.

jurisdiction has been advocated, among others, by the European Union.[105] In 2011, the Ad Hoc Open-ended Informal Working Group recommended the General Assembly to initiate a process, "with a view to ensuring that the legal framework for the conservation and sustainable use of marine biodiversity in areas beyond national jurisdiction effectively addresses those issues by identifying gaps and ways forward, including through the implementation of existing instruments and the possible development of a multilateral agreement under the [LOSC]".[106] In response, the General Assembly decided to initiate such a process within the Ad Hoc Open-ended Informal Working Group.[107]

In summary, first, area-based management tools have been advocated in a fisheries context for use by RFMO/As. Second, multi-purpose MPAs could affect fisheries management by RFMO/As through the potential coordinated designation of MPAs in an integrated manner, whose purpose might include the regulation of fisheries even where the regulation of activities in a given area is not necessary from a fisheries management perspective. Third, multi-purpose MPAs have received mixed reactions from states, and, if they are to be established in the future, it is uncertain which institution takes on the coordinating or leading role in the establishment of such MPAs.

At the Rio+20, the commitment concerning area-based management tools has been renewed, in particular taking into account the developments within the CBD. The outcome document refers to the reaffirmation of "the importance of area-based conservation measures, including [MPAs], consistent with international law and based on best available scientific information, as a tool for conservation of biological diversity and sustainable use of its components" and notes CBD Decision X/2, in particular Target 11 of the Aichi Biodiversity Targets.[108] The difference in terms (i.e., 'area-based management tools' and 'area-based conservation measures') should not be given too much importance. The Rio+20 commitment in this regard draws heavily upon the wording of the Aichi Biodiversity Targets; CBD Decision X/2 uses both of these two terms, without explaining any difference. In addition, in its recommendations to the UNGA in 2011, the Ad Hoc Open-ended Informal Working Group still referred to "area-based management tools, including marine protected areas".[109]

---

105 See, e.g., Green Paper: Towards a future Maritime Policy for the Union: A European vision for the oceans and seas, at p. 17. See also EU Presidency Statement – Oceans and Law of the Sea and Sustainable Fisheries at the UNGA in 2006, available at <http://www.eu-un.europa.eu/articles/en/article_6575_en.htm>. For a succinct overview of this issue, see Morgera, "Competence or Confidence? The Appropriate Forum to Address Multi-Purpose High Seas Protected Areas", at pp. 9–10.

106 A/66/119, at p. 2, para. 1(a).

107 UNGA Resolution 66/231, para. 167.

108 The Future We Want, para. 177.

109 UNGA Resolution 66/231, Annex.

3.4 *Concluding Remarks*

A growing awareness of the need to address the conservation and management of DHSFS by incorporating new approaches found in the FSA is discernible at various global fora. While the Working Group of the CBD did not reach agreement on the application of the principles of the FSA in discussing options for cooperation for the establishment of MPAs beyond national jurisdiction, agreement was secured in all the other fora on the extended application of certain principles of the FSA to DHSFS, including the UNGA and the Review Conference.[110] In this regard, two points should be noted. First, a careful reading of the recommendations shows that states are recommended to recognize that some principles of the FSA *should* be applicable to DHSFS. In other words, hortatory calls are further qualified by an auxiliary verb. Second, references to the application of FSA general principles at the global level are always directed at discrete high seas 'fish' stocks. In fact, the FSA excludes sedentary species as defined in Article 77 of the LOSC from the definition of 'fish' in its application. It is therefore not clear, at least from practice at the global level, whether the exploitation of sedentary species beyond areas of national jurisdiction is also governed by the same principles.

Which principles of the FSA should be applied to DHSFS? In the documents concerned, principles or provisions of the FSA to be applied to discrete stocks are qualified by adjectives such as 'fundamental', 'general', and 'relevant', while their content has yet to be clarified. As a minimum, principles applicable to DHSFS should include Articles 5 and 6 as well as annexes I and II of the FSA.[111] Provisions relating to RFMO/As, non-flag enforcement, and dispute settlement may also be applied by RFMO/As or included in a new global instrument.[112]

110 However, note the following comments: "there is reportedly resistance to extending the FSA to include [stocks that are located entirely in the high seas]". Background Paper on the Conservation Status of Migratory Sharks and Possible Options for International Cooperation under the Convention on Migratory Species, UNEP/CMS/MS/4, 23 March 2007, at p. 47. This background paper was prepared in the context of an initiative launched under the Convention on Migratory Species (CMS) to negotiate a global instrument dealing with migratory sharks. The paper, which was prepared on behalf of the CMS Secretariat, briefly touched upon the issue of DHSFS in the context of the CMS: it referred to the possibility to discuss the problem of "high seas migratory species that probably never or only rarely cross administrative boundaries between the high seas and EEZs", despite the fact that the CMS is exclusively concerned with species that cross a boundary between the maritime area of two or more states or between the maritime area under national jurisdiction and the high seas (pp. 4 and 49).

111 See, for example, Chairman's background paper on possible initiatives at the fourth meeting of states parties to the FSA (Report of the Fourth ICSP, at p. 20). See also Molenaar, "Addressing Regulatory Gaps in High Seas Fisheries", at p. 564.

112 Hayashi, suggesting a new global agreement for high seas fisheries, argues that the main parts of such an agreement could be modelled on the FSA and a number of provisions of the FSA,

The provision concerning compatibility between conservation and management measures for the high seas and those for the EEZ would be, *mutatis mutandis*, applicable when a discrete high seas stock occurs in areas within the competence of different RFMO/As (in other words, 'straddling' DHSFS) or when two or more RFMO/As have overlapping competence in relation to certain discrete stocks in the same area.[113] An examination of the practice by RFMO/As would help to understand which principles should be applied and in what manner they should be applied (e.g., whether or not in the same manner as straddling and highly migratory fish stocks).

A remaining question regarding DHSFS would be whether and how to develop a legal or non-legal instrument governing the exploitation of these stocks.[114] No single view on how best to proceed has been agreed at the international level.[115]

Second, the UNGA and the FAO have been addressing the management of deep-sea fisheries at the global level. Through its annual Resolutions, the General Assembly has been recommending actions to be taken by states and RFMO/As, while the FAO has taken on the technical work including the development of common criteria and standards for the identification of VMEs and SAIs under the International Guidelines on deep-sea fisheries. Instead of imposing a blanket moratorium on bottom trawling on the high seas, the UNGA called for bottom fisheries on the high seas to be addressed at the regional level in a precautionary manner in accordance with the guidance by the UNGA. Two points are of interest. First, the UNGA designated RFMO/As as a principal vehicle to carry out the management of deep-sea fisheries; individual flag states have a complementary

---

including most of the provisions in parts III to VII, in addition to Articles 5 and 6 in part II and annexes I and II, could be used with minor adjustment. Hayashi, "Global Governance of Deep-Sea Fisheries", at pp. 296–297. In addition, an FAO information paper states that the recommendation in UNGA Resolution 60/31 to recognize that the general principles of the FSA should also apply to DHSFS enhances the role of high seas RFMOs, suggesting that such principles should include provisions concerning RFMOs. COFI/2007/9 Rev.1, at p. 5.

113 However, Molenaar argues that the choice of provisions to be applicable to discrete stocks should be based on whether a provision is *not* suitable for discrete fish stocks, and on that understanding that Article 7 on the compatibility of conservation and management measures is not applicable. Molenaar, "Addressing Regulatory Gaps in High Seas Fisheries", at p. 565.

114 On advantages and disadvantages of legally binding instruments, see D.A. Balton and D.C. Zbicz, "Managing Deep-Sea Fisheries: Some Threshold Questions", 19 *International Journal of Marine and Coastal Law* (2004), at pp. 252–255. See also Oral, "Protection of Vulnerable Marine Ecosystems in Areas Beyond National Jurisdiction: Can International Law Meet the Challenge?" at pp. 107–108 (stating, in the context of the conservation and management of high seas and deep ocean VMEs, "[t]here is no reason to attempt to create a new set of rules for the high seas as the existing ones do provide the necessary foundation").

115 For discussions on possible future options, see further Chapter 6 below.

role.[116] Neither UNGA Resolutions nor the FAO Guidelines explicitly articulate the potential role of coastal states with regard to deep-sea fisheries on the high seas above the continental shelf. While, in theory, the jurisdiction of coastal states is safeguarded by the non-prejudice clause in the Guidelines and UNGA Resolution 63/112, the primary role of RFMO/As complemented by flag states virtually means that the exercise of coastal state jurisdiction remains a last resort that may entail a risk of being accused of violating the law of the sea. Second, measures for deep-sea fisheries need to be more precautionary than those for pelagic fisheries under the UNGA recommendations and the FAO Guidelines. The UNGA Resolution, developed by the Guidelines, emphasizes the requirement of prior impact assessment. The prohibition of bottom fisheries where SAIs are likely, as recommended in UNGA Resolution 61/105, is precautionary in nature in its own right. Furthermore, the FAO Guidelines fill a potential gap of uncertainty by the further application of the precautionary approach. The burden of proof is reversed: in cases of substantial uncertainty, deep-sea fisheries need to proceed in a cautious manner.

Practice regarding deep-sea fisheries at the global level has not led to a modification of the LOSC regime governing fisheries on the high seas, and this will continue to be the case, at least in the near future. However, some developments should be noted as potentially interesting trends. Among others, no distinction is made by international organizations between fisheries for straddling stocks and those for DHSFS in addressing deep-sea fisheries. In fact, as will be seen in the next chapter, stock distribution is often controversial and inconclusive, and it is difficult to determine which stock is straddling and which stock is not. Therefore, to a certain extent, a possible dichotomy between the regime for DHSFS and regimes for transboundary stocks on the high seas is blurred by recent practice concerning deep-sea fisheries. This does not, however, imply that the same thing is usually true of fisheries other than deep-sea fisheries.

Third, recent calls for MPAs in the high seas as a new management tool shows that there are proponents of taking a cross-sectoral, all-encompassing approach to various activities on the high seas, including high seas fisheries, by coordinating the regulation of activities in different sectors. This may impact the regulatory framework for high seas fisheries because the designation of high seas MPAs by a global international organization may be initiated by motives distinct

[116] The relevant recommendations in UNGA Resolution 61/105 also testify to the claim that the international community is not eager to accept an argument that corals are mineral resources and the International Seabed Authority has (concurrent) competence to regulate bottom trawling on the high seas above the Area. For discussions on this issue, see Section 2.1.4 above.

from fisheries management.[117] In this regard, legal and policy issues arise. In legal terms, this concerns a balance between freedom of high seas fishing and the protection of the marine environment. The LOSC itself does not give a clear answer since, generally speaking, the LOSC provides each zone with a set of rules governing it even though more than one part may govern certain activities. In the policy domain, one of the main issues is the leading or coordinating role to be played by a particular institution or institutions with regard to MPAs. At the global level, this legal/policy uncertainty has led to the discussions on the need for an implementing agreement to the LOSC. Considering the uncertainty associated with the issue of multi-purpose MPAs, it is concluded that the LOSC/FSA regulatory framework attributing the primary role in fisheries management to RFMO/As will continue in the foreseeable future, but initiatives such as the UNGA biodiversity working group might lead to reform by assigning a supervisory role to a particular global organization concerning the establishment of high seas MPAs.

[117] The danger of potential conflicts and the need for coordination are more imminent at the regional level. The next chapter describes the relationship between RFMOs and other international organizations with regard to area-based management tools.

Chapter Four

Existing Regional Fisheries Management Organizations

Chapter 3 observed that states have been addressing the contemporary challenges with regard to high seas fisheries in three ways. First, states have been shaping a regulatory framework to manage fisheries for DHSFS at the global level building upon the framework of the LOSC and the principles of the FSA. Second, the UNGA has been playing a leading role in addressing deep-sea fisheries on the high seas, assigning a central role to RFMO/As in the implementation of its recommendations on the regulation of high seas deep-sea fisheries. Third, area-based management tools have been employed by RFMOs and have proven to be a useful tool in fisheries management; coordination and cooperation in establishing and operating integrated MPAs might have impacts on fisheries management.

This chapter investigates how RFMOs with the competence to manage straddling fish stocks and DHSFS have addressed fisheries for discrete high seas stocks and deep-sea fisheries on the high seas and how they have used area-based management tools and coordinated their actions with other regional organizations. The scope of the competence of RFMOs is determined by the provisions of their constitutive instruments and their practice. The constitutive instruments of RFMOs dealing with highly migratory species, anadromous stocks and catadromous species define their competence by listing the species concerned. By definition, DHSFS do not fall within the competence of these RFMOs.[1] The constitutive instruments of RFMOs dealing with straddling stocks define the scope of their competence by delimiting areas geographically. Therefore, as a point of departure, it may be assumed that RFMO/As dealing with straddling stocks are also competent with respect to discrete high seas stocks unless certain species are excluded from the scope of these RFMO/As. In fact, the Review Conference on the FSA found that RFMO/As which are competent to regulate straddling fish stocks have the necessary competence to conserve and manage DHSFS.[2] On the

---

[1] For the potential competence of the Western and Central Pacific Fisheries Commission, see Section 5.4.4 below.

[2] See Outcome of the Review Conference, para. 16.

basis of this assumption, each section of this chapter first examines whether five RFMOs with the competence to manage straddling fish stocks also have competence with regard to discrete high seas stocks.[3]

After confirming the competence to deal with fisheries for DHSFS, each section deals with the following issues: the similarity and difference in the approaches of the RFMOs concerned in managing fisheries for DHSFS and straddling fish stocks, including the application of the precautionary and ecosystem approaches; conservation measures for deep-sea fisheries, including bottom trawling, from the perspective of implementing the UNGA fisheries Resolutions as well as the influence on the actions at the global fora; the use of area-based management tools and cooperation and coordination with other organizations aimed at the protection of the marine environment in the same region.

The sections of this Chapter are organized following the chronological order of the date of the latest amendment to constitutive instruments and, if no amendment has been made to a constitutive instrument, the original date of the adoption of the constitutive instrument. By organizing its content in this way, the chapter intends to reflect trends in discussions and to illustrate interrelations in the negotiations on establishing or updating RFMOs.

4.1 *Commission for the Conservation of Antarctic Marine Living Resources (CCAMLR)*

At the Special Consultative Meetings in 1978 and 1980, Antarctic Treaty Consultative Parties negotiated a draft treaty dealing with the conservation of Antarctic marine living resources. The Convention on the Conservation of Antarctic Marine Living Resources (CCAMLR Convention) was adopted in 1980 and entered into force in 1982.[4]

The Convention applies to marine living resources in the area south of 60° south latitude and those in the area between that latitude and the Antarctic

---

[3] The regime dealing with Pollock stocks in the Central Bering Sea is also competent to deal with straddling stocks. Section 5.3 below briefly touches upon this regime in the context of the establishment of an RFMO for the North Pacific Ocean.

[4] See, e.g., D.G.M. Miller *et al.*, "Managing Antarctic Marine Living Resources: The CCAMLR Approach", 19 *International Journal of Marine and Coastal Law* (2004), at pp. 317–363; R. Herr, "The International Regulation of Patagonian Toothfish: CCAMLR and High Sea Fisheries Management", in O.S. Stokke (ed.), *Governing High Seas Fisheries: The Interplay of Global and Regional Regimes* (2001), at pp. 303–328; E.J. Molenaar, "CCAMLR and Southern Ocean Fisheries", 16 *International Journal of Marine and Coastal Law* (2001), at pp. 465–499.

Convergence which form part of the Antarctic marine ecosystem.[5] Thus, the Convention covers the Antarctic marine ecosystem as a whole.[6]

While only a few islands exist below 60° south, a number of islands exist in the sub-Antarctic area between that latitude and the Antarctic Convergence (referred to as 'sub-Antarctic islands'). At the Conference which adopted the CCAMLR Convention in 1980, the Chairman made a statement concerning these islands. The Statement explicitly recognized the French claim to sovereignty over Kerguelen and Crozet Islands and its jurisdiction over waters adjacent to these islands. It would be open to France whether to agree on the application of any specific conservation measure to waters adjacent to these islands or to indicate the exclusion of its application thereto. France would be bound by any conservation measures of the Commission only with its participation as far as these islands are concerned.[7] Moreover, the same "understandings [...] also apply to waters adjacent to the islands within the area to which [the] Convention applies over which the existence of State sovereignty is recognized by all Contracting Parties".[8] Unless a reservation is made by coastal states to conservation measures, the analysis in this section does not distinguish among waters in the entire Convention Area.[9]

5 CCAMLR Convention, Article I(1).

6 Under the Antarctic Treaty System, claims to sovereignty or title thereto over Antarctica are frozen: nothing in the Treaty is interpreted as a renunciation of asserted rights or claims. Antarctic Treaty, Washington, 1 December 1959, Article IV. See also CCAMLR Convention, Articles III–IV (prescribing that both contracting and non-contracting parties to the Antarctic Treaty are also bound by Articles IV of the Treaty in their relations with other contracting parties to the CCAMLR Convention). Claimant states have restrained from exercising enforcement jurisdiction over sections of the continent and islands or waters adjacent thereto in relation to third-state nationals. See Molenaar, "CCAMLR and Southern Ocean Fisheries", at pp. 477–478; F. Orrego Vicuña, "The Law of the Sea and the Antarctic Treaty System: New Approaches to Offshore Jurisdiction", in C.C. Joyner and S.K. Chopra (eds.), *The Antarctic Legal Regime* (1988), at p. 108. For the purposes of this study, waters adjacent to the continent and the islands south of 60° south latitude are treated like high seas. See Antarctic Treaty, Article VI. See also Molenaar, "CCAMLR and Southern Ocean Fisheries", at p. 481.

7 See paras. 2 and 3 of the Statement.

8 See ibid., para. 5.

9 As regards islands whose sovereignty is not recognized by all contracting parties, see a description concerning South Georgia and the South Sandwich Islands over which the governments of the UK and Argentina agreed in their joint statement that measures adopted by CCAMLR were valuable but there was a need to strengthen the existing arrangements, and agreed to renew their efforts in the context of CCAMLR. E. Meltzer, "Global Overview of Straddling and Highly Migratory Fish Stocks: The Nonsustainable Nature of High Seas Fisheries", 25 *Ocean Development and International Law* (1994), at p. 278.

Commercially attractive species fished in this area include krill and toothfish.[10] There are very few stocks that straddle both EEZs of the sub-Antarctic islands and high seas waters within the Convention Area.[11] This study analyzes conservation measures for all marine living resources within the Antarctic Convergence unless a certain species to which the measure is applied is determined as a straddling stock.

4.1.1 *The Convention*

The objective of the Convention is the conservation of Antarctic marine living resources, which, for the purposes of the Convention, includes rational use.[12] These living resources are defined as "the populations of fin fish, molluscs, crustaceans and all other species of living organisms, including birds, found south of the Antarctic Convergence".[13] Thus, living resources covered by the Convention include discrete high seas stocks and straddling stocks.

The principles of conservation applicable to activities in the area concern not only harvested populations but also dependent and related populations and the Antarctic marine ecosystem.[14] The latter means the complex of relationships of Antarctic marine living resources with each other and with their physical environment, thus including both associated or dependent species and habitats.[15] Provisions concerning the maintenance of the ecological relationships and the restoration of depleted populations to the designated levels make it possible for the Commission to take measures for bycatch reduction both to ensure sustainable fisheries for the target species and to seek the conservation of non-harvested species as such.

The Commission formulates, adopts and revises conservation measures on the basis of the best scientific evidence available.[16] Even though the Convention does not explicitly employ the term 'precautionary approach', it is clear

---

10 Despite the intention of the drafters of the Convention, who aimed to conserve Antarctic marine living resources as a whole by extending its area of application to the outer limit of the Antarctic Convergence, species such as Patagonian toothfish occur not only within the Convention area but also in adjacent areas under national jurisdiction outside the Convention Area. On the stock distribution of Patagonian toothfish, see Molenaar, "CCAMLR and Southern Ocean Fisheries", at pp. 471–472; Herr, "The International Regulation of Patagonian Toothfish: CCAMLR and High Sea Fisheries Management", at notes 58–61.

11 M.L. Mooney-Seus and A.A. Rosenberg, Regional Fisheries Management Organizations (RFMOs): Progress in Adopting Precautionary Approach and Ecosystem-Based Management, 10 February 2007, at p. 6.

12 CCAMLR Convention, Article II(1) and (2).

13 Ibid., Article I(2).

14 Ibid., Article II(3).

15 Ibid., Article I(3).

16 Ibid., Article IX(1)(f).

that the Convention bases itself on the concept of precaution. For example, any harvesting and associated activities in the Convention Area shall be conducted in accordance with the principles of conservation, including the prevention of changes which are potentially irreversible in the long term (i.e., over two or three decades) and the minimization of the risk thereof, taking into account the state of available knowledge concerning the direct and indirect impact of harvesting.[17] This approach, together with the protection of the ecosystem, has been fully implemented in the practice of the Commission as stated below.

CCAMLR cooperates with other international organizations. The Commission shall take full account of any relevant measures or regulations of ATCMs. The same applies to measures of other fisheries commissions responsible for species which may enter the CCAMLR Convention Area, in order to ensure consistency between the rights and obligations under the measures and regulations of these organizations and CCAMLR conservation measures.[18] Moreover, it has been argued that there is a corresponding need for other RFMOs to respect the efforts of the Commission.[19]

4.1.2 *Implementation of the Precautionary Approach*

CCAMLR has been praised by many commentators with regard to its application of the precautionary approach and ecosystem approaches.[20] Regarding the measures applying the precautionary approach, first, an innovative and precautionary characteristic of management by the Commission is that a new fishery is only possible when prior notification is made and certain conditions are complied with.[21] This applies to a fishery for a new target species, in a new fishing ground, or with new fishing gear.[22] Following a year of fishing as a new fishery, the fishery becomes an exploratory fishery.[23] This classification is intended to deal with a fishery which still lacks critical management information.[24] An exploratory fishery may become a regulated fishery when data for management purposes

17 Ibid., Article II(3)(c).

18 Ibid., Article IX(5). See also Article XXIII.

19 Molenaar, "CCAMLR and Southern Ocean Fisheries", at p. 471.

20 See, e.g., Mooney-Seus and Rosenberg, Regional Fisheries Management Organizations (RFMOs): Progress in Adopting Precautionary Approach and Ecosystem-Based Management, at p. 3 *et seq.*

21 See CCAMLR, Schedule of Conservation Measures in Force, CM 21-01 (2010), paras. 3–11.

22 Ibid., para. 1. See also Miller *et al.*, "Managing Antarctic Marine Living Resources: The CCAMLR Approach", at p. 324.

23 See CCAMLR Conservation Measures, CM 21-02 (2010), para. 1. See also Mooney-Seus and Rosenberg, Regional Fisheries Management Organizations (RFMOs): Progress in Adopting Precautionary Approach and Ecosystem-Based Management, at p. 9 ("a new fishery lasts for one year unless no catch is taken at which time it retains its classification").

24 CCAMLR Conservation Measures, CM 21-02, para. (1)(ii). See also Miller *et al.*, "Managing Antarctic Marine Living Resources: The CCAMLR Approach", at p. 325.

are sufficient to allow a regulated fishery.[25] Second, the Commission adopts 'precautionary' catch limits for determining conservation measures for various species such as toothfish, icefish, krill and crab.[26] This implies that reference points are set at levels which are more conservative than those capable of achieving the MSY, ensuring that uncertainty associated with the lack of knowledge is taken into consideration and risks of irreversible changes are abated.

4.1.3 *Protection of Vulnerable Marine Ecosystems from Bottom Fisheries*

In the CCAMLR Convention Area, fishing vessels have been widely deploying bottom fishing gear: virtually all Patagonian toothfish fisheries on the high seas have been conducted by bottom longlines, while bottom trawl fishing on the high seas of the Southern Ocean has been rather limited.[27] The Commission has adopted various bycatch limits and a number of technical measures dealing with fishing gear.

Recently, CCAMLR started addressing bottom fisheries from the perspective of the protection of VMEs from destructive fishing practices.[28] Initially, it only took measures for gillnet and bottom trawl, but later decided to address all bottom fishing gear. Below, first, measures for bottom trawl and gillnet are analyzed and then the latest measure to address all bottom fisheries is described.

Bottom trawling is to a great extent restricted in the CCAMLR Convention Area. As early as 1999, a measure was introduced to limit fishing for several species by trawling in Statistical Division 58.4.2 to new and exploratory fisheries by Australian-flagged vessels, limiting these fisheries to the mid-water trawl method only.[29] In addition, when the 25th meeting of CCAMLR was held, bottom trawl

---

25 Miller *et al.*, "Managing Antarctic Marine Living Resources: The CCAMLR Approach", at pp. 326–327.

26 The title of the conservation measures is not uniform. Sometimes the term 'catch limits' is used while on other occasions the term 'precautionary catch limits' is employed. In the former case, sometimes the word 'precautionary' is used in the text of the conservation measures. Therefore, the title of a given measure does not indicate whether the precautionary approach is applied.

27 See M. Gianni, *High Seas Bottom Trawl Fisheries and their Impacts on the Biodiversity of Vulnerable Deep-Sea Ecosystems: Options for International Action* (2004), at p. 43.

28 In 2005, the Commission requested the Scientific Committee to begin considering the issue of destructive fishing practices, including the consideration of the benthic impacts of bottom trawling, by identifying vulnerable deep-sea habitats, including deep-sea corals, which may require protection from fishing. Report of the Twenty-fourth Meeting of the Commission (Hobart, Australia, 24 October to 4 November 2005), at para. 10.23.

29 CCAMLR Conservation Measures, CM 186/XVIII, para. 1. Under this Measure, no direct fishery by trawl was allowed except for those in paragraph 1 (paragraph 6(i)). The Measure for these species was followed by CM 212/XIX, 237/XX and 43-04, which allow research programmes on shallow-water bottom trawling in accordance with research and data collection plans annexed to them requiring sampling to assess benthos.

fishing had been prohibited in some areas adjacent to the Antarctic Continent,[30] and there were no plans to initiate new bottom trawl fishing.[31]

At the same Commission meeting, the US proposed (1) to immediately freeze the footprint of bottom trawl fishing (i.e., no expansion into new areas or in existing areas), unless CCAMLR determines that such expansion would not cause significant adverse harm to VMEs; and (2) to end all bottom trawling by 2009, unless CCAMLR determines that its continuation would not cause significant adverse harm. The US also wished to establish rules based on sound science to enhance sustainable fishing practices and to phase out destructive fishing practices.[32] Most members supported freezing the footprint and limiting future expansion to operations which would not cause significant adverse harm to VMEs, while some members requested that the environmental impact of bottom trawl fishing be further considered by the Scientific Committee.[33] The Commission agreed that for any proposal to conduct bottom trawling on the high seas within the Convention Area to be approved, the Commission needs to be satisfied that the fishery would not have a significant impact on VMEs, including benthos and benthic communities, in the area of proposed fishing activities.[34] Such a proposal needs to be notified in accordance with the notification procedure for new fisheries, and any future notifications of new or exploratory fisheries using bottom trawl gear would need to provide information on the known and anticipated impacts of the gear on VMEs, including benthos and benthic communities.[35] Furthermore, the Commission requested the Scientific Committee to review the use of bottom trawling gear in high-seas areas of the Convention Area, including with respect to relevant criteria for determining what constitutes significant harm to benthos and benthic communities in the Convention Area.[36] The adopted conservation measure was valid for fishing seasons 2006/07 and 2007/08.[37] The measure did not apply to the use of bottom trawling gear in

30 Ibid., CM 41-05 (2005) and 41-11 (2005).

31 Report of the Twenty-fifth Meeting of the Commission (CCAMLR-XXV Report) (Hobart, Australia, 23 October to 3 November 2006), at para. 11.29.

32 Ibid., at paras. 11.27–11.28.

33 Ibid., at paras. 11.32–11.33.

34 Ibid., at para. 11.36.

35 Ibid., at para. 12.18. See also measures adopted in 2006. CCAMLR Conservation Measures, CM 21-01 (2006) and 21-02 (2006).

36 CCAMLR-XXV Report (2006), at para. 11.37. CCAMLR Conservation Measures, CM 22-05 (2006), para. 2.

37 The measure was reviewed by the Commission in 2007 based on the best scientific evidence available, as stipulated in CM 22-05. The Commission decided that the Measure would remain in force for 2007/08. Report of the Twenty-sixth Meeting of the Commission (CCAMLR-XXVI Report) (Hobart, Australia, 22 October to 2 November 2007), at para. 13.3.

scientific research in the Convention Area.[38] The measure was revised in 2008 to apply for an indefinite period.[39]

Second, deep-water gillnet fishing is mostly prohibited. The Scientific Committee agreed that gillnets, including trammel nets, are non-selective fishing devices and, if not utilized correctly, could take mobile species indiscriminately, and that gillnets may have adverse impacts if dragged along the bottom and have the potential of ghost fishing.[40] Endorsing the advice from the Scientific Committee, the Commission agreed to introduce an interim ban on the use of deep-water gillnets for purposes other than scientific research in the Convention Area, unless and until the Commission would agree otherwise on the basis of advice from the Scientific Committee.[41] The use of gillnets for scientific research in waters shallower than 100 m shall be permitted subject to the requirements of Conservation Measure 24–01, while its use for scientific research in waters deeper than 100 m shall be notified in advance to the Scientific Committee and shall be approved by the Commission before such research can commence.[42]

In 2007, the Commission addressed the implementation of UNGA Resolution 61/105 by considering the impacts of all bottom fisheries (not limited to bottom trawling and gillnetting) on VMEs and adopted CM 22-06.[43] The Commission endorsed the view of the Scientific Committee that the existing practices can be used and tasked it with developing pragmatic and flexible guidelines for identifying VMEs and defining actions taken by vessels which may encounter evidence of VMEs during the course of fishing; the guidelines may take account, where appropriate, of relevant international technical guidelines.[44] The Commission further noted that some assemblages may be readily classified as vulnerable when they are characterized by slow growing, habitat-forming, sessile species, including cold-water coral communities, sponge communities, hydrothermal vents and other communities associated with seamounts.[45]

Conservation Measure 22-06 seeks to implement UNGA Resolution 61/105 by addressing each of the recommendations of that Resolution. The Measure applies to areas in the Convention Area south of 60 degrees south, to the rest

38 CCAMLR Conservation Measures, CM 22-05 (2006), para. 3.

39 See ibid., CM 22-05 (2008).

40 CCAMLR-XXV Report (2006), at para. 11.25.

41 CCAMLR Conservation Measures, CM 22-04 (2006); CCAMLR-XXV Report (2006), at paras. 11.25–11.26.

42 CCAMLR-XXV Report (2006), at para. 12.26.

43 CM 22-06 was subsequently revised several times. For the latest, see CCAMLR Conservation Measures, CM 22-06 (2010).

44 CCAMLR-XXVI Report (2007), at paras. 5.14–5.18.

45 Ibid., at para. 5.19. The term 'vulnerable marine ecosystem' in the CCAMLR context includes 'seamounts, hydrothermal vents, cold water corals and sponge fields'. See CCAMLR Conservation Measures, CM 22-06 (2010), para. 3.

of the Convention Area with the exception of sub-areas and divisions where an established fishery was in place in 2006/07 with a catch limit greater than zero, and to Division 58.4.1 north of 60 degrees south.[46] The Measure addresses the use of any gear that interacts with the bottom.[47] All bottom fishing activities commenced on 1 December 2008 or thereafter shall be subject to assessment by the Scientific Committee.[48]

The Scientific Committee assesses, based on the best available scientific information, whether individual bottom fishing activities would contribute to having SAIs on VMEs, taking account of the history of bottom fishing in the areas proposed.[49] Each contracting party proposing to participate in bottom fishing shall submit information and a preliminary assessment of the known and anticipated impacts of its bottom fishing activities on VMEs, including benthos and benthic communities, and the submissions shall include the proposed mitigation measures.[50] In assessing the fishing proposal, the Scientific Committee may use additional information available to it, including information from other fisheries in the region or similar fisheries elsewhere.[51] The Commission shall, taking account of advice and recommendations provided by the Scientific Committee concerning bottom fishing activities, adopt conservation measures to prevent SAIs on VMEs.[52] The measure would, *inter alia*, allow, prohibit or restrict bottom fishing activities within particular areas and/or with certain gear types and/or require specific mitigation measures.[53] Although the phrase 'taking account of' might give rise to fears that the Commission disregards advice and recommendations by the Scientific Committee, such a fear does not seem to be justified as the Commission has generally followed the advice of the Scientific Committee.[54] In 2008, the Commission noted that only five members had submitted preliminary assessments out of 11 members who had submitted notifications for exploratory longline fisheries and, in other exploratory fisheries, data were insufficient to assess and review the potential impacts on VMEs or possible mitigation requirements. It confirmed that all members notifying bottom fisheries covered by Conservation Measure 22-06 must submit preliminary assessments of the potential

46 CCAMLR Conservation Measures, CM 22-06 (2010), paras. 1–2.
47 Ibid., para. 4.
48 Ibid., paras. 5 and 7.
49 Ibid., para. 7 *chapeau*.
50 Ibid., para. 7(i).
51 Ibid., para. 7(ii).
52 Ibid., para. 7(iii).
53 Ibid.
54 See Mooney-Seus and Rosenberg, Regional Fisheries Management Organizations (RFMOs): Progress in Adopting Precautionary Approach and Ecosystem-Based Management, at p. 21.

for their proposed bottom fishing activities.[55] In 2009, with regard to late notifications from Russia and the Republic of Korea, the Commission agreed on a process which would satisfy the assessment requirements of CM 22-06.[56]

When evidence of VMEs is encountered, contracting parties shall require their vessels to cease bottom fishing activities, in the absence of site-specific or other conservation measures to prevent SAIs on VMEs.[57] The Measure adopted in 2007 did not in itself specify the minimum distance the vessel concerned needs to keep from the encountered site.

The encounter protocol was elaborated in Conservation Measure 22-07 adopted in 2008.[58] The measure applies to the same area as Conservation Measure 22-06.[59] This means that the measure applies to exploratory fisheries for toothfish and the exploratory fishery for crab in Subarea 48.2.[60] First, members shall require their vessels, if 10 or more VME indicator units are recovered in one line segment, to complete hauling any lines intersecting with the Risk Area without delay and not to set any such lines further.[61] Upon receipt of a notification of this, the Secretariat shall, within one working day of receipt, notify all fishing vessels in the relevant fishery and their flag states that the Risk Area is closed and all vessels shall immediately cease setting any such lines further.[62] The Risk Area remains closed for any fishery until reviewed by the Scientific Committee and management actions are determined by the Commission, although scientific research as agreed by the Scientific Committee shall be allowed in that Area. Second, upon the receipt of five notifications made under paragraph 5 (i.e., communications

---

55 Report of the Twenty-seventh Meeting of the Commission (CCAMLR-XXVII Report) (Hobart, Australia, 27 October–7 November 2008), at paras. 5.8 and 5.10.

56 See Report of the Twenty-eighth Meeting of the Commission (CCAMLR-XXVIII Report) (Hobart, Australia, 26 October–6 November 2009), at paras. 8.26–8.34 and 12.18–12.21.

57 CCAMLR Conservation Measures, CM 22-06 (2010), para. 9.

58 CCAMLR-XXVII Report (2008), at para. 5.25. CM 22-07 was amended in 2009 and 2010. It is expected to be reviewed in 2012. Report of the Twenty-ninth Meeting of the Commission (CCAMLR-XXIX Report) (Hobart, Australia, 25 October–5 November 2010), at paras. 5.7 and 12.13.

59 CCAMLR Conservation Measures, CM 22-07 (2008), para. 1.

60 CCAMLR-XXVII Report (2008), at para. 13.24.

61 CCAMLR Conservation Measures, CM 22-07 (2010), para. 4. The term 'VME indicator unit' is defined as either one litre of those VME indicator organisms that can be placed in a 10-litre container or one kilogram of those VME indicator organisms that do not fit into a 10-litre container. See para. 2(iii). The term 'Risk Area' is defined as an area where 10 or more VME indicator units are recovered within a single line segment, having a radius of 1 nautical mile from the midpoint of the line segment from which the VME indicator units are recovered. See para. 2(v). The Commission considered the size of the buffer zone, noting the Scientific Committee's advice (1 nautical mile), existing bycatch move-on rules in Conservation Measure 33–03 (5 nautical miles) and standards agreed in other fora such as the EC Regulation (5 nautical miles). CCAMLR-XXVII Report (2008), at para. 5.26.

62 CCAMLR Conservation Measures, CM 22-07 (2010), para. 6(ii).

that five or more VME indicator units are recovered within one line segment) within a single fine-scale rectangle, the Secretariat shall, within one working day of receiving the fifth notification, notify all fishing vessels in the relevant fishery and their flag states of the coordinates of the fine-scale rectangle, indicating that VMEs may occur within that area but, as opposed to the situation described above, vessels may continue to fish in the area consistent with paragraphs 4 and 5 of the measure.[63] The Commission noted that the presence of VME taxa or indicators of VMEs in catches from fishing gears would be evidence that VMEs could be present while their absence in the catches did not necessarily represent an absence of VMEs.[64]

By the Commission meeting in 2009, the Secretariat had received 30 VME indicator notifications, of which seven notifications consisted of at least 10 VME indicator units and this resulted in seven Risk Areas being declared in Subareas 88.1 and 88.2.[65] The Commission agreed that VMEs reported in Subarea 48.2 be given protection in the experimental harvest regime for crabs in that subarea by closing experimental blocks A, C and E.[66]

On the basis of the 30 notifications of encounters with VMEs during the course of research surveys, the Commission agreed that these VMEs be recorded in the Secretariat's VME Register in 2009.[67] The Commission endorsed the inclusion into the VME Register of two new sites identified during a fishery-independent trawl survey in Subarea 48.2 in 2010.[68]

In 2011, the Commission adopted Conservation Measure 22-9 to protect registered VMEs notified in accordance with Conservation Measure 22-06.[69] All bottom fishing activities are prohibited within the defined areas of VMEs (i.e., circles with the radius of 1.25 nautical miles) with the exception of scientific research activities agreed by the Commission for monitoring or other purposes.[70]

Conservation Measure 22-06 goes beyond the requirement of UNGA Resolution 61/105 in stipulating that each vessel carries at least one CCMALR-designated scientific observer to collect data.[71] The data collection plans for bottom fisheries will be developed by the Scientific Committee and included in conservation measures.[72] It should also be noted that CCAMLR members have conducted

63 Ibid., paras. 5 and 7.
64 CCAMLR-XXVII Report (2008), at para. 5.5(i).
65 CCAMLR-XXVIII Report (2009), at para. 5.3.
66 Ibid., at para. 5.4. See also para. 11.17.
67 Ibid., at para. 5.4.
68 CCAMLR-XXIX Report (2010), at para. 5.3.
69 Report of the Thirtieth Meeting of the Commission (CCAMLR-XXX Report) (Hobart, Australia, 24 October–4 November 2011), at para. 12.18. See also para. 5.1.
70 CCAMLR Conservation Measures, CM 22-09 (2011), paras. 3–4.
71 Ibid., CM 22-06 (2010), para. 11(ii).
72 See ibid., para. 11(iii).

extensive scientific research on marine ecosystems. The accumulated outcome of that research has been reflected in the bioregionalization project.

One important feature of Conservation Measure 22-06 is the treatment of bottom fishing for scientific research purposes. In the case of small-scale research taking less than 50 tonnes of finfish and meeting the requirement of Conservation Measure 24-01 Annex 24-01/B, the scientific bottom fishing research activities shall be notified in accordance with paragraph 2 of Conservation Measure 24-01, proceed according to that Measure, and be undertaken with due regard to potential impacts on VMEs.[73] Paragraph 2 of Conservation Measure 24-01 stipulates that any member planning such research activities shall notify the Secretariat and, as opposed to paragraph 3 of that Measure, does not need to provide the opportunity for other members to review and comment on its research plan. On the other hand, those research activities going beyond the above-mentioned criteria in Conservation Measure 24-01 shall be treated in accordance with paragraph 9 of Conservation Measure 22-06, rather than procedures in Conservation Measure 24-01.[74] Information regarding the location and the type of any VME encountered in the course of scientific bottom fishing research activities shall be reported to the Secretariat.[75] The relevant paragraph does not explicitly require research vessels notifying under paragraph 2 of Conservation Measure 24-01 to cease scientific bottom fishing activities where they encounter evidence of VMEs. If these activities also need to cease in these circumstances, this may lead to a situation where conservation measures cannot be taken on the basis of scientific information because research cannot be undertaken on the VME site.

In addition, the Commission agreed that a prohibition on bottom fishing in depths shallower than 550 m, as applied in Divisions 58.4.1 and 58.4.2, should be generally applied to bottom fisheries covered by Conservation Measure 22-06. Exploratory fisheries for toothfish in depths shallower than 550 m in Divisions 48.6, 58.4.1, 58.4.2, 58.4.3a, 58.4.3b, 88.1 and 88.2 were prohibited to protect benthic communities.[76]

4.1.4 *Area-Based Management Tools*

CCAMLR has adopted a number of area-based management measures. The prohibition of directed fishing for certain species is imposed in a number of measures dealing with a variety of areas.[77] In particular, Sub-areas 48.1 and 48.2 are

---

[73] See ibid., para. 14; CM 24-01 (2011).

[74] Ibid., CM 22-06 (2010), para. 14. Paragraph 9 is a provision concerning VME encounters. Note that the original measure adopted in 2007 referred to paragraph 7, which concerned assessments by the Scientific Committee and the adoption of conservation measures by the Commission.

[75] Ibid.

[76] See CCAMLR Conservation Measures, CM 41-04 to 41-07 and 41-09 to 41-11.

[77] See ibid., CM 32-04 to 32-17.

closed for all finfish fisheries.[78] Basically, the moratorium will be continued until the Commission has decided otherwise on the basis of the advice of the Scientific Committee.[79] The Scientific Committee will act in a precautionary manner in its evaluation of the stock assessment, and the Commission takes decisions by consensus.[80] Therefore, once a moratorium is set, it seems difficult to lift it. Area closures are supplemented by seasonal closures. CCAMLR has adopted a variety of measures to limit fishing in specific seasons.[81] Such measures are intended for the protection of spawning stocks. Furthermore, as noted in the previous sub-section, registered VME sites are protected from the impacts of bottom fisheries.

In 2009, the Commission adopted Conservation Measure 91-03 (2009) on the Protection of the South Orkney Islands southern shelf. The designation was intended to contribute towards the conservation of marine biodiversity in Subarea 48.2.[82] All types of fishing activities with the exception of scientific fishing research activities agreed by the Commission shall be prohibited within the defined area.[83] One member commented that it accepted the proposal because the definition of the area was amended to the effect that the area where fishing activity was carried out had been excluded so as to avoid restricting the fishery and further noted that "the amended MPA constitutes a good precedent for future deliberations with respect to the establishment of MPAs by this organisation".[84]

Given the frequent use of area-based management tools in CCAMLR, there is a potential for an overlap of competence between CCAMLR and other components of the Antarctic Treaty System. In particular, Antarctic Specially Protected Areas (ASPAs) as well as Antarctic Specially Managed Areas (ASMAs) have been established under Annex V of the 1991 Environmental Protocol (Madrid Protocol) to the Antarctic Treaty, replacing the earlier categories of specially protected areas.[85] Annex V stipulates that prior approval by CCAMLR is required for the

78 Ibid., CM 32-02 and 32-03.

79 Conservation measures often use the following wording: "[t]his prohibition shall apply until at least such time that a survey [...] is carried out, its results reported to and analysed by the Working Group on Fish Stock Assessment and a decision that the fishery be reopened is made by the Commission based on the advice of the Scientific Committee".

80 CCAMLR Convention, Article XII(1).

81 CCAMLR Conservation Measures, CM 41-02 to 41-11, 42-01 and 42-02, and 43-02 and 43-03, 51-01 to 51-03, 52-01 and 61-01.

82 CCAMLR Conservation Measures, CM 91-03, para. 1.

83 Ibid., para. 2. Discharges and dumping of waste by fishing vessels or transhipment activities involving fishing vessels are not allowed. See paras. 3–4.

84 CCAMLR-XXVIII Report (2009), at para. 7.4 (Japan). A significant number of members rejected the notion expressed by Japan that MPAs and fishing activities should be mutually exclusive.

85 Currently, 71 ASPAs and 7 ASMAs had been established. Information available at the Secretariat of the Antarctic Treaty, "Status of Antarctic Specially Protected Area and Antarctic Specially Managed Area Management Plans", <http://ats.aq/documents/ATCM34/WW/atcm34_ww003_e.pdf>.

establishment of ASPAs and ASMAs in the marine area of the Antarctic Treaty area.[86] Thus, formally speaking, there is no potential discrepancy or divergence in the designation and management of protected areas under the CCAMLR Convention and the Antarctic Treaty.

Nevertheless, it is possible that the criteria and standards employed to designate ASPAs or ASMAs in areas other than the marine area might affect the management of fisheries by CCAMLR. This also applies to proposals of ASPAs and ASMAs for the marine area. In addition, given the acknowledgement of "the special obligations and responsibilities of the Antarctic Treaty Consultative Parties for the protection and preservation of the environment of the Antarctic Treaty area",[87] it would be appropriate to conclude that cooperation has to be established with other bodies in the Antarctic Treaty System, in particular the Committee for Environmental Protection.[88] Thus, once a particular area is designated as an ASPA or an ASMA under Annex V of the Madrid Protocol, CCAMLR has to cooperate in this regard.

Discussions are ongoing with regard to the elaboration of a representative system of MPAs. For example, Australia introduced a proposal on an overarching measure to support the establishment of a representative system of MPAs.[89] The Commission acknowledged that an overarching measure may help facilitate the designation of MPAs in the future and therefore it was important that such a measure reflected appropriate conservation and policy objectives.[90] In 2011, the Australian-proposed conservation measure providing a general framework for the establishment of CCAMLR MPAs was adopted (Conservation Measure 91-04).[91] At the same Commission meeting, proposals for MPAs in the Ross Sea region and in the East Antarctica were discussed but members did not agree on these proposals; controversial issues included: the definition of 'rational use', the

86 Protocol on Environmental Protection to the Antarctic Treaty, Annex V, Article 6(2). Areas for ASPAs and ASMAs may be proposed both by the Committee for Environmental Protection established under this Protocol and by CCAMLR (Article 5(1)). For example, in 2010, CCAMLR endorsed the revised management plan for ASPA No. 149, Cape Shirreff and San Telmo Islands and the plan would be taken forward for consideration by the Committee for Environmental Protection. See CCAMLR-XXIX Report (2010), at para. 7.6.

87 CCAMLR Convention, Article V(1).

88 Molenaar, "CCAMLR and Southern Ocean Fisheries", at p. 473.

89 CCAMLR-XXIX Report (2010), at para. 7.12.

90 Ibid., at para. 7.18. The Commission noted that in 2005 the Scientific Committee had identified a number of conservation objectives that could be achieved through the establishment of MPAs and "[w]hile representativeness was one such objective, other objectives include those identified in SC-CAMLR-XXIV, paragraphs 3.54(i), (iii), (iv-b) and (iv-c)". Ibid., at para. 7.17.

91 CCAMLR-XXX Report (2011), at paras. 12.38–12.39.

protection commensurate with the objectives of the MPA concerned and the appropriateness of the size and delineation of the proposed MPA.[92]

4.2 *General Fisheries Commission for the Mediterranean (GFCM)*

The General Fisheries Council for the Mediterranean was established within the framework of the FAO under the Agreement for the Establishment of the General Fisheries Council for the Mediterranean. The Agreement was amended several times. The major issues in the most recent amendment in 1997 included allowing regional economic integration organizations (REIOs) to participate, establishing an autonomous budget and incorporating certain aspects of modern fisheries management.[93] The organization was renamed as the 'General Fisheries Commission for the Mediterranean'.[94]

The GFCM carries out its functions and responsibilities in the Mediterranean and the Black Sea and connecting waters.[95] No part of the Mediterranean is beyond 200 miles from the baselines of the coastal states. However, as many coastal states in the region have not declared the EEZ to the full extent, there remains a substantial area of high seas. The high seas areas of the Mediterranean are likely to remain for the foreseeable future.

The Mediterranean is considered as a semi-enclosed sea, and the legal regime applied to this area is not necessarily the same as that for other areas of the high seas. For example, the provisions of Part IX of the LOSC for enclosed and semi-enclosed seas are applicable to the Mediterranean.[96] But provisions of the LOSC, including the provisions of Part VII of the LOSC, are also applicable irrespective

92 Ibid., at paras. 7.10–7.30.

93 As of 2012, only one out of 24 members has not yet deposited its instrument of acceptance of the 1997 revision of the Agreement. "Summary of Recommendations of Regional Fishery Bodies Established within the FAO Framework", COFI/2012/SBD.4, at p. 8. The GFCM agreed to hold an extraordinary session to amend the Agreement in 2012 in response to the report of the Task Force on this issue (ibid.).

94 See GFCM Agreement, Article I(1). Strictly speaking, the original Agreement in its title uses 'Council' and the title of the one adopted in 1997 uses 'Commission'. However, for the purpose of this study, they are treated as a single instrument amended later.

95 Ibid., preamble and Article IV.

96 For the purposes of the LOSC, 'enclosed or semi-enclosed sea' means a gulf, basin or sea "surrounded by two or more States and connected to another sea or the ocean by a narrow outlet" or "consisting entirely or primarily of the territorial seas and exclusive economic zones of two or more coastal States". LOSC, Article 122. The Mediterranean satisfies, at least, the first requirement, while the latter requirement might be satisfied in the future if coastal states decide to declare the 200-mile EEZ. See also Nandan and Rosenne (eds.), *United Nations Convention on the Law of the Sea 1982: A Commentary*, vol. III, at p. 352; *I.C.J. Reports 1985*, at p. 40, para. 47.

of the applicability of Part IX.[97] Article 123 does not grant additional rights or jurisdiction to coastal states.[98]

4.2.1 *The Agreement*

The purpose of the GFCM is, among other things, to promote the development, conservation, rational management and best utilization of 'living marine resources'.[99] The GFCM Agreement does not expressly exclude any species from such resources. Discrete high seas stocks, whether fish or other organisms, are therefore covered by the Agreement.

The 1997 Agreement has introduced some of the modern approaches to fisheries management as evidenced by reference in its preamble to the LOSC, Agenda 21 and the FAO Code of Conduct for Responsible Fisheries. In particular, Article III(2) stipulates that the Commission shall take into account the best scientific evidence available and shall apply the precautionary approach.

The GFCM Agreement does not distinguish between target species and non-target species or related ecosystems in providing for the functions of the Commission. Thus, it could be argued that the Agreement implicitly allows the GFCM to adopt measures for the conservation of non-target species and marine ecosystems, whether or not for the purpose of the conservation of target species.

4.2.2 *Practice*

In the Agreement area of the GFCM, bottom-dwelling (demersal) fish account for around 40% of catches.[100] The GFCM started the management of deep-sea

97 See Nandan and Rosenne (eds.), *United Nations Convention on the Law of the Sea 1982: A Commentary*, vol. III, at p. 365. But at least one author casts doubt on the proposition that the LOSC is applicable to the conservation of living resources of the high seas in the Mediterranean. See M.C. Maffei, "The Protection of Endangered Species of Animals in the Mediterranean Sea", in E.L. Miles and T. Treves (eds.), *The Law of the Sea: New Worlds, New Discoveries* (1993), at p. 256 note 5.

98 Nandan and Rosenne (eds.), *United Nations Convention on the Law of the Sea 1982: A Commentary*, vol. III, at p. 365; Miovski, "Central Bering Sea Overfishing", at pp. 558–560; A.G. Oude Elferink, "Fisheries in the Sea of Okhotsk High Seas Enclave: The Russian Federation's Attempts at Coastal State Control", 10 *International Journal of Marine and Coastal Law* (1995), at p. 15. In fact, the text of Article 123 employs hortatory language such as 'should cooperate' and 'shall endeavour', and other interested states or international organizations are invited 'as appropriate', rather than obliged, to cooperate. Nandan and Rosenne (eds.), *United Nations Convention on the Law of the Sea 1982: A Commentary*, vol. III, at p. 366.

99 GFCM Agreement, Article III(1).

100 FAO press release, "Mediterranean fisheries: as stocks decline, management improves", 27 July 2005, available at <http://www.fao.org/newsroom/en/news/2005/105722/index.html>. For a comprehensive study on the ecosystem impacts of fisheries in the Mediterranean, including those caused by bottom-trawling, see S. Tudela, Ecosystem Effects of Fishing in the

fisheries by introducing the regulation of certain fishing techniques in 2005. It was agreed to prohibit the use of towed dredges and trawlnet fisheries at depths of more than 1000 m.[101]

In 2006, upon a proposal by the European Community, the GFCM adopted a recommendation entitled "Establishment of Fisheries Restricted Areas in order to Protect the Deep Sea Sensitive Habitats" to prohibit fishing with towed dredges and bottom trawl nets in three sites: 'Lophelia reef off Capo Santa Maria di Leuca', 'The Nile delta area cold hydrocarbon seeps' and 'The Eratosthemes Seamount'.[102] The cold water coral reef off Capo Santa Maria di Leuca is located in international waters.[103] The site is located on the continental shelf of Italy in the Ionian Sea.

The main fishery operating around the cold water coral reef off Capo Santa Maria di Leuca is that by vessels from Gallipoli. Most of them fish by means of trawling, targeting, *inter alia*, deep-water shrimps, Norway lobster and Hake, while some vessels fish with bottom longlines targeting, *inter alia*, hake, bluntnose sixgill shark and silver scabbard fish.[104] Nothing in the Recommendation indicates that either sedentary species or bottom-dwelling (non-sedentary species) fish are excluded from its scope. In fact, dredges are used to target sedentary species such as clams, scallops and oysters.[105]

Mediterranean: An Analysis of the Major Threats of Fishing Gear and Practices to Biodiversity and Marine Habitats, GFCM Studies and Reviews No. 74.

101 GFCM, Recommendation GFCM/29/2005/1, para. 2.

102 GFCM, Recommendation GFCM/30/2006/3, para. 1.

103 A change to the existing coordinates for this site was suggested in a workshop meeting on MPAs in 2007. The proposed coordinates straddle the Italian territorial sea and the high seas. Report of the Transversal Workshop on Marine Protected Areas (MPAs) (GFCM & RAC/SPA), Salammbô, Tunisia, 24 and 25 May 2007, at para. 26.

104 S. Tudela, Proposal for a representative network of protected deep-sea sensitive habitats in the Mediterranean (2005), at p. 13. When the Sub-Committee on Marine Environment and Ecosystems considered the protection of this site as part of the network of deep-sea sensitive habitats, it recommended the banning of demersal fishery practices over the area. Report of the Sixth Session of the Sub-Committee on Marine Environment and Ecosystems (SCMEE), Rome, Italy, 27–30 September 2005, GFCM:SAC8/2005/Dma.1, at p. 3.

105 A. Freiwald *et al.*, *Cold-water Coral Reefs: Out of Sight – No Longer Out of Mind* (2004), at p. 39. The present study is without prejudice to the question whether deep-water shrimps targeted by this fishery belong to sedentary species. Note that the classification of some crustaceans as sedentary species has been disputed in some cases. See, e.g., Agreement concerning shrimp between Brazil and the United States, Brasília, 9 May 1972, preambular para. 2; J.M. Van Dyke, "Modifying the 1982 Law of the Sea Convention: New Initiatives on Governance of High Seas Fisheries Resources: The Straddling Stocks Negotiations", 10 *International Journal of Marine and Coastal Law* (1995). Article (1)(c) of the FSA stipulates that fish includes "crustaceans except those belonging to sedentary species". On the other hand, see A.-K. Westberg, "Governance and Management of Living Marine Resources and Fisheries on the Continental Slope and in the Deep-sea: A Legal Framework and Some Points of Departure", in R. Shotton (ed.), *Deep Sea*

The above-mentioned GFCM recommendation further stipulates that members shall call the attention of the appropriate authorities in order to protect these areas from the impact of any other activity jeopardizing the conservation of the features that characterize these particular habitats.[106]

In addition, it was also agreed to increase the selectivity of demersal trawlnets by the immediate implementation of at least a 40 mm mesh size opening for the whole demersal trawl cod-end. Members are also required to explore and implement additional measures to this end.[107] This measure aims at both allowing small, juvenile fish to escape so as to conserve breeding stocks and reducing accidental catches of non-target species. In 2009, the GFCM adopted a further recommendation on the minimum mesh size in the cod-end of demersal trawlnets (40-mm square mesh cod-end or a diamond mesh size of at least 50 mm for all trawling activities exploiting demersal stocks).[108]

In 2009, it adopted a recommendation to establish a fisheries restricted area in the Gulf of Lions to protect spawning aggregations and deep-sea sensitive habitats: the recommendation stipulates that the fishing effort for demersal stocks of vessels using towed nets, bottom and mid-water longlines and bottom-set nets in the fisheries restricted area shall not exceed the level applied in 2008.[109] As suggested in the preamble to the Recommendation,[110] it was not certain whether the spatial extent of the fisheries restricted area was adequate; the GFCM requested the Scientific Advisory Committee to undertake studies on

---

2003: Conference on the Governance and Management of Deep-sea Fisheries: Part 1: Conference Reports. Queenstown, New Zealand, 1–5 December 2003 (2005), at p. 713 ("typical demersal species such as skates, some flatfish and perhaps wolf-fish, and certain benthic sharks could [...] be included in the definition of sedentary species [under Article 77(4) of the LOSC]").

106 GFCM, Recommendation GFCM/30/2006/3, para. 2. The EU transposed this Recommendation to Council Regulation 1967/2006 of 21 December 2006. Other parties, however, reported no action taken at the national level on this Recommendation. See Report of the Thirty-First Session, Rome, 9–12 January 2007, GFCM Report 31, at pp. 73–75 (Appendix F, Annex C).

107 GFCM, Recommendation GFCM/29/2005/1, para. 1. In 2007, a new recommendation allowing a derogation from Article 1 of this Recommendation was adopted under Article V of the GFCM Agreement. According to the new recommendation, the members of the GFCM may continue authorizing, until 31 May 2010, the use of mesh size smaller than 40 mm to operate in certain local and seasonal demersal trawl fisheries exploiting non-shared demersal stocks only if such fishing activities are already formally authorized by the GFCM members. GFCM, Recommendation GFCM/31/2007/1, paras. 1–2. At the same time, a non-binding resolution was adopted that members of the GFCM implement on a voluntary basis at least the 40 mm square mesh cod-end in bottom trawling. GFCM, Resolution GFCM/31/2007/3, para. 1.

108 Recommendation GFCM/33/2009/2.

109 Recommendation GFCM/33/2009/1.

110 Ibid., preambular para. 7.

the protection of spawning aggregations on the continental shelf edge and slope of the northern western Mediterranean.[111]

In addition, the GFCM adopted a resolution on the reduction of bottom trawling fishing efforts by 10% in all GFCM areas.[112] The Scientific Advisory Committee advised to convert the Resolution into a binding Recommendation under Article V of the GFCM Agreement.[113]

In 2011, the GFCM adopted a recommendation requiring Contracting Parties and Cooperating Non-contracting Parties (CPCs) to prohibit the use of any kind of towed gear to exploit red coral and the use of Remotely Operated underwater Vehicles (ROVs) for the exploitation of red coral in the GFCM Competence Area and to ensure the prohibition of the exploitation of red coral populations at a depth of less than 50 m until scientific studies indicate otherwise, with the possibility for a derogation as specified in the recommendation.[114]

As opposed to other RFMOs reviewed in this chapter, the GFCM has not fully addressed the UNGA Resolutions on Sustainable Fisheries with regard to bottom fisheries. In 2010, it was suggested by the Chairperson of the Scientific Advisory Committee that the Committee should investigate the likely consequences of changing the current framework for the management of bottom trawl fisheries to one in which specific areas for trawling would be designated.[115]

4.3 *South East Atlantic Fisheries Organization (SEAFO)*

Fisheries on the high seas of the South-East Atlantic Ocean were managed by the International Commission on South East Atlantic Fisheries since 1971.[116] Following Namibia's independence in 1990, it became inoperative and, in practical terms, no management body dealt with high seas fisheries in this region except for tuna fisheries during the 1990s.[117]

---

111 Report of the Thirty-Third Session of the General Fisheries Commission for the Mediterranean, Tunis, 23–27 March 2009, at p. 10 and Appendix H.

112 Resolution GFCM/33/2009/1.

113 Report of the Thirty-Fourth Session of the General Fisheries Commission for the Mediterranean, Athens, 12–17 April 2010, at p. 11.

114 Recommendation GFCM/35/2011/2. Further measures have been taken in 2012. See Recommendation GFCM/36/2012/1.

115 Report of the Thirty-Fourth Session of the General Fisheries Commission for the Mediterranean, Athens, 12–17 April 2010, at p. 11.

116 Convention on the Conservation of the Living Resources of the Southeast Atlantic, Rome, 3 October 1969. The Protocol of Termination was adopted in 1990, although not ratified. No action has been taken since then and the Commission is considered to be inoperative.

117 Meltzer, "Global Overview of Straddling and Highly Migratory Fish Stocks" (1994), at pp. 293–304; Miller, "Management and Governance Conventions and Protocols: SEAFC, WCPFC and

In 1995, Namibia approached South Africa to discuss the need for an RFMO in the South East Atlantic Ocean. Namibia was concerned that unsustainable fishing practices on the high seas could undermine management efforts within its EEZ for orange roughy fisheries. Having agreed upon the need to establish an RFMO, Namibia and South Africa invited other coastal states of the South East Atlantic. Several consultations among the coastal states in 1997 resulted in adopting a coastal states' draft for the SEAFO Convention, largely based on the FSA. Following this, negotiations started in December 1997 between coastal states and distant water fishing nations identified as having an interest in the fisheries (the EC, Japan, Norway, Russia and the US). The participants in the latter category were later enlarged to include Ukraine, Iceland, the Republic of Korea and Poland. Following six sessions of negotiations as well as one technical consultation during 1998–2000, the Conference on the South East Atlantic Fisheries Organization adopted the Convention on the Conservation and Management of Fishery Resources in the South East Atlantic Ocean (SEAFO Convention).[118]

The Convention Area, where the Convention applies, includes waters beyond areas of national jurisdiction in the South East Atlantic Ocean as defined in Article 4. This is almost identical to FAO Statistical Area 47 exclusive of areas subject to national jurisdiction.[119]

Participants in the process were conscious that the Convention would set up the first RFMO to manage straddling fish stocks following the FSA and would set an important international precedent. Consequently, the status and implementation of the FSA, rather than the actual fisheries, became a focal point during the negotiations.[120] One commentator argues that, as opposed to the coastal

---

SADC", at p. 597; Rayfuse, *Non-Flag State Enforcement in High Seas Fisheries*, at pp. 308–309; A.K. Sydnes, "New Regional Fisheries Management Regimes: Establishing the South East Atlantic Fisheries Organisation", 25 *Marine Policy* (2001), at p. 354. Sydnes notes that there is no relation between the SEAFO process and its predecessor. Sydnes, "New Regional Fisheries Management Regimes: Establishing the South East Atlantic Fisheries Organisation", at p. 354.

118 For a detailed account of the entire process, see Sydnes, "New Regional Fisheries Management Regimes: Establishing the South East Atlantic Fisheries Organisation", at pp. 355–359.

119 In order to accommodate Angola's preference, participants in the SEAFO process met a day ahead of signing the SEAFO Convention to sign a declaration stating that the parties agreed to consider "to extend the boundary of the Convention Area northward to include those areas of high seas adjacent to all waters under Angolan jurisdiction [...] subject to the cooperation and agreement of other coastal states concerned". Ibid., at p. 359.

120 Ibid., at pp. 356–357. Some commentators even argued that part of the driving force behind the establishment of SEAFO in spite of the low level of fishing activities was the desire of certain distant water fishing nations to put the FSA into practice so as to establish an international precedent for regional fisheries management. Henriksen *et al.*, *Law and Politics in Ocean Governance*, at p. 164. In addition to the FSA, it has been pointed out that the SEAFO Convention has been modelled on the CCAMLR Convention and ICCAT Convention. Rayfuse, *Non-Flag State Enforcement in High Seas Fisheries*, at p. 309.

states' draft, the negotiations gradually departed from attempting to fully reflect the provisions of the FSA and a main characteristic of the SEAFO Convention is its 'limited' application of the FSA.[121] Although the above comments appear to relate mainly to provisions on enforcement, this issue is a particularly important factor for the legal regime of discrete high seas stocks. Since the FSA as such does not apply to DHSFS, the constitutive instruments and practice of RFMOs are likely to exert more influence on the regime for DHSFS. The following analysis first attempts to investigate to what extent the provisions of the SEAFO Convention reflect the FSA with regard to the principles of conservation. Then, it analyzes the practice of SEAFO in respect of, *inter alia*, area closures around seamounts to protect benthic ecosystems and the setting of the TAC.

4.3.1 *The Convention*

The objective of the Convention is to ensure the long-term conservation and sustainable use of the fishery resources in the Convention Area.[122] 'Fishery resources' mean resources of fish, molluscs, crustaceans and other sedentary species within the Convention Area, and excludes sedentary species subject to the fishery jurisdiction of coastal states pursuant to Article 77 of the LOSC and highly migratory species listed in Annex I of the LOSC.[123] Thus, the Convention applies not only to straddling stocks but also to discrete high seas stocks.[124] In fact, one of the participants in the negotiations commented in 2000 in retrospect as follows: the "first challenge is the design of an agreement to provide for conservation and management measures applicable to the [DHSFS] that occur in the proposed SEAFO area".[125]

121 Sydnes, "New Regional Fisheries Management Regimes: Establishing the South East Atlantic Fisheries Organisation", at pp. 361–362. See also Orrego Vicuña, "The Law Governing High Seas Fisheries: In Search of New Principles", at p. 390 ("some definitions of the [FSA] seem to be generously interpreted so as to accommodate historical fishing"). Cf. Convention on the Conservation and Management of Highly Migratory Fish Stocks in the Western and Central Pacific Ocean (WCPFC Convention), Honolulu, 5 September 2000.

122 SEAFO Convention, Article 2.

123 Ibid., Article 1(l). Since sedentary species subject to coastal state jurisdiction are excluded from the category of sedentary species with which SEAFO has competence, by referring to "other sedentary species", the SEAFO Convention recognizes the concept of sedentary species also in the context of deep seabed, and SEAFO has the competence to ensure their long-term conservation and sustainable use.

124 Fisheries resources in the SEAFO Convention Area include deep-water species such as alfonsinos, horse mackerel, orange roughy, armourhead, deep-sea red crab, octopus and squid, Patagonian toothfish, hake and oreodories. Ibid., Annex, Section Two.

125 A. Jackson, "Developments in the Southeast Atlantic, 1997–1999: Meetings of Coastal States and Other Interested Parties on a Fisheries Management Organization for the South East Atlantic (the SEAFO Process)", in M.H. Nordquist and J.N. Moore (eds.), *Current Fisheries Issues and the Food and Agriculture Organization of the United Nations* (2000), at p. 61. He identifies two key

Contracting parties are to adopt measures based on the best scientific evidence available and to apply the precautionary approach.[126] The Commission is to take into account recommendations and advice from the Scientific and Compliance Committees, in particular on biological unity and other biological characteristics of the stocks in formulating its decisions,[127] while the Commission is to apply the precautionary approach widely to the conservation and management and exploitation of fishery resources in order to protect those resources and preserve the marine environment.[128] Although the Commission is to be more cautious when information is uncertain, unreliable or inadequate, the absence of adequate scientific information shall not be used as a reason for postponing or failing to take conservation and management measures.[129] The Commission is to take cognizance of best international practices regarding the application of the precautionary approach, including Annex II of the FSA and the Code of Conduct.[130]

The preamble to the Convention expresses the commitment of the contracting parties to safeguard the environment and marine ecosystems in which living marine resources of the South-East Atlantic Ocean occur.[131] Contracting parties apply the provisions relating to fishery resources, taking due account of the impact of fishing operations on ecologically related species such as seabirds, cetaceans, seals and marine turtles.[132] Measures are to be adopted not only for target species but also for species belonging to the same ecosystems as well as dependent or associated species.[133] Due account of the need to minimize harmful impacts on living resources as a whole shall also be taken.[134] Contracting parties are to protect biodiversity in the marine environment.[135]

---

areas where the test of how far states are ready to bind themselves to apply the provisions of the FSA to all fishing on the high seas, and considers one area as dispute settlement measures in Part VIII of the FSA, including those relating to the application of generally accepted standards for the conservation and management of living marine resources. Jackson, "Developments in the Southeast Atlantic, 1997–1999", at pp. 61–62. See also Sydnes, "New Regional Fisheries Management Regimes: Establishing the South East Atlantic Fisheries Organisation", at p. 362.

126 SEAFO Convention, Article 3(a) and (b).

127 Ibid., Article 6(6).

128 Ibid., Article 7(1).

129 Ibid., Article 7(2).

130 Ibid., Article 7(3).

131 Ibid., preambular para. 1. In 2007, the Commission agreed to the principle of the protection of the marine environment during fishing activities in the Convention Area and that this should be encompassed in the SEAFO conservation measures. Report of the 4th Annual Meeting of the Commission, 2007, at p. 6.

132 SEAFO Convention, Article 3(c).

133 Ibid., Article 3(d).

134 Ibid., Article 3(e). The term 'Living marine resources' means all living components of marine ecosystems including seabirds. Ibid., Article 1(n).

135 Ibid., Article 3(f).

The Convention explicitly refers to conservation and management measures such as the TAC, area and season closure, size and sex restrictions, fishing gear and method restrictions, as well as a fishing effort restriction.[136] In the case of straddling stocks, the interests of coastal developing states are to be taken into account in determining fishing opportunities,[137] while there is no comparable provision for discrete high seas stocks.

The SEAFO Convention provides that SEAFO is to cooperate with other relevant fisheries management organizations and to take account of their conservation and management measures which affect living marine resources in the Convention Area and to seek consistency with such measures.[138] SEAFO shares its border with CCAMLR. In addition, the South African Community Protocol on Fisheries applies to living aquatic resources of the high seas as may be considered to be of interest to states parties to the Protocol.[139] The parties to the Protocol do not purport to compete with SEAFO in high seas fisheries management; rather, they attempt to complement SEAFO through coordinated activities under the Protocol.[140] The Protocol stipulates that the states parties shall endeavour to establish common positions and undertake coordinated and complementary actions with regard to, *inter alia*, SEAFO.[141]

4.3.2 *Practice*

SEAFO has taken several measures to conserve and manage fisheries as well as to protect vulnerable deep-water habitats and ecosystems, including closed areas and catch limits.

In 2006, the Scientific Committee recommended closing 13 seamounts in the Convention Area.[142] Five of them (numbered as 3–7) appear to be located on the outer continental shelf of Namibia and South Africa, while the other eight

136 Ibid., Article 6(8).

137 Ibid., Article 20(d).

138 Ibid., Articles 6(12) and 18(4).

139 SADC Protocol, Article 2(b).

140 See Miller, "Management and Governance Conventions and Protocols: SEAFC, WCPFC and SADC", at pp. 620–628, particularly p. 628.

141 SADC Protocol, Article 6(1)(b) and Appendix 2. Furthermore, the Protocol provides that the States Parties undertake, among other things, to "collaborate in the establishment of common positions and policies with regard to the effective management of the high seas living aquatic resources" and to "support the activities of international organisations which conserve and manage living aquatic resources on the high seas, and which act in non-discriminatory manner in relation to State Parties". Article 11(c)–(d).

142 Report of SEAFO Scientific Committee, 2006, at pp. 16–18. In the same meeting, the Commission also endorsed another Scientific Committee recommendation concerning the protection of benthic ecosystems, i.e., the collection of data by the scientific observers on benthic organisms such as sponges and corals in their sampling protocols. Report of the 3rd Annual Meeting of the Commission, 2006, at p. 4.

are beyond the outer limit of the continental shelf. Conservation Measure 06/06 stipulates that all fishing activities for species covered by the SEAFO Convention Area shall be prohibited from 1 January 2007 to 31 December 2010 in the 10 seamount areas defined in Annex 1.[143] Three seamount areas (Nos 3 and 4 on the Namibian continental shelf and No. 13 in the Area) were excluded from the area closures by the Commission.[144] Contracting Parties shall communicate to the Executive Secretary information relating to fishing activities, including those by trawlers and longliners, undertaken in 2004–2006 in areas identified in Annex 1.[145]

In 2010, the Commission revised the Measure by agreeing to close 11 seamounts, as recommended by the Scientific Committee: a number of new areas (mostly along the Mid-Atlantic Ridge) were added and many of the existing closures were reduced in terms of size.[146] It should be noted that the new Measure does not specify the expiry date or the possibility for limited access: the only operative paragraph stipulates that "All fishing activities for fisheries resources covered by the SEAFO Convention shall be prohibited from 1 January 2011 in the areas defined in Annex 1 of this Conservation Measure".

Conservation Measure 06/06 stipulated that the Commission would consider at its 2007 Annual Meeting according access to a small-scale and restricted exploratory fishery from 1 January 2008 for an area not exceeding 20% of the fishable area of each seamount as follows.[147] The Commission at its Annual meeting in 2007 decided to maintain the closure in Conservation Measure 06/06, taking into account the Scientific Committee recommendation.[148] The Commission stipulated processes to be followed before the resumption of fishing in the closed area. First, VMEs will have been identified and mapped and an assessment of the impact of any resumption on such VMEs will have been made, and that information shall be submitted to the Scientific Committee.[149] Second, contracting

143 SEAFO, Conservation Measure 06/06 on the Management of Vulnerable Deep Water Habitats and Ecosystems in the SEAFO Convention Area, para. 1. Note the similarity between this measure and the NAFO Conservation and Enforcement Measure, Article 12, paragraphs 5–10, adopted in September 2006.

144 While the Scientific Committee again recommended closing these three seamounts next year, the Commission did not accept the proposal; parties raised concerns that if all areas are closed, there is no other way that data will be obtained. Report of the 4th Annual Meeting of the Commission, 2007, at p. 4.

145 Conservation Measure 06/06, para. 4.

146 SEAFO, Conservation Measure 18/10 on the Management of Vulnerable Deep Water Habitats and Ecosystems in the SEAFO Convention Area. For the discussions on the revision of the Measure at the Scientific Committee, see Report of SEAFO Scientific Committee 2010, at pp. 7–21.

147 Conservation Measure 06/06, para. 2.

148 SEAFO, Conservation Measure 11/07 laying down conditions for the resumption of fishing activities in areas subject to closure through conservation measures 06/06, para. 1.

149 Ibid., para. 2(a).

parties shall submit Research Fishing Plans for evaluation by the Scientific Committee on the impact of the proposed fisheries on the sustainability of the fisheries resources and on vulnerable marine habitats and the Scientific Committee shall submit its recommendation to the Commission for a decision on any reopening of the area to fishing.[150] In 2008, the Commission agreed that all proposals for scientific surveys in SEAFO closed areas be submitted to the Scientific Committee for scrutiny and comments before consideration for approval by the Commission and also approved that exploratory fishing should not be allowed in existing closed areas.

The Scientific Committee recommended that all forms of trawling and gillnet fishing should be banned in view of, among other things, the likely impact of trawls on vulnerable habitats on seamounts in the SEAFO Convention Area.[151] The parties did not accept the recommendation, stating that they felt that all types of fishing gears have an impact on vulnerable habitats and there is no ground for differentiating one from the others; in addition, the parties observed that as gillnet fishing was not conducted in the Convention Area, it was a 'non-issue'.[152] However, in 2009, the Commission, informed of the banning of gillnets in the CCAMLR area, approved that gillnets be banned in the SEAFO Convention Area until such time that more information became available.[153]

The Scientific Committee has repeatedly urged the Commission to control fishing opportunities, in view of the lack of complete and reliable data. In 2005, the Scientific Committee recommended the freezing of current fishing efforts. The Commission, however, deferred the consideration of the proposal to the following year, stating that since only one party had conducted fisheries in 2004, the freezing of current fishing efforts could discriminate against the possibility for other parties to deploy vessels in the area.[154] In 2006, the Scientific Committee further recommended a considerable reduction of fishing pressure in existing fisheries as well as the establishment of relevant indicators of the status of stocks and fishing pressure, combined with precautionary catch limits or effort limitations, in developing new fisheries or expanding existing fisheries to new areas.[155]

In 2007, the Commission adopted catch limits for the Patagonian Toothfish and red crab fisheries in the Convention Area in 2008 and 2009.[156] No national quotas were agreed. When the catch limits are deemed to be exhausted, the

---

150 Ibid., para. 2(b).

151 Report of SEAFO Scientific Committee, 2007, at pp. 7–8.

152 Report of the 4th Annual Meeting of the Commission, 2007.

153 Report of the 6th Annual Meeting of the Commission, 2009, at p. 6.

154 Report of the 2nd Annual Meeting of the Commission, 2005, at p. 5.

155 Report of the 3rd Annual Meeting of the Commission, 2006, at pp. 3–4.

156 SEAFO, Conservation Measure 10/07 fixing catch limits and related conditions for the Patagonian Toothfish and red crab fisheries in the SEAFO Convention area in 2008 and 2009, paras. 1–2. This measure was maintained for 2009. See Report of the 5th Annual Meeting of the Commission,

Executive Secretary is mandated to close the fisheries.[157] Two points are of interest here. First, in formulating catch limits for toothfish in the entire SEAFO Area, the Scientific Committee took account of CCAMLR Conservation Measure 41/04 relating to toothfish.[158] Second, fisheries for deep-sea species are managed on the basis of the TAC, rather than fishing effort limitations. As elaborated in the Conclusions, this is different from NEAFC, where effort limitations were first introduced and followed by catch limitations. In 2008, the Commission also approved the TAC for orange roughy of 100 tonnes and for alfonsino of 200 tonnes for 2009.[159] To the contrary, the Commission could not reach a decision on the TAC for armourhead in 2011, and the issue was referred to the Scientific Committee for more analysis in 2012.[160]

In 2008 the Commission agreed with the recommendation of the Scientific Committee to ban directed fisheries for deep-water shark in the SEAFO Convention Area.[161]

At the annual meeting in 2008, SEAFO adopted a conservation measure relating to bottom fishing activities in all existing and new bottom fishing areas outside SEAFO closed areas.[162] The Measure provides for, *inter alia*, the mapping of existing bottom fishing areas within the Convention Area, the assessment of bottom fishing activities and the operational procedures for handling encounters with VMEs.[163] In particular, all bottom fishing activities in new bottom fishing areas or with bottom gear not previously used in the area concerned shall be considered as exploratory fisheries and shall be conducted in accordance with an Exploratory Bottom Fisheries Protocol to be adopted by the Commission as

2008, at p. 4. The TAC for toothfish was changed to 200 tonnes for 2010 and 230 tonnes for 2011 for 2012. The TAC for red crab was maintained for 2010 and for 2011 and 2012.

157 SEAFO, Conservation Measure 10/07, para. 3.

158 Report of SEAFO Scientific Committee, 2007, at p. 8. CCAMLR Conservation Measures CM 41-04 (2006), *inter alia*, limited toothfish fisheries in Statistical Subarea 48.6 to exploratory fisheries using longlines by vessels from four parties (Japan, Korea, New Zealand and Norway) and set a precautionary catch limit of 455 tonnes north of 60 degrees south and 455 tonnes south of 60 degrees south.

159 Report of the 5th Annual Meeting of the Commission, 2008, at p. 4. The TAC for orange roughy was reduced to 50 tonnes for the whole SEAFO Convention Area for 2010. In 2010, it was decided to set a zero TAC for Sub-division B1 and the 50 tonnes TAC for the remainder of the Convention Area for 2011 and 2012. The TAC for alfonsino was maintained for 2010 as well as for 2011 and 2012.

160 Report of the 8th Annual Meeting of the Commission, 2011, at p. 6.

161 Report of the 5th Annual Meeting of the Commission, 2008, at p. 4.

162 SEAFO, Conservation Measure 12/08 on Bottom Fishing Activities in the SEAFO Convention Area.

163 Ibid., Articles 2–5 and Annex 3.

soon as possible.[164] Before exploratory bottom fishing can take place, a detailed proposal shall be submitted by the Contracting Party to the Scientific Committee for scrutiny and the Scientific Committee will provide a recommendation to the Commission, which will decide if the exploratory fishing may proceed.[165] Exploratory bottom fishing activities shall be subject to the assessment procedure set forth in Article 4, with the understanding that particular care shall be taken in the evaluation of risks of the SAI on VMEs, in line with the precautionary approach.[166] The Scientific Committee shall undertake an assessment and provide advice to the Commission as to whether the proposed bottom fishing activity would have SAIs on VMEs and, if so, whether mitigation measures would prevent such impacts.[167] The Commission shall adopt, taking account of advice and recommendations provided by the Scientific Committee, conservation and management measures to prevent SAIs on VMEs, including prohibiting bottom fishing activities.[168]

The Commission, in 2009, amended Conservation Measure 12/08 by revising the reporting format and the threshold levels as 60 kg live corals and 800 kg live sponges.[169] It is particularly noted that the decision to revise the threshold levels was taken after the Commission was informed of the recent development within CCAMLR and NAFO to reduce encounter threshold levels.[170] No instance was reported in 2010 and 2011 of bycatches exceeding the current VME threshold values.[171] In view of the outcome of the UNGA Workshop held in September 2011, the Scientific Committee recommended that an adapted version of the CCAMLR encounter protocols be applied in the SEAFO Convention Area, but it could not reach consensus on VME threshold levels and gave two options to the Commission; the Commission adopted a compromised proposal on threshold levels with the less strict threshold level for the trawl tow in Annex 5 of Conservation Measure 22/11.[172]

In 2011, the Commission adopted the fishing footprint (i.e., the existing bottom fishing areas) with the reference period of 1987 to July 2011.[173] In addition, it was made clear that an impact assessment, as well as a detailed proposal, shall be

164 Ibid., Article 3(1). Until its adoption, the interim protocol attached as an annex to the measure applies to exploratory bottom fisheries.
165 Ibid., Article 3(2).
166 Ibid.
167 Ibid., Article 4(3)(iii).
168 Ibid., Article 4(4).
169 SEAFO, Conservation Measure 17/09 on Bottom Fishing Activities in the SEAFO Convention Area.
170 Report of the 6th Annual Meeting of the Commission, 2009, at p. 5.
171 Report of the 8th Annual Meeting of the Commission, 2011, at p. 4.
172 Ibid., at pp. 4 and 6.
173 Ibid., at p. 6.

submitted by the contracting party before exploratory bottom fishing can take place and that the Scientific Committee shall undertake an evaluation of the impact assessment.[174]

4.4 *North East Atlantic Fisheries Commission (NEAFC)*

NEAFC was established under the 1959 North-East Atlantic Fisheries Convention, which entered into force in 1963. Following the extension of coastal state fisheries jurisdiction to 200 miles, the 1959 Convention was replaced by a new convention adopted in 1980, which entered into force in 1982.[175]

The NEAFC Convention was further amended in 2004 and 2006, and the contracting parties have agreed to voluntarily use the new convention on a provisional basis, pending ratification of the new convention by the parties to the 1980 Convention.[176] The amendments to the NEAFC Convention are primarily concerned with recommendations for procedures for dispute settlement (Article 18 bis) and Articles 1, 2 and 4.[177] Although the amendments to the NEAFC Convention as such remain minor, a number of modern approaches to fisheries management have been reflected in documents adopted by NEAFC, including the Scheme of Control and Enforcement. Given the fact that contracting parties agreed to voluntarily apply the new convention pending the ratification procedures, this section refers to the provisions of the new convention unless it is necessary to compare the texts of the new and 1980 conventions.

The Convention applies to the areas of the Northeast Atlantic and Arctic Oceans, exclusive of the Baltic Sea and the Mediterranean Sea, as provided for in the Convention.[178] Therefore, the Convention Area includes at least three high

---

174 SEAFO, Conservation Measure 22/11 on Bottom Fishing Activities in the SEAFO Convention Area, Articles 3–4. Compare these provisions with those in Conservation Measure 17/09.

175 For the history of NEAFC, see also the Performance Review Panel Report of the North East Atlantic Fisheries Commission, NEAFC, at pp. 10–11.

176 For information, see Report of the 24th Annual Meeting of the North-East Atlantic Fisheries Commission, 14–18 November 2005, at pp. 32–34. At the 30th Annual Meeting in November 2011, it was reported that the EU and Norway were the only contracting parties having finalized the process of ratifying the amended NEAFC Convention. See 30th Annual Meeting of the North-East Atlantic Fisheries Commission, 7–11 November 2011, Executive Summary of the Report.

177 See Declaration on the Interpretation and Implementation of the Convention on the Future Multilateral Cooperation in North-East Atlantic Fisheries, London. In addition, 'fisheries jurisdiction' was replaced by 'jurisdiction' in Articles 5, 6, 8, 9, 12, 13, 15, 18 and 20. See also Molenaar, "Current Legal and Institutional Issues Relating to the Conservation and Management of High Seas Deep-Sea Fisheries", at p. 125 note 91.

178 New NEAFC Convention, Article 1(a).

seas areas:[179] the Reykjanes Ridge-Azores Area, part of the Norwegian Sea (also known as the 'Banana Hole')[180] and the so-called 'Loophole' of the Barents Sea.

Fisheries management in the Loophole is a mixture of coastal state regulation and a multilateral approach. On the one hand, fisheries in the Loophole are not excluded from the NEAFC regulation, and NEAFC has in fact adopted measures for some fisheries in this area.[181] On the other hand, outside the framework of NEAFC, the two coastal states (i.e., Norway and Russia) have managed to restrain fishing by distant water fishing nations such as the EC, Greenland and the Faroe Islands as well as Iceland in the high seas part of the Barents Sea through the conclusion of access agreements to their EEZ.[182] Commentators vary in their evaluation of the regime governing the management of fisheries in the Barents Sea.[183] In any case, the two coastal states whose EEZ surrounds the Loophole

179 For the applicability of the NEAFC Convention to the high seas part of the Arctic Ocean, see Section 5.4.1 below.

180 Some authors also refer to this area as the 'Doughnut Hole' (or 'Donut Hole'). See, e.g., Henriksen *et al., Law and Politics in Ocean Governance*, at p. 101; Meltzer, "Global Overview of Straddling and Highly Migratory Fish Stocks" (1994), at pp. 280–282.

181 For example, conservation and management measures for pelagic redfish in ICES Sub-areas I and II apply to the Loophole (i.e., ICES Sub-area Ia).

182 For the EC, Greenland and Faroe Islands, see Meltzer, "Global Overview of Straddling and Highly Migratory Fish Stocks" (1994), at p. 281. The conflict between the coastal states and Iceland was resolved in 1999 by a multilateral agreement among the three states concerned and two protocols between Iceland and each of the coastal states. See Agreement between the Government of Iceland, the Government of Norway and the Government of the Russian Federation concerning Certain Aspects of Cooperation in the Area of Fisheries (Barents Sea Loophole Agreement), St. Petersburg, 15 May 1999; Protocol between the Government of Norway and the Government of Iceland under the Agreement between the Government of Norway, the Government of Iceland and the Government of the Russian Federation concerning Certain Aspects of Cooperation in the Area of Fisheries, St. Petersburg, 15 May 1999; Protocol between the Government of Iceland and the Government of the Russian Federation under the Agreement between the Government of Iceland, the Government of Norway and the Government of the Russian Federation concerning Certain Aspects of Cooperation in the Area of Fisheries, St. Petersburg, 15 May 1999. See also R.R. Churchill, "The Barents Sea Loophole Agreement: A "Coastal State" Solution to a Straddling Stock Problem", 14 *International Journal of Marine and Coastal Law* (1999), at pp. 468–475; Henriksen *et al., Law and Politics in Ocean Governance*, at p. 143. Note that the matter was referred to NEAFC by Russia and there was a brief discussion in the 1994 annual meeting. For an account of the reason why the involvement of NEAFC did not take place in NEAFC thereafter, see Churchill, "The Barents Sea Loophole Agreement", at pp. 479–480.

183 Compare, e.g., O.S. Stokke, "The Loophole of the Barents Sea Fisheries Regime", in O.S. Stokke (ed.), *Governing High Seas Fisheries: The Interplay of Global and Regional Regimes* (2001), at p. 286 (arguing that the regime for the Barents Sea is clearly characterized as a regional fisheries management arrangement); Churchill, "The Barents Sea Loophole Agreement", at pp. 476–477, 480 and 482 (arguing that the regime is characterized as 'direct cooperation' between interested states in the sense of Article 7(1)(a) of the FSA); Örebech *et al.*, "The 1995 United Nations

have succeeded in persuading other interested states to refrain from fishing in that area for some species.[184] Stocks which may fall under the category of deep-sea species or DHSFS are not regulated by the above-mentioned fisheries management regime between the coastal states and the high seas fishing states.[185]

In the Norwegian Sea, NEAFC is involved in the management of a straddling stock of spring-spawning herring in the high seas areas. Coastal states first set the TAC for Norwegian spring-spawning herring for the EEZ, and NEAFC then sets its total catch for the high seas.[186]

In the light of the above observations, no separate section needs to be devoted to fisheries management in the Barents Sea 'Loophole' or the Norwegian Sea 'Banana Hole'.

4.4.1 *The Convention*

The objective of the new NEAFC Convention is to ensure the long-term conservation and optimum utilization of the fishery resources in the Convention Area, providing sustainable economic, environmental and social benefits, and the Commission shall perform its functions in order to fulfil this objective.[187]

The new NEAFC Convention defines fishery resources as "resources of fish, molluscs, crustaceans and including sedentary species, excluding, in so far as they are dealt with by other international agreements, highly migratory species listed in Annex I of [the LOSC], and anadromous stocks".[188] A major difference between the new and the 1980 convention is that while, under the 1980 convention, sedentary species are excluded from the scope of fishery resources, the

Straddling and Highly Migratory Fish Stocks Agreement: Management, Enforcement and Dispute Settlement", at p. 124 (stating that the Norway-Russia Fisheries Commission is, unless expanded, no longer a valid arrangement for the management of Norwegian Arctic cod since it would be contrary to the FSA for two states to continue to cooperate within the framework of a bilateral arrangement, excluding others, if the stock distribution includes the high seas, and observing that the appropriate organization or arrangement would be NEAFC).

184 It could be argued that Iceland may conduct fisheries for species not regulated by the Loophole Agreement. Churchill observes that this could happen, while this is contrary to the Norwegian view and any such fishing would not be of real significance in practice. Churchill, "The Barents Sea Loophole Agreement", at p. 473.

185 In the preamble to the Loophole Agreement, it is recognized that "a considerable part of the living marine resources of the northern Norwegian and Barents Seas represent a single regulated biological stock system". Barents Sea Loophole Agreement, preambular paragraph 5. See also Churchill, "The Barents Sea Loophole Agreement", at p. 468.

186 See Churchill, "The Barents Sea Loophole Agreement", at pp. 480–481. He characterizes the regime as a regional fisheries arrangement utilizing an RFMO.

187 New NEAFC Convention, Articles 2 and 4(1). Under the 1980 NEAFC Convention, the Commission would perform its functions in the interests of the conservation and optimum utilization of the fishery resources of the Convention Area. 1980 NEAFC Convention, Article 4(1).

188 New NEAFC Convention, Article 1(b).

new convention explicitly includes sedentary species in the definition of fishery resources.[189]

4.4.1.1 *Precautionary Approach*

The Commission shall ensure that recommendations are based on the best scientific evidence available.[190] In addition, contracting parties shall furnish any available scientific and statistical information needed for the purposes of the Convention at the request of the Commission, and the Commission may also make recommendations to collect additional statistical information relating to fisheries beyond areas under the jurisdiction of the contracting parties.[191]

While the 1980 Convention did not expressly refer to the precautionary approach, most contracting parties were not opposed to the application of the precautionary approach before the new convention was adopted in 2006. In practice, through the 2003 Memorandum of Understanding (MoU) between NEAFC and ICES, NEAFC accepted the application of the precautionary approach, including the ICES format.[192] In 2005, it was agreed to amend the convention, and the new convention provides that the Commission shall apply the precautionary approach.[193] The article concerned does not define the 'precautionary approach'. In this regard, recourse could be had to the FSA and the FAO Code of Conduct for Responsible Fisheries, both of which are referred to in the preamble to the new convention.[194]

4.4.1.2 *Ecosystem Approach*

The 1980 NEAFC Convention only stipulated the incorporation of ecosystem considerations concerning species interrelationships. The Commission shall seek to ensure consistency between its recommendation and coastal states' measures and decisions for fisheries for straddling stocks conducted within the area under their jurisdiction if that would have an effect on those stocks through species interrelationships. Moreover, the relevant contracting parties and the Commission shall promote the coordination of such recommendations, measures and

189 1980 NEAFC Convention, Article 1(2).

190 New NEAFC Convention, Article 4(2)(a). The 1980 NEAFC Convention merely provides that, in performing its function, the Commission shall take into account the best scientific evidence available to it. 1980 NEAFC Convention, Article 4(1).

191 New NEAFC Convention, Articles 9(1) and 16(2).

192 Memorandum of Understanding 2003–2006 between the North-East Atlantic Fisheries Commission and the International Council for the Exploration of the Sea, reproduced in NEAFC Review Panel Report, at pp. 77–87 (Appendix IX).

193 New NEAFC Convention, Article 4(2)(b).

194 Henriksen *et al.*, *Law and Politics in Ocean Governance*, at p. 118. They point out that there are some variations between the ICES reference points and Annex II of the FSA. Henriksen *et al.*, *Law and Politics in Ocean Governance*, at pp. 111–112.

decisions.[195] Thus, when a measure for discrete high seas stocks has impacts on straddling stocks, relevant national measures are taken into consideration.

Before the amendment of the Convention, there were examples in practice, albeit limited, that showed ecosystem considerations. One such initiative relates to the inclusion of ecosystem considerations in the MoU between NEAFC and ICES.[196] Since the latter provides scientific advice to NEAFC, the way in which it supplies the advice impacts upon management by NEAFC. An MoU for 2004–2006 was unanimously adopted by NEAFC in 2003.[197] The MoU stipulated that NEAFC requested ICES to advise on impacts that fisheries have on the ecosystem, and potential imbalances caused by fisheries and reference points as guidance for management purposes in an ecosystem context. The MoU expired at the end of 2006, and a draft for a new MoU developed in the Permanent Committee on Management and Science (PECMAS)[198] was adopted, with a minor amendment proposed by the EU at the Extraordinary Meeting in June 2007.[199]

The new convention gives NEAFC a clearer mandate with respect to the ecosystem approach and biodiversity protection. It provides that the Commission shall take due account of the impact of fisheries on other species and marine ecosystems, and in doing so adopt, where necessary, conservation and management measures that address the need to minimize harmful impacts on living marine resources and marine ecosystems, and shall take due account of the need to conserve marine biological diversity.[200]

In addition, the Commission shall now provide a forum for consultation and the exchange of information on the state of the fishery resources in the Convention Area and on the management policies, including an examination of the overall effects of such policies on the fishery resources and, as appropriate, other living marine resources and marine ecosystems.[201]

---

195 This provision remains unchanged under the new convention. New NEAFC Convention, Article 5(2).

196 For a discussion, see Henriksen *et al., Law and Politics in Ocean Governance*, at pp. 114–115. For closed areas to protect VMEs, see Section 4.4.2.2 below.

197 Report of the Extraordinary Meeting of the North-East Atlantic Fisheries Commission, 14–15 May 2003, at pp. 10–12; Report of the 22nd Annual Meeting of the North-East Atlantic Fisheries Commission, 10–14 November 2003, at p. 28.

198 Permanent Committee on Management and Science, PECMAS, of the North-East Atlantic Fisheries Commission, Report of the Second Meeting 19–21 February 2007, at pp. 6–7.

199 Report of the Extraordinary Meeting of the North-East Atlantic Fisheries Commission, 13–14 June 2007, at pp. 15–16.

200 New NEAFC Convention, Article 4(2)(c)–(d). The new convention defines the term 'living marine resources' as "all living components of marine ecosystems". New NEAFC Convention, Article 1(c). It also defines the term 'marine biological diversity' in an almost identical manner as the CBD. The only difference is that the new NEAFC Convention is only concerned with marine aspects. See New NEAFC Convention, Article 1(d).

201 New NEAFC Convention, Article 4(3).

4.4.2 *Practice*

NEAFC has adopted a number of measures for fisheries on the high seas. The rest of this sub-section examines such measures to the extent that they may have implications for new challenges in high seas fisheries such as fisheries for DHSFS, deep-sea fisheries and area-based management tools.

Recently, discussions have taken place with respect to the stock classification of oceanic redfish in the NEAFC Convention Area. However, for the time being, it is unlikely that one of its stocks proves to be, or will become, a discrete high seas fish stock. Targeted DHSFS are thus deep-sea species. Therefore, the following parts first deal with deep-sea fisheries and then analyze area-based management tools (i.e., NEAFC closed areas and cooperation with the OSPAR Commission).

4.4.2.1 *Deep-sea Fisheries*

Some deep-sea species such as roundnose grenadier have been fished in the Regulatory Area of NEAFC for a long time.[202] While the issue of deep-sea fisheries was discussed and ICES advice was sought on several occasions before 2002,[203] intensive discussions only started after the 20th Annual Meeting in 2001, where the parties agreed to establish the Working Group on the Appraisal of Regulatory Measures for Deep-sea Species.[204] Since then, as predicted by the EC, the management of deep-sea fisheries has become a "major issue for NEAFC to solve".[205] Measures for deep-sea species have been taken in five ways: information gathering; the regulation of fishing effort; the prohibition on new fisheries; the regulation of fishing gear; and a prohibition and control of fisheries for certain species.

The first sort of measure relates to the collection of catch and effort data and other information. The lack of information was a major source of difficulties in determining management measures for deep-sea fisheries.[206] Discussions took place, especially at the early stage of management, regarding the scientific basis of the measures. For example, at the 18th Annual Meeting in 1999, when the EU suggested the freezing of efforts at current levels, all parties agreed that more input from ICES was needed before progress could be made, and the president of NEAFC stated that the meeting should encourage the compilation of catch data

202 Mooney-Seus and Rosenberg, Regional Fisheries Management Organizations (RFMOs): Progress in Adopting Precautionary Approach and Ecosystem-Based Management, at p. 129.

203 For an overview, see NEAFC Review Panel Report, at p. 70.

204 Report of the 20th Annual Meeting of NEAFC, 5–9 November 2001, at pp. 11–12.

205 Summary Report, North East Atlantic Fisheries Commission, Nineteenth Annual Meeting: 21–24 November 2000, at p. 13.

206 For the first time in 1997, parties reported on catches of deepwater species. Report of the Sixteenth Annual Meeting, London, November 1997, at p. 14.

on these species on a voluntary basis.[207] Even in 2002, while all parties agreed on the need to establish management measures for deep-sea species at the outset of the discussion, Russia claimed that such measures could not be taken before more information was available.[208]

The reporting regime was already in place as part of the Control and Enforcement Scheme. However, in relation to deep-sea fisheries, the existing regime was problematic not only because data was not split between the Regulatory Area and coastal zones prior to 2006,[209] but also because ICES statistical areas were too large for these species due to their often limited local distribution.[210] The Commission adopted a special regime for compiling information on national measures and the submission of scientific information on deep-sea fisheries, including the development of sampling plans for deep-sea species.[211] The measures use ICES statistical rectangles as the reporting unit, which are smaller and more suited to the management of deep-sea fisheries.

The second sort of measure concerns fishing effort regulation: freezing and a reduction of fishing efforts. At the 21st Annual Meeting in 2002, the capacity of the deep-sea fleet was established and then frozen. The original Norwegian-EU proposal (AM2002/41) attempted to establish the fishing effort using the period 1998–2001 as a baseline. However, in the face of Russian and Danish[212] opposition, the proposal was amended to use as the reference period the highest level in previous years, and the measure thus deviated from the original intent to freeze efforts at *current* levels.[213] Due to the disagreement on the scope of application of effort reduction and the definitions of deep-sea fishing activities, NEAFC failed to establish conservation and management measures in 2003.[214]

---

207 Summary Report, North East Atlantic Fisheries Commission, Eighteenth Annual Meeting: 22–25 November 1999 at p. 10. Similarly, the need for more information for managers to be able to fully comprehend the advice needed was expressed by the EU at the 19th Annual Meeting in 2000. NEAFC 2000 Annual Meeting Report, at p. 13.

208 Report of the 21st Annual Meeting of the North-East Atlantic Fisheries Commission, 12–15 November 2002, at p. 25.

209 NEAFC Review Panel Report, at p. 29.

210 This was pointed out by ICES as early as at the 19th Annual Meeting in 2000. NEAFC 2000 Annual Meeting Report, at p. 13.

211 Originally adopted in 2005 and amended in 2006. See NEAFC 2005 Annual Meeting Report, at p. 24; Report of the 25th Annual Meeting of the North-East Atlantic Fisheries Commission, 13–17 November 2006, at p. 94.

212 Although Denmark is a member of the European Union and has transferred the competence to regulate high seas fisheries to the EU, the Faroe Islands and Greenland are excluded from the application of the Common Fisheries Policy and therefore the Danish government participates on their behalf in NEAFC and other RFMOs of which it is member.

213 NEAFC 2002 Annual Meeting Report, at pp. 25–29.

214 NEAFC 2003 Extraordinary Meeting Report, at pp. 3–7; NEAFC 2003 Annual Meeting Report, at pp. 20–22.

At the 23rd Annual Meeting, the reduction of fishing efforts by 30% was finally agreed upon. Despite strong EU insistence, the scope of the recommendation was limited to the Regulatory Area, and was only concerned with directed fishing for deep-sea species.[215] At the 25th Annual Meeting, a further 10% reduction in fishing efforts was proposed by the EU, while Russia preferred to stick to 70%. Upon a suggestion by the President, it was agreed by consensus that the effort shall not exceed at most 65% in 2007 of the highest level put into deep-sea fishing in previous years.[216] A roll-over of this measure was agreed in subsequent years.[217]

Third, the control of new fisheries has been attempted. The EU proposed the prohibition of new fisheries without proof of sustainability, and this proposal was referred to the Deep-Sea Working Group in 2005.[218] The Working Group recommended limiting new fisheries by adopting either of the following options: (1) fishing for deep-sea species shall be prohibited except in designated ICES rectangles; (2) fishing for deep-sea species shall be prohibited in designated ICES rectangles.[219] The proposal was referred to PECMAS by consensus at the Commission in 2006.[220] The PECMAS meeting in February 2007 did not reach agreement on this issue. Two parties indicated that they did not consider the first approach (i.e., first close everything, and then open up) to be acceptable. It was noted that there are other measures that control new or current fisheries, especially the cap on effort in deep-sea fisheries.[221] PECMAS agreed to propose to the Commission that PECMAS start a process of clarifying the objectives of and developing criteria for closures.[222]

Fourth, limitations on the use of certain fishing gear have been adopted. At the 24th Annual Meeting, a proposal on an interim ban on the use of gillnets, entangling nets and trammel nets was adopted by consensus.[223]

---

215 Report of the 23rd Annual Meeting of the North-East Atlantic Fisheries Commission, 8–12 November 2004, at pp. 31–32. The EU declared that it would apply the measure of a 30% reduction of fishing efforts throughout the entire Convention Area, while Iceland stated it could only accept that that measure applied to the Regulatory Area (p. 31). This measure was maintained for 2006. NEAFC 2005 Annual Meeting Report, at p. 24.

216 NEAFC 2006 Annual Meeting Report, at p. 93.

217 See, e.g., Recommendation VI:2010.

218 NEAFC 2005 Annual Meeting Report, at p. 25.

219 Report, NEAFC Working Group on Deep Sea Species, 20–21 June 2006, at p. 19.

220 NEAFC 2006 Annual Meeting Report, at p. 93.

221 PECMAS Second Meeting Report, at p. 8.

222 Ibid., at pp. 4 and 8. See also discussions below in the context of the regulation of bottom fishing in the NEAFC Regulatory Area.

223 NEAFC 2005 Annual Meeting Report, at p. 25.

Fifth, fisheries for certain species were regulated in terms of catch limits.[224] At the outset, it should be noted that it has been repeatedly emphasized that setting the TAC does not suit the management of deep-sea fisheries and effort regulation should be employed.[225] However, when ICES proposed the introduction of catch limits for the first time in 2006, the contracting parties in the Working Group on Deep Sea Species discussed the feasibility of setting catch levels. It was noted that the parties are still unable to give specific advice on individual stocks and, even if possible, there is the problem of mixed fisheries, which probably need to be regulated by effort measures. In addition, Norway pointed out that a limit in effort could lead to the required catch levels, without introducing catch limits, while the EU noted that this may not be the case where there are aggregations of fish.[226]

At the 25th Annual Meeting, the Commission discussed the proposal by the Working Group on Deep Sea Species that fishing for orange roughy should be prohibited. In response to the Danish view that there was not sufficient time to consider the proposal, the president suggested making a preliminary regulation and revising it later. The proposal to ban the fishery for the first half of 2007 was adopted by consensus, with the understanding that it would be reconsidered at the Extraordinary Meeting in June 2007.[227] At the Extraordinary Meeting in 2007, the Danish proposal for management measures for orange roughy in 2007 and 2008 was opposed by other contracting parties, including Norway, Iceland and the EU due the reason that the ICES advice recommended no directed fisheries for this species. It was decided to extend the interim prohibition on directed orange roughy fisheries in the NEAFC Regulatory Area until 31 December 2007.[228]

At the 26th Annual Meeting in 2007, Denmark made a proposal that directed fisheries for orange roughy in the NEAFC Regulatory Area in ICES Sub-areas V–VII (i.e., the Regulatory Area adjacent to waters under the national jurisdiction of EU member states) are prohibited while the directed fisheries for orange roughy in other areas of the Regulatory Area are allowed under precautionary conditions, including: fishing activities shall be restricted to vessels of contracting parties having participated in the fishery for orange roughy in the NEAFC Regulatory Area other than Sub-areas V–VII prior to 2005; and total catches of any contracting party shall not exceed 150 tonnes.[229] The EU opposed this proposal, submit-

224 Apart from orange roughy discussed below, the Commission banned directed fisheries for deep-sea sharks for 2012. Report of the 30th Annual Meeting of the North-East Atlantic Fisheries Commission, 7–11 November 2011, at pp. 19–22; Recommendation 7: 2012.

225 See, e.g., NEAFC 2003 Annual Meeting Report, at p. 22 (Norway and the EU indicating that it would be difficult to advise on TACs for deep-sea species).

226 NEAFC Working Group 2006 Report, at p. 3.

227 NEAFC 2006 Annual Meeting Report, at pp. 93–94.

228 NEAFC 2007 Extraordinary Meeting Report, at pp. 5–6.

229 See AM 2007/18 rev, paras. 2.1–2.2.

ting its own proposal to maintain the prohibition of fishery for orange roughy in the NEAFC Regulatory Area for 2008.[230] The Danish proposal was adopted by four votes against one.[231] Three features of the recommendation are worth commenting upon. First, it restricted the fishing opportunities to states with a fishing record.[232] Second, it adopted catch limits, rather than fishing effort limits. Third, it allocated 150 tonnes equally to all qualified contracting parties.

Apart from the measures mentioned above, it is interesting to note a reference to DHSFS in the measure on the collection of information adopted in 2006 (Recommendation X: 2007), which reads as follows:

> Among the general principles of the [FSA] the states party to this agreement undertake to; [*sic*] adopt measures to ensure long-term sustainability of straddling fish stocks and highly migratory fish stocks and to promote the objective of their optimum utilization; to ensure that such measures are based on the best scientific evidence available and to collect and share, in a timely manner, complete and accurate data concerning fishing activities. The Review Conference on the UN Fish Stocks Agreement held in May 2006 agreed that these general principles should also apply to [DHSFS].

This may shed some light on which principles of the FSA should be applied to DHSFS, since the reference singles out principles such as long-term sustainability, optimum utilization, the basis of best scientific evidence available, and data collection and sharing.

Furthermore, the same measure stipulates that measures that apply to deep-sea species in waters under national jurisdiction shall not undermine those established for the Regulatory Area. Since NEAFC manages deep-sea fisheries on a species basis, rather than a stock basis, the compatibility between measures for areas under national jurisdiction and those for the high seas becomes even more important.[233]

---

230 See AM 2007/57.

231 Report of the 26th Annual Meeting of the North-East Atlantic Fisheries Commission, 12–16 November 2007, at p. 13. Subsequently, an objection to this recommendation was received from the EU on 29 November 2007. The measure was discontinued at the annual meeting in 2008. Report of the 27th Annual Meeting of the North-East Atlantic Fisheries Commission, 10–14 November 2008, at pp. 17–19. In 2009, the Commission adopted another Danish proposal for 2010 and 2011, which was substantially identical to the 2007 measure. Recommendation IX:2010. In 2011, the Commission was again unable to agree on measures for orange roughy; it was confirmed that the general measures applicable to deep-sea species would apply during 2012. NEAFC 2011 Annual Meeting Report, at pp. 21–23.

232 NEAFC 2007 Annual Meeting Report, at p. 13. This entails the issue of new entrants. In fact, the observer from the Cook Islands criticized the recommendation, invoking Article 300 of the LOSC. NEAFC 2007 Annual Meeting Report, at p. 13.

233 Compare with the management of bottom fisheries by NAFO in Section 4.5.3 below.

In addition to separate measures for deep-sea species, NEAFC adopted comprehensive measures for bottom fishing to address the implementation of UNGA Resolution 61/105 at the Extraordinary Meeting in July 2008.[234] They provide for the mapping of existing bottom fishing areas, the assessment of exploratory fisheries in new bottom fishing areas, the identification of VMEs and unexpected encounters with VMEs.[235] In particular, from 1 January 2009, all bottom fishing activities in new bottom fishing areas or with bottom gear not previously used in the area concerned, shall be considered as exploratory fisheries and shall be conducted in accordance with an Exploratory Bottom Fisheries Protocol to be adopted by the Commission as soon as possible.[236] The exploratory bottom fishing activities shall be subject to the assessment procedure set forth in Article 5, with the understanding that particular care shall be taken in the evaluation of risks of the SAI on VMEs, in line with the precautionary approach.[237] On the basis of an assessment made by ICES, PECMAS shall provide advice to the Commission as to whether the proposed bottom fishing activity would have SAIs on VMEs and, if so, whether mitigation measures would prevent such impacts.[238] Taking account of the advice and recommendations provided by ICES and considered by PECMAS, the Commission shall adopt conservation and management measures to prevent SAIs on VMEs, including prohibiting bottom fishing activities.[239] Prior to commencing new bottom fishing activities based upon the results of exploratory fisheries conducted in the prior two years, the Commission shall review the assessments undertaken in accordance with Article 5 and the results of the fishing protocols and shall either authorize these bottom fishing activities to proceed and establish conservation and management measures to prevent SAIs on VMEs and to ensure the long-term sustainability of deep-sea fish stocks or not authorize these bottom fishing activities to proceed.[240]

At the 2008 Annual Meeting, NEAFC further elaborated the measures addressing UNGA Resolution 61/105 by adopting the protocol dealing with encounters with VMEs.[241] The new measure defined the threshold levels (i.e., a catch per

234 NEAFC, Recommendation by the North-East Atlantic Fisheries Commission in Accordance with Article 5 of the Convention on Future Multilateral Cooperation in North-East Atlantic Fisheries at Its Extraordinary Meeting on 1–2 July 2008 to Adopt the Following recommendation on Bottom Fishing Activities in the NEAFC Regulatory Area.

235 Ibid., Articles 3–6.

236 Ibid., Article 4(1).

237 Ibid., Article 4(2).

238 Ibid., Article 5(3)(iii). See also Article 5(2).

239 Ibid., Article 5(4).

240 Ibid., Article 4(5).

241 NEAFC 2008 Annual Meeting Report, at p. 19. The original proposal, which was adopted after some amendments, was to a large extent based on a recommendation adopted in NAFO in September 2008 (p. 10). On the NAFO recommendation, see Section 4.5.2 below.

set of more than 100 kg of live coral or 1000 kg of live sponge) for an encounter with VMEs and provided for the requirement of the 2-mile moving-away from the suggested location of encounters.[242]

At the same meeting, PECMAS reported to the Commission that, because areas where VMEs were identified were already closed to fishing, the risk of encounters with corals and sponges is minimal in areas open for commercial fisheries.[243] In fact, as of November 2011, no encounters have been reported and no impact assessments have been submitted by contracting parties.[244] Confirming no encounters having been reported, the PECMAS Chair pointed out that NEAFC has put a great effort into closing the areas where encounters would be likely and not having encounters in the open areas could be seen as a positive situation that constitutes a success of the area closures exercise.[245] To the contrary, the EU suggested a link between the lack of reporting any encounters and the high threshold levels.[246]

In 2011, the Commission agreed to undertake an extensive review of measures to protect VMEs during 2012.[247]

4.4.2.2 *Closed Areas*

Closely related to the issue of deep-sea fisheries is the protection of marine ecosystems through the closure of certain areas to fisheries. To protect deep-sea species and vulnerable habitats from trawl fishing, the closure of several seamounts (i.e., the Hecate and Faraday seamounts and a section of the Reykjanes Ridge, the Altair seamounts and the Antialtair seamounts) for bottom trawling and fishing with static gear was agreed at the 23rd Annual Meeting for the period from 1 January 2005 to 31 December 2007.[248] All of these seamounts are entirely located

---

242 Recommendation XIII: 2009, Recommendation by the North-East Atlantic Fisheries Commission in Accordance with Article 5 of the Convention on Future Multilateral Cooperation in North-East Atlantic Fisheries at Its Annual Meeting on 10–14 November 2008 to Adopt the Following Recommendation on Operational Procedures for Fishing in Existing and New Bottom Fishing Areas. In 2009, the threshold levels for corals and sponges were lowered to 60 kg of live coral and 800 kg of live sponge, respectively. Recommendation XI: 2010. At the 30th Annual Meeting, the EU proposal to change the current threshold levels in harmony with the measures in place for NAFO was not adopted. NEAFC 2011 Annual Meeting Report, at pp. 10–12 and 26–27.

243 NEAFC 2008 Annual Meeting Report, at p. 10.

244 NEAFC 2011 Annual Meeting Report, at p. 11.

245 Ibid., at pp. 11–12.

246 Ibid., at p. 26.

247 Ibid., at pp. 27–28.

248 NEAFC 2004 Annual Meeting Report, at pp. 39–41. Also note that NEAFC closed an area on the Western slope of Rockall Plateau to all fisheries except longlines in 2001, in order to protect juvenile haddock. See Report of the Extraordinary Meeting of the North-East Atlantic Fisheries

in the high seas. Subsequently, the interim measure was prolonged by consensus for 2008 at the 26th Annual Meeting in 2007.[249]

In a postal vote in April 2009, NEAFC adopted a measure to close five areas on the Mid-Atlantic Ridge (MAR) for bottom fisheries with a view to protecting VMEs from SAIs in force until 31 December 2015.[250] The area covered by the new measure included all five areas closed in 2004; the size of the total area covered by the new measure was estimated at 333,000 km² covering a large portion of the MAR while the remaining part of the MAR was regarded as a 'new fishing area'.[251] While the EU proposed to extend the closed areas in 2009 and 2010, the contracting parties did not agree on this proposal.[252]

Furthermore, in response to ICES advice presented at the 25th Annual Meeting, the EU and Russia tabled proposals for closures of the Hatton Bank, North West Rockall, Logachev Mounds, South Rockall, West Rockall Mounds and South West Rockall.[253] The EU eventually accepted the Russian proposal. Bottom trawling and fishing with static gear, including fishing with bottom gillnets and longlines, were prohibited in areas on the Hatton Bank, and Rockall Bank in the parts of areas that fall within the NEAFC Regulatory Area (i.e., North-West Rockall, Logachev Mounds and West Rockall Mounds), as delimited in the proposal, for the period 1 January 2007–31 December 2009.[254] The North-West Rockall Bank and the Logachev Mounds straddle the EEZ and the high seas, while the Hatton Bank is entirely located in the high seas.

At the 26th Annual Meeting in 2007, based on new data and advice by ICES, the EU and Russian proposal for the protection of cold-water corals was adopted by consensus. A new area was added to the existing closed area of the Hatton Bank, the existing boundary of the North-West Rockall closed area was adjusted, and a new closed area was established in the South-West Rockall (Empress of

---

Commission, 22–23 March 2001, at p. 5; A/61/154, at p. 33. See also NEAFC 2011 Annual Meeting, at p. 19.

249 NEAFC 2007 Annual Meeting Report, at p. 13.

250 Recommendation by the North East Atlantic Fisheries Commission in accordance with Article 5 of the Convention on Future Multilateral Cooperation in North East Atlantic Fisheries on the Protection of Vulnerable Marine Ecosystems from significant adverse impacts in the NEAFC Regulatory Area, adopted on 27 April 2009 (hereinafter '2009 MAR Closure Recommendation').

251 See NEAFC, "NEAFC closes large areas to bottom fisheries on the Mid-Atlantic Ridge to protect Vulnerable Marine Ecosystems in the High Seas of the North East Atlantic", Press Release, 27 April 2009, available at <http://www.neafc.org/system/files/vmes_press_rel_april2009.pdf>.

252 Report of the 29th Annual Meeting of the North-East Atlantic Fisheries Commission, 8–12 November 2010, at p. 7. On this issue, especially from the perspective of coordination between NEAFC and OSPAR, see *infra*.

253 For previous discussions on these areas, see NEAFC 2004 Annual Meeting Report, at p. 40; NEAFC 2005 Annual Meeting Report, at p. 26.

254 NEAFC 2006 Annual Meeting Report, at pp. 94–95.

Britain Bank).[255] It was noted that the recommendation was in full conformity with UNGA Resolution 61/105.[256] Subsequently, the closed areas of the Hatton Bank, Rockall Bank, Logachev Mounds and West Rockall Mounds were maintained with minor modifications.[257]

In 2009, NEAFC also closed the blue ling spawning area in the high seas area south of the Icelandic EEZ (i.e., ICES Division XIV).[258] The measure prohibited all fishing with bottom contacting gear in the period 15 February to 15 April, applicable for 2010–2012.

The accelerated progress on the establishment of closed areas from 2006 could arguably be attributed to the decision of the Commission to use the new NEAFC Convention on a provisional and voluntary basis as of 2006. Under the 1980 Convention, it was argued that the NEAFC competence to close areas outside the fisheries jurisdiction of contracting parties for fishing activities is limited to closure related to 'fishery resources' and the Commission does not have the right to establish closed areas in order to protect seamounts, hydrothermal vents or cold-water corals as such.[259] It is conceivable that the provisions of the new Convention allowing a wider scope of measures have facilitated the proposals.

Apart from the area closure as such, it is interesting to note that the measure explicitly indicates that closures are "without prejudice to any sovereign rights of Coastal States over the continental shelf [...] including sovereign rights of Coastal States to exploit sedentary species on the continental shelf".[260] The insertion of this paragraph in 2009 [i.e., after the 2008 UNGA Resolution on sustainable fisheries] suggests that NEAFC parties were influenced by the global development in this respect. However, it is noted that the measure does not refer to coastal state 'jurisdiction'.

4.4.2.3 *Cooperation with the OSPAR Commission*

The Commission may establish working arrangements with any other international organization which has "related objectives".[261] At the regional level,

255 NEAFC 2007 Annual Meeting Report, at p. 14.

256 Ibid.

257 See Recommendation VIII: 2010; Recommendation 14: 2011; Recommendation 8: 2012.

258 Recommendation X: 2010.

259 See the Paper presented by Norway to the 23rd Annual Meeting on "Deep-water Habitats Vulnerable to Fishing Activities: Closing of Areas for Trawling in the Regulatory Area", at p. 5. In that paper, Norway took the position that the establishment of protected areas in the high seas would be in conflict with the prohibition of the LOSC Articles 89 and 137(3), without amending the LOSC.

260 2009 MAR Closure Recommendation, para. 6. This caveat is also found in the recommendations on other closures. Recommendation VIII: 2010, para. 5; Recommendation 14: 2011, para. 5; Recommendation 8:2012, para. 5.

261 New NEAFC Convention, Article 14(3).

interactions may occur between NEAFC and several kinds of international organizations: (1) an organization dealing with the protection of the marine environment in the region; (2) an RFMO regulating straddling stocks in the neighbouring area of the high seas (i.e., NAFO); (3) another RFMO or organization regulating marine living resources in the same area (e.g., the North Atlantic Salmon Conservation Organization (NASCO)); (4) a scientific organization (i.e., ICES). The rest of this part focuses on (1).[262]

The Convention for the Protection of the Marine Environment of the North-East Atlantic (OSPAR Convention)[263] established a Commission as the executive body of the Convention (hereinafter 'OSPAR Commission'). The purpose of the Convention is to provide a framework for the comprehensive protection of the marine environment of the North-East Atlantic.[264]

The OSPAR Commission does not have competence to regulate fisheries. The penultimate recital of the preamble to the OSPAR Convention recognizes that "questions relating to the management of fisheries are appropriately regulated under international and regional agreements dealing specifically with such questions". Annex V to the OSPAR Convention also provides that "no programme or measure concerning a question relating to the management of fisheries shall be adopted under this Annex".[265] The phrase 'questions relating to the management of fisheries' includes the question of which action can be taken under instruments constituting, *inter alia*, NEAFC, "whether or not such action has been taken".[266]

Nevertheless, the OSPAR Commission could affect fisheries management in the North-East Atlantic in two ways. First, where the OSPAR Commission considers that action is desirable in relation to such a question, the Commission is to draw that question to the attention of the authority or international body

262 Note also that NEAFC adopted the Terms of an Agreement of Co-operation between the IMO and NEAFC. See NEAFC 2008 Annual Meeting Report, at p. 24.

263 Convention for the Protection of the Marine Environment of the North-East Atlantic, Paris, 22 September 1992. The OSPAR Convention is exceptional in the sense that most of the other regional conventions on the protection of the marine environment are only applicable to areas within the limits of the national jurisdiction of the contracting parties (and do not have the competence to address the establishment of MPAs in the Area). See Oude Elferink, "The Regime of the Area", at pp. 171–172.

264 L. de La Fayette, "The OSPAR Convention Comes into Force: Continuity and Progress", 14 *International Journal of Marine and Coastal Law* (1999), at p. 247.

265 OSPAR Annex V, Article 4(1).

266 OSPAR Agreement on the Meaning of Certain Concepts in Annex V to the 1992 OSPAR Convention on the Protection and Conservation of the Ecosystems and Biological Diversity of the Maritime Area (Reference number: 1998–15.2).

competent for that question.[267] In the context of high seas fisheries (for species other than salmon or tuna) in the North-East Atlantic, this is NEAFC. In fact, as part of the commitment to establish a network of MPAs in the OSPAR maritime areas by 2010, the OSPAR Commission has already drawn the attention of NEAFC to the need to protect corals on the western slopes of the Rockall Bank.[268] Second, the establishment of MPAs by the OSPAR Commission in the North-East Atlantic might indirectly affect similar undertakings by NEAFC. It is recalled that the unsuccessful EU proposal for the extension of the MAR closures was partly based on the call for coordination between NEAFC and OSPAR.[269]

As noted earlier, NEAFC and the OSPAR Commission have already engaged in dialogue for several years.[270] However, cooperation is not yet sufficient: the relationship and future links between NEAFC and the OSPAR Commission were identified in the NEAFC Performance Review Report as the only area for potential improvement in the relationship with other organizations.[271] The Working Group on the Future of NEAFC has acknowledged this and recommended the Commission to develop an MoU or other working arrangements.[272] The MoU was signed by both the OSPAR Commission and NEAFC in 2008.[273] In the MoU, NEAFC and the OSPAR Commission reached the understanding to promote mutual cooperation towards the conservation and sustainable use of marine biological diversity in a number of areas. In particular, NEAFC and the OSPAR Commission would work together to develop a common understanding of the application of the

267 OSPAR Annex V, Article 4(1). According to this article, where action within the competence of the Commission is desirable to complement or support action by those authorities or bodies, the Commission shall endeavour to cooperate with them.

268 See Meeting of the OSPAR Commission, Stockholm, 26–30 June 2006, at Annex 6 (Briefing on OSPAR's Work on the Protection of the Marine Environment of the High Seas), para. 19. As a response, NEAFC instructed the Secretary to answer the letter from the OSPAR Commission by drawing attention to the area already closed for bottom trawling, for different reasons (i.e., to protect juvenile haddock), on the western slopes of the Rockall Bank. NEAFC 2004 Annual Meeting Report, at p. 39.

269 See NEAFC 2009 Annual Meeting Report, at p. 14.

270 See NEAFC 2004 Annual Meeting Report, at p. 39; NEAFC 2005 Annual Meeting Report, at p. 38; NEAFC 2006 Annual Meeting Report, at p. 101.

271 NEAFC Review Panel Report, at p. 57. In the past, NEAFC members were reluctant to cooperate with the OSPAR Commission more actively: NEAFC noted that "the cooperation has to start on a national level and spill over to the RFMOs and [regional seas programmes] from there". NEAFC 2005 Annual Meeting Report, at p. 38.

272 Report of the Meeting of the Working Group on the Future of the North-East Atlantic Fisheries Commission, 30–31 January and 1 February 2007, at pp. 10–11. This was agreed by the Commission. NEAFC 2007 Extraordinary Meeting Report, at p. 14.

273 Press release of NEAFC and the OSPAR Commission, "Fisheries and Environment bodies join forces to strengthen protection of the North-East Atlantic", available at <http://www.neafc.org/system/files/22092008_mou_neafc_osparv3.pdf>.

precautionary approach/principle and would cooperate regarding marine spatial planning and area management.[274] These provisions clearly indicate that standards and criteria for fisheries management by NEAFC will be affected by the work of the OSPAR Commission (and *vice versa*) in the future.

In 2010, the OSPAR Commission took part in a session within the NEAFC PECMAS meeting held in September/October 2010.[275] The NEAFC Secretariat was invited to participate in the OSPAR Biodiversity Committee in April 2011.[276]

Further cooperation between NEAFC and OSPAR is likely to take place in respect of area-based management. In fact, the NEAFC and OSPAR Secretariats, together with the ISA Secretariat, attended a meeting on options for the management of the proposed Charlie-Gibbs Fracture Zone MPA in March 2010.[277] In this respect, divergent views were presented within NEAFC on the issue of a proposal for a collective arrangement between competent authorities on the management of MPAs in areas beyond national jurisdiction.[278]

The OSPAR Commission established six MPAs in 2010: four MPAs in waters superjacent to areas subject to submission by Portugal to the CLCS and two MPAs in areas beyond national jurisdiction.[279] It further developed the draft collective arrangement between competent authorities on the management of selected areas in areas beyond national jurisdiction within the OSPAR Maritime Area.[280] The arrangement states that:

> The competent authorities decide within the framework of their respective mandate and competence, to cooperate in the development and implementation of appropriate measures for the conservation and management of these areas, informed, where appropriate, by conservation objectives established for these areas.[281]

---

274 Memorandum of Understanding between the North East Atlantic Fisheries Commission (NEAFC) and the OSPAR Commission, September 2008, reproduced in Report of the Extraordinary Meeting of the North-East Atlantic Fisheries Commission, 1 and 2 July 2008, volume II – Annexes, at pp. 23–25.

275 See Permanent Committee on Management and Science – PECMAS, of the North-East Atlantic Fisheries Commission, 30 September–1 October 2010, Report, at pp. 11–18.

276 Ibid., at p. 18.

277 Ibid., at pp. 17–18.

278 Ibid., at p. 18.

279 OSPAR 10/23/1-E, at p. 30, para. M5.2(c). The seabed of the northern part of the Charlie Gibbs Fracture Zone is subject to the submission of information to the CLCS by Iceland. The members of the OSPAR Commission were not in a position to resolve issues concerning the establishment of a water column only MPA for the northern part of that Zone and decided to work within the framework of the OSPAR Commission to resolve any outstanding issues with regard to the high seas of the northern part of the originally proposed Charlie Gibbs Fracture Zone MPA by 2012. OSPAR 10/23/1-E, at p. 13, para 5.16(b). In 2012, the Charlie Gibbs North High Seas MPA was established. OSPAR 12/22/1-E, at pp. 13–14, paras. 5.15–5.20.

280 OSPAR 11/20/1-E, at pp. 13–14, paras. 4.15–4.19.

281 Ibid., Annex 15, para. 2.

It further states that the competent authorities should exchange information, cooperate on, *inter alia*, environmental impact assessments and strategic environmental assessments and consult annually and cooperate to obtain a better knowledge of the areas concerned.[282] In addition, the Arrangement provides that the cooperation of competent authorities on the management of selected areas in areas beyond national jurisdiction in the North East Atlantic should be based on the joint principles of competent authorities on the management of human activities in selected areas in areas beyond national jurisdiction in the North East Atlantic, which aim to set out a common understanding between competent authorities on the management of human activities in selected areas in areas beyond national jurisdiction in the North East Atlantic, enshrined in existing international instruments and comprising the nine general principles specified in the document.[283]

The Second Informal Meeting of Competent Authorities (Madeira II) was held in Paris in January 2012 to discuss the management of selected areas in areas beyond national jurisdiction.[284] The Marine Environment Protection Committee of the IMO at its 63rd session held in February–March 2012 discussed the information document containing the proposed Arrangement and agreed that any proposal to amend maritime traffic in the North East Atlantic must be made by IMO member governments to the appropriate IMO body and requested the Secretariat to keep it informed of any future developments in this regard.[285] The ISA and NEAFC are expected to consider the draft Arrangement in July 2012 and October 2012, respectively.[286] It still remains to be seen how this process of the consideration of the Arrangement among the competent authorities will develop and what outcome it will yield.

In addition, as part of the cooperation with the OSPAR Commission, the NEAFC held the joint NEAFC/OSPAR/CBD workshop on the identification of ecologically or biologically significant marine areas in the North East Atlantic.[287]

---

282 Ibid., para. 3.

283 These are: (a) Ecosystem Approach; (b) Obligation to protect and preserve the marine environment as in LOSC Article 192; (c) Precautionary Principle; (d) Sustainable Use of Natural Resources; (e) Use of Best Available Scientific Advice; (f) Application of Environmental Impact Assessments and Strategic Environmental Assessments; (g) Polluter Pays Principle; (h) Public Availability of Information; (i) Application of Best Available Technique / Best Environmental Practice. OSPAR 11/20/1-E, Annex 15 (para. 1 and Annex 2).

284 OSPAR 12/22/1-E, at p. 14, para. 5.21.

285 IMO Doc. MEPC/63/23, at p. 59, paras. 9.3–9.4.

286 OSPAR 12/22/1-E, at p. 14, para. 5.22.

287 Discussions took place whether or not to send the report to the CBD. See NEAFC 2011 Annual Meeting, at pp. 50–52. The ecologically or biologically significant areas are not management areas but rather very large scientific areas containing separate VMEs; not all parts of an ecologically or biologically significant area are VMEs (p. 9).

4.5 *Northwest Atlantic Fisheries Organization (NAFO)*

Following the extension of national jurisdiction by coastal states in the late 1970s, states agreed in 1978 to establish a new organization, the Northwest Atlantic Fisheries Organization (NAFO), under a new convention, which entered into force in 1979. The Organization is composed of a General Council, a Scientific Council, a Fisheries Commission and a Secretariat.[288]

In 2005, the contracting parties to the NAFO Convention agreed to initiate a reform process within NAFO. The General Council and the Fisheries Commission established an *ad hoc* Working Group on NAFO Reform to review and, where appropriate, develop recommendations to modify and/or complete the provisions of the NAFO Convention.[289] The contracting parties to the 1978 Convention adopted a new convention at the annual meeting in 2007. In 2008, the contracting parties adopted a resolution providing for guidance with regard to NAFO's commitment to modern fishery management approaches as reflected in the amended Convention, pending the ratification process for the New NAFO Convention.[290] Under the new Convention, NAFO will consist of a Commission, a Scientific Council, and a Secretariat.[291]

The NAFO Convention applies to the waters of the Northwest Atlantic Ocean as defined in the Convention.[292] The Regulatory Area is the part of the Convention Area beyond areas under national jurisdiction.[293]

Under the 1978 Convention, the object of NAFO is to contribute through consultation and cooperation to the optimum utilization, rational management and conservation of the fishery resources of the Convention Area.[294] The Fisheries Commission is responsible for the management and conservation of the fishery resources of the Regulatory Area, and its proposals are to be designed to achieve the optimum utilization of the fishery resources of the Regulatory Area.[295] The objective of the new NAFO Convention is to ensure the long-term conservation and sustainable use of the fishery resources in the Convention Area and, in so

288 NAFO Convention, Article II(2).

289 Meeting Proceedings of the General Council and Fisheries Commission for 2005/2006, at p. 170 (Reform of NAFO (GC Doc. 05/2), para. 1).

290 Report of the General Council, 30th Meeting, 22–26 September 2008, Vigo, Spain, at p. 5. Paragraphs (a)–(g) reiterate Article III(b)–(h) of the New NAFO Convention.

291 New NAFO Convention, Article V(2).

292 Ibid., Article IV(1).

293 Ibid., Article I(p). In a previous draft, it was defined as the part "which lies beyond the areas in which coastal states exercise fisheries jurisdiction". E.g., Draft New NAFO Convention (Revision 3, Corr.), Article I(1)(p). Now, it is clear that the outer continental shelf is not included in the Regulatory Area.

294 NAFO Convention, Article II(1).

295 Ibid., Article XI(1) and (2).

doing, to safeguard the marine ecosystems in which these resources are found.[296] The new convention also provides that contracting parties shall promote the optimum utilization and long-term sustainability of fishery resources.[297]

The term 'fishery resources' is defined in the new convention as all fish, molluscs and crustaceans within the Convention Area, excluding sedentary species over which coastal states may exercise sovereign rights consistent with Article 77 of the LOSC and, in so far as they are managed by other international conventions or agreements, anadromous and catadromous stocks as well as highly migratory species listed in Annex I of the LOSC.[298] Discrete high seas *fish* stocks, therefore, are clearly included in the species coverage of the new convention. The absence of any reference to 'other sedentary species', unlike the SEAFO Convention, may be interpreted to imply the exclusion of sedentary species other than molluscs and crustaceans beyond the outer limit of the continental shelf.[299]

The new Convention is still pending ratification by members.[300] In addition, some of the new approaches adopted in the new Convention have already been introduced in practice under the 1978 Convention. For these reasons, the rest of the present section analyzes the provisions of the new Convention and the 1978 Convention as well as practice under the 1978 Convention.

4.5.1 *Precautionary Approach*

Under the new convention, contracting parties shall adopt measures based on the best scientific advice available to ensure that fishery resources are maintained at or restored to levels capable of producing the MSY, and to rebuild fishery resources to the said levels[301] and they shall apply the precautionary approach

296 New NAFO Convention, Article II.

297 Ibid., Article III(a).

298 Ibid., Article I(f). The 1978 Convention applied to all fishery resources of the Convention Area except for, *inter alia*, sedentary species of the continental shelf. NAFO Convention, Article I(4).

299 Note that NAFO has used the term 'fish' in a narrow sense, as seen in the practice of distinguishing between fish and invertebrates. See, e.g., Meeting Proceedings of the General Council and Fisheries Commission for 2005/2006, at p. 165. Molenaar considers that the new NAFO Convention 'explicitly' includes sedentary species of the Area. E.J. Molenaar, Current Legal and Institutional Issues Relating to the Conservation and Management of High Seas Deep Sea Fisheries, at p. 13 note 88 and its accompanying text.

300 For the amendment procedures, see NAFO Convention, Article XXI. At the 2011 Annual Meeting, it was reported that the amended NAFO Convention had been ratified by three contracting parties (Norway, Canada and the EU). Report of the General Council, 33rd Annual Meeting, 19–23 September 2011, Halifax, Nova Scotia, Canada, at p. 5.

301 Under the 1978 Convention, the Commission only needed to "take into account any relevant information or advice provided to it by the Scientific Council". NAFO Convention, Article XI(2).

in accordance with Article 6 of the FSA.[302] In that provision, the term 'advice', rather than 'evidence', is employed. This may reflect recent calls for RFMOs to base their conservation measures on the advice of their scientific bodies, and this terminology probably allows members to avoid a controversy associated with the term 'evidence' (i.e., whether information is sufficiently proven).

In addition, contracting parties shall prevent or eliminate overfishing and excess fishing capacity, and ensure that levels of fishing effort do not exceed those commensurate with the sustainable use of the fishery resources.[303]

In practice, NAFO has already adopted the precautionary approach for several years. As early as 1997, the Precautionary Approach framework ('PA framework') was developed by the Scientific Council, although it was not formally adopted by the Fisheries Commission.[304] In 1999, the Fisheries Commission adopted a resolution "to apply a precautionary approach widely for stocks under NAFO purview" and, to achieve this goal, agreed, *inter alia*, to determine precautionary reference points for stocks where sufficient information exists, and to determine provisional precautionary reference points, whenever possible, and a precautionary approach for other stocks.[305]

The 1997 framework was revised and recommended by the Scientific Council, and the revised framework was finally adopted by the Fisheries Commission in 2004.[306] The new PA framework adopts five zones (Safe Zone, Overfishing Zone, Cautionary Fishing Zone, Danger Zone, and Collapse Zone), according to which management strategies and courses of action are determined.[307] In the new framework, the fishing mortality limit should be no higher than the MSY, whereas there should be a low probability that the fishing mortality exceeds the limit reference point.[308] Thus, in this framework, while the concept of MSY is not discarded, the adoption of new reference points substantially changed the traditional approach to fisheries management. The three sets of reference points (i.e., limit reference points, buffer reference points and target reference points) used in the 2004 framework seem to correspond to those of Annex II of the FSA, although the 2004 framework does not advise

302 New NAFO Convention, Article III(b)–(c).

303 Ibid., Article III(f).

304 Concerns were expressed by fisheries managers over the 1997 framework. Precautionary Approach Framework (PA Framework), at pp. 1–2.

305 Resolution 2/99 of the Fisheries Commission of NAFO, adopted on 17 September 1999 to guide implementation of the Precautionary Approach within NAFO.

306 Meeting Proceedings of the General Council and Fisheries Commission, September 2004–August 2005, at p. 97.

307 PA framework, at pp. 2–3.

308 Ibid., at p. 4.

adopting pre-agreed conservation and management actions to be taken where reference points are approached or exceeded.[309]

Canada proposed to test the framework on two stocks i.e., yellowtail flounder in Division 3LNO (a straddling stock) and shrimp in Division 3M (a discrete high seas stock) before applying it to all regulated stocks and to request the Scientific Council also to provide a description of how the advice using the PA framework differs from advice provided in the traditional manner. It was stated that yellowtail flounder in Division 3LNO was a data-rich stock in good health and managed by TAC/Quota, while shrimp in Division 3M was a data-poor stock in good condition and managed by effort controls. This proposal was adopted by the Fisheries Commission.[310] Its request in 2004 for scientific advice on management in 2006, noting the 2004 PA framework, requested information concerning: the limit and precautionary reference points as described in Annex II of the FSA indicating areas of uncertainty; the stock biomass and fishing mortality trajectory over time; information regarding the current Zone within which the stock is located as well as proposals regarding possible harvest strategies to move the resource to (or maintain it in) the Safe Zone including medium-term considerations and associated risk or probabilities which will assist the Commission in developing the management strategies described in paragraphs 4 and 5 of Annex II in the FSA; and a description of the advice using the Framework which is different from advice provided in the traditional manner.[311]

In 2005, Canada proposed to extend the PA framework to other stocks by requesting the Scientific Council to determine biological reference points for all outstanding stocks, but the US cautioned against making a 'blanket' request to the Scientific Council. The Fisheries Commission encouraged the continuing work of the Scientific Council in determining reference points for the PA framework.[312] In the end, noting the PA framework, the Fisheries Commission requested that the Scientific Council provide information related to the implementation of the PA framework for *all* stocks under its responsibility, including the limit and precautionary reference points.[313]

309 Henriksen *et al.*, *Law and Politics in Ocean Governance*, at p. 78. In addition, at p. 79, they point to the Fisheries Commission disregarding much of the information on stocks.

310 Meeting Proceedings of the General Council and Fisheries Commission, September 2004-August 2005, at pp. 97 and 140.

311 Ibid., at p. 155.

312 Meeting Proceedings of the General Council and Fisheries Commission for 2005/2006, at p. 121.

313 Ibid., at pp. 166–167 (especially, paragraph 4).

4.5.2 *Ecosystem Approach*

The Fisheries Commission under the 1978 Convention was only required to take into account very limited ecosystem considerations.[314] It is questionable whether the 1978 Convention provided a firm basis for the conservation of marine ecosystems.[315] The new convention explicitly provides that the contracting parties are "[committed] to applying an ecosystem approach to fisheries management".[316] Under the new convention, contracting parties shall take due account of the impact of fishing activities on other species and marine ecosystems and, in doing so, adopt measures to minimize the harmful impact on living resources and marine ecosystems.[317] In collaboration with the Scientific Council, the Commission is to assess the impact of fishing and other human activities on living resources and marine ecosystems.[318] The Commission may refer to the Scientific Council any question pertaining to the scientific basis for the decisions it may need to take concerning "fishery resources, the impact of fishing activities on living resources, and the safeguarding of the ecosystem in which these resources are found".[319] Furthermore, contracting parties shall take due account of the need to minimize pollution and waste originating from fishing vessels as well as minimize discards, catch by lost or abandoned gear, catch of species not subject to a directed fishery and impacts on associated or dependent species, in particular endangered species.[320] In addition, contracting parties shall take due account of the need to preserve marine biodiversity.[321]

While the introduction in the Convention text of an ecosystem approach is new, existing NAFO measures already reflect this approach.[322] In particular, NAFO has addressed the protection of VMEs from bottom fisheries.

314 The Fisheries Commission shall take into account, among other things, an effect through species interrelationships. NAFO Convention, Article XI(3).

315 Henriksen and others argue that "it is more questionable whether the Fisheries Commission may ban all fishing in a specific area by establishing a marine protected area, or prohibit bottom trawling in areas to protect the environment". Henriksen *et al.*, *Law and Politics in Ocean Governance*, at pp. 75 and 79–80.

316 New NAFO Convention, preambular para. 8.

317 Ibid., Article III(d). See also Article VI(8)(b). In Article I(k), the term 'living resources' is defined as "all living components of marine ecosystems".

318 Ibid., Article VI(6)(c).

319 Ibid., Article VI(7).

320 Ibid., Article III(i).

321 Ibid., Article III(e). The term 'marine biological diversity' is defined in an almost identical way to the NEAFC Convention, echoing the definition in the CBD. Ibid., Article I(l).

322 NAFO Annual Report 2006, at p. 4. See Fisheries Commission Report at the 28th Annual Meeting in 2006, at p. 9. See also NAFO Annual Report 2005, at pp. 10–11. The 2006 Annual Report noted that despite the formal acceptance of the ecosystem approach to fisheries (EAF) by NAFO, single species assessments are still necessary for the management of the commercial

In 2004, during the discussion concerning implications for NAFO resulting from the JPOI, delegates were especially interested in an application of the ecosystem approach within NAFO and mentioned the ecological impact of fishing gear (bottom trawling), seamounts, marine protected areas, and bycatch problems as possible areas of attention for NAFO.[323] In fact, the regulation of the impacts of bottom trawl on marine ecosystems is an important topic for NAFO, given the fact that "[t]he main gear type used in the [NAFO Regulatory Area] is the bottom trawl".[324]

In 2005, the Fisheries Commission adopted measures entitled "Ecosystem Approach to Fisheries Interim Measures" as proposed by Canada. The Fisheries Commission asked the Scientific Council for advice on the development of criteria for determining areas of marine biological and ecological significance and on the identification of such areas in the NAFO Regulatory Area. In this context the Fisheries Commission also decided to amend Article 20 of the Conservation and Enforcement Measures (CEM) to invite vessels engaged in fishing in four areas in the NAFO Regulatory Area (i.e., Orphan Knoll, Corner Seamounts, Newfoundland Seamounts and New England Seamounts), where possible, to provide for the collection of biological data.[325] By subjecting any fishing activity on known seamounts found in the NAFO Regulatory Area to extensive data collection, it was expected that the measure would provide additional data for the Scientific Council to consider and eventually assist the Fisheries Commission to prioritize the areas of ecological and biological significance.[326]

In 2006, the Fisheries Commission adopted a proposal by Canada on "Precautionary Closure to Four Seamount Areas based on the Ecosystem Approach to Fisheries".[327] Three of the areas subject to this measure are located entirely in the Regulatory Area and the other one straddles the high seas and the EEZ.[328] First,

---

target fish and this continues to form the backbone of the work of the Scientific Council. NAFO Annual Report 2006, at p. 10.

323 Meeting Proceedings of the General Council and Fisheries Commission, September 2004-August 2005, at p. 97.

324 See NAFO, "Fishing in the NAFO Area", available at <http://www.nafo.int/fisheries/frames/fishery.html>.

325 See Meeting Proceedings of the General Council and Fisheries Commission for 2005/2006, at pp. 115–116 and 140. See also FSA Review Conference Preparatory Report, at p. 37.

326 Meeting Proceedings of the General Council and Fisheries Commission for 2005/2006, at pp. 115–116. See also the Fisheries Commission's request for scientific advice, especially paragraph 9 (p. 168).

327 Fisheries Commission Report at the 28th Annual Meeting in 2006, at pp. 9 and 48.

328 In an FAO map, the southwest portion of New England Seamounts area appears to straddle the NAFO Regulatory Area and Bermudian EEZ. The map is available at FAO, "Marine Protected Areas in the High Seas", <http://www.fao.org/fishery/topic/16204>. But that area under national jurisdiction does not appear in the map of NAFO, "NAFO CONVENTION AREA", available at <http://www.nafo.int/image/maps/nafo_map_hr.jpg>. While the New England

it decided to close the above-mentioned four seamounts to all fishing activities involving demersal fishing gears until 31 December 2010.[329] Second, the Fisheries Commission was to consider providing access to small-scale and restricted exploratory fishery, effective as of 1 January 2008, not to exceed 20% of the fishable area of each seamount at the 2007 Annual Meeting.[330] These representative areas would be recommended by the Scientific Council based on existing surveys and commercial data from these seamount areas. Recommendations would be made on areas that could be fished on each seamount and on a protocol for the collection of the data required to assess these seamounts.[331] Vessels may only fish in the defined areas in accordance with the protocol established by the Scientific Council and adopted by the Fisheries Commission.[332] If vessels fishing in the areas encounter hard corals, notification of the location of the coral area is to be provided to the Executive Secretary who will implement an immediate temporary closure of that area to all contracting parties pending a Fisheries Commission decision at the next Annual Meeting.[333] In 2007, the Scientific Council recommended, among other things, that any research survey in the closed areas should first be reviewed by the Scientific Council before proceeding.[334]

In 2007, Canadian and EU proposals concerning bottom fishing restrictions to protect VMEs were discussed. Consensus was not reached on either of them; instead, interim measures were adopted to establish a Coral Protection Zone closing all fishing activity involving bottom contact gear in Division 3O from

---

Seamounts area encircled by the coordinates specified in the measure straddles the high seas and the EEZ, the seamounts as such appear to be located within the NAFO Regulatory Area. Cf. International Seabed Authority, "GIS Interactive Resource Maps", available at <http://www.mapserver.isa.org.jm/GIS/>.

329 See Fisheries Commission Report at the 28th Annual Meeting in 2006, at p. 48. Originally, the proposal was a closure to all fishing activities, but Russia expressed its view that the closure should be limited only to bottom fishing gears. The proposal was revised by incorporating the term 'demersal' to accommodate the position of Russia, and the revised proposal was adopted with this understanding (p. 9).

330 This provision was deleted from the seamount closure as extended until 2014, which is subject to a review at the Annual Meeting in 2012. See Report of the Fisheries Commission and its Subsidiary Body (STACTIC), 32nd Annual Meeting, 20–24 September 2010, Halifax, Nova Scotia, Canada, at p. 10.

331 Fisheries Commission Report at the 28th Annual Meeting in 2006, at p. 48. See also NAFO Annual Report 2006, at p. 7. According to the measure, Contracting parties shall provide the Executive Secretary, in advance of the June 2007 Scientific Council meeting, with all existing data from survey and commercial fisheries that have taken place in these seamount areas. Fisheries Commission Report at the 28th Annual Meeting in 2006, at p. 49.

332 Fisheries Commission Report at the 28th Annual Meeting in 2006, at p. 49. In addition to the protocol, vessels fishing in the areas shall have a scientific observer on board.

333 Ibid.

334 Report of the Fisheries Commission, 29th Annual Meeting, 24–28 September 2007, Lisbon, Portugal, at p. 6.

1 January 2008 until 31 December 2012.[335] This closed area straddles the Canadian EEZ and the Regulatory Area.[336]

The intersessional Meeting held in May 2008 discussed a strategy for the protection of VMEs. The meeting agreed on measures to implement UNGA Resolution 61/105, including: to promptly identify VMEs in the North-West Atlantic, assess existing bottom fishing areas and their impact on sensitive habitats and adopt conservation and management measures to prevent SAIs on such habitats.[337] A new Ad Hoc Working Group of Fishery Managers and Scientists on VMEs was established to advise the Fisheries Commission on adequate measures for the protection of VMEs.[338] Under the measures, from 1 January 2009 all bottom fishing activities in new fishing areas or with bottom gear not previously used in the area concerned shall be considered as exploratory fisheries and shall be conducted in accordance with an exploratory fisheries protocol to be adopted by the Fisheries Commission in 2008.[339] The exploratory bottom fishing shall be subject to the assessment procedure set forth in Article 4bis, with the understanding that particular care will be taken in the evaluation of risks of the SAIs on VMEs, in line with the precautionary approach.[340] The Scientific Council shall undertake an assessment and provide advice to the Fisheries Commission as to whether the proposed bottom fishing activity would have SAIs on VMEs and, if so, whether mitigation measures would prevent such impacts.[341] The Fisheries Commission shall adopt, taking account of advice and recommendations provided by the Scientific Council and the *ad hoc* Working Group of scientists and managers, conservation and management measures to prevent SAIs on VMEs, including prohibiting bottom fishing activities.[342] Prior to commencing new bottom fishing activities based upon the results of exploratory fisheries conducted in the prior two years, the Fisheries Commission shall review the assessments undertaken in accordance with Article 4bis and the results of the fishing protocols

335 Ibid., at pp. 10–11.

336 See ibid., Annex 24.

337 See Report of the Fisheries Commission Intersessional Meeting, 30 April–7 May 2008, Montreal, Quebec, Canada, at pp. 8 and 48–52. In 2010, Article 5bis was revised to enhance reporting requirements. In addition, the entire Chapter I bis of the NAFO Conservation and Enforcement Measures was updated (without substantial change) in 2010. See Fisheries Commission and STACTIC Report at the 32nd Annual Meeting in 2010, at p. 10.

338 See 2008 Fisheries commission Intersessional Meeting Report, at p. 8. In 2010, it was agreed that the working group should continue, and the Fisheries Commission removed the ad hoc nature of the working group and defined new terms of reference. Fisheries Commission and STACTIC Report at the 32nd Annual Meeting in 2010, at p. 10 and Annex 13.

339 Article 3bis(1).

340 Article 3bis(2).

341 Article 4bis(3)(iii). See also Article 4bis(2).

342 Article 4bis(5).

and shall either establish conservation and management measures to prevent SAIs on VMEs and to ensure the long-term sustainability of deep-sea fish stocks or not authorize these fishing activities to proceed.[343]

At the annual meeting in 2008, the Scientific Council chair informed the meeting that, drawing on the criteria given by the FAO International Guidelines and the best available scientific information, the Scientific Council identified eight areas as potential candidates for VMEs but left open what type of mitigation measures might be appropriate.[344]

The Fisheries Commission, based on the Ad Hoc Working Group report, observed that there appears to be little interaction between species of corals and fishing activity in the Regulatory Area and this situation reflects decades of sustained fishing.[345] It also observed that, in general, the probability of significant adverse impacts on VMEs is higher in new fishing areas than in existing fishing areas.[346] It revised the CEM by referring to the definition and criteria for VMEs and SAIs agreed in the FAO International Guidelines and adopted mitigation measures, including the extension of the current seamount closures to include the Fogo Seamounts (part of the eight VMEs identified by the Scientific Council) and the adoption of an Interim Exploratory Protocol and the interim Encounter Provisions for Deep Sea VMEs.[347] The Encounter Provision was inserted into the CEM as Article 5bis, providing that the vessel master shall cease fishing and move away at least 2 nautical miles from the endpoint of the tow. The Provision provisionally defined an encounter with a VME as that above the threshold levels with indicator species (i.e., a catch per set of more than 100 kg of live coral and/or 1000 kg of live sponges), excluding an encounter with a VME indicator species less than the defined amount or a mere detection of the presence of an element itself as insufficient to identify a VME. In discussing the provision, Iceland expressed its concerns about the inclusion of sponges in the Encounter Provisions as premature and also commented that there is no basis for the 2 nautical miles move-away clause.[348]

In accordance with Article 2bis of the CEM, the Fisheries Commission identified the existing bottom fishing areas. It was observed that the preliminary map shows the presence of anomalous bottom locations and this was likely due to errors in the data. Therefore, the Chair requested the contracting parties

---

343 Article 3bis(5).

344 Report of the Fisheries Commission, 30th Meeting, 22–26 September 2008, Vigo, Spain, at p. 6.

345 Ibid., at p. 43.

346 Ibid., at p. 44.

347 Ibid.

348 Ibid., at p. 10.

to submit or resubmit their respective footprint data.[349] In 2009, the Fisheries Commission supported the draft footprint and further instructed the ad hoc Working Group of Fishery Managers and Scientists to further review the draft footprint and report back to the Fisheries Commission in 2010.[350] In 2010, the Fisheries Commission adopted the map of existing fishing areas.[351]

In 2009, the Fisheries Commission adopted a number of measures for the management of bottom fisheries and VME protection. First, it closed several areas of high concentration of corals or sponges near the Flemish Cap and Nose of the Grand Banks (coral protection zones in Divisions 3L and 3M and sponge protection zones in Divisions 3L, 3M and 3N) to all bottom fishing activities until 31 December 2011.[352] Second, it modified the encounter threshold levels to 60 kg of live coral and/or 800 kg of live sponges.[353] Third, it adopted the Exploratory Fishery Data Collection Form, which was intended for use during exploratory fishery in accordance with Article 5bis(2) (b) of the NAFO Conservation and Enforcement Measures.[354] In this connection, it endorsed the NAFO Coral Identification Guide to be used to accomplish the Form.[355]

In 2010, the Fisheries Commission refined its provisions to protect VMEs from SAIs of bottom fisheries. First, it agreed to extend the closure of 6 seamounts for the next 4 years (i.e., until 31 December 2014).[356] Second, it agreed to revise Article 4bis of the CEM on impact assessments of bottom fishing: this provides for the requirement of the submission of information and a preliminary assessment of the known and anticipated impacts on VMEs if the proposed bottom fishing is outside the NAFO fishing footprint, if new scientific information indicates the existence of a VME in a given area or if significant changes occur in fishing conduct or technology.[357] In 2011, the existing coral and sponge closures were extended until 31 December 2014 to synchronize with the seamount closure.[358]

349 Ibid.

350 Ibid., at p. 10 and Annex 13.

351 Fisheries Commission and STACTIC Report at the 32nd Annual Meeting in 2010, at p. 10 and Annex 11.

352 Report of the Fisheries Commission and its Subsidiary Body (STACTIC), 31st Annual Meeting, 21–25 September 2009, Bergen, Norway, at p. 9 and Annexes 9–10. The boundary of the Northeast Flemish Cap was extended to the deeper boundary up to the 2500 metre-contour. Report of the Fisheries Commission, 33rd Annual Meeting, 19–23 September 2011, Halifax, Nova Scotia, Canada, at p. 10.

353 Fisheries Commission and STACTIC Report at the 31st Annual Meeting in 2009, at p. 9 and Annex 11.

354 Ibid., at p. 9 and Annex 12. The Exploratory Fishery Data Collection Form was revised in 2010. Fisheries Commission and STACTIC Report at the 32nd Annual Meeting in 2010, at p. 10.

355 Fisheries Commission and STACTIC Report at the 31st Annual Meeting in 2009, at p. 9.

356 Fisheries Commission and STACTIC Report at the 32nd Annual Meeting in 2010, at p. 10.

357 Ibid., at p. 10 and Annex 12.

358 Fisheries Commission Report at the 33rd Annual Meeting in 2011, at p. 10.

The threshold level for sponge was reduced to 400 kg and 600 kg in new and existing fishing areas, respectively.[359]

In 2011, NAFO decided to reassess the likely impacts of NAFO managed fisheries on known or likely VME by 2016 and every five years thereafter.[360] Such a reassessment will concern existing bottom fishing areas.

4.5.3 *Discrete High Seas Stocks*

NAFO has taken conservation and management measures for several DHSFS in Division 3M: cod, American plaice, redfish and shrimp.[361]

With regard to cod and American plaice, in response to the Scientific Council's advice to allow no directed fishery as well as to limit bycatch to the lowest level possible, the Fisheries Commission has agreed to continue moratoria (i.e. no directed fishery) for the cod stock and the American plaice stock in Division 3M as well as the bycatch provisions of the CEM in 2003–2004, 2005–2006 and 2007–2008.[362] In 2008, it was decided that there will be no directed fishery of cod in Division 3M in 2009 and the Fisheries Commission adopted bycatch provisions concerning this stock. It also decided that there will be no directed fishery applicable in the years 2009–2011 and the bycatch provisions as stipulated in Article 11(1)(b) of the CEM shall apply.[363]

In 2009, the Fisheries Commission decided to reopen directed fishery for Atlantic cod on the Flemish Cap (in Division 3M) with the TAC of 5500 tonnes being allocated using pre-moratoria values, with precautionary measures for bycatch limitations.[364] The decision was not taken unanimously, in particular because

---

359 Ibid.

360 Ibid.

361 The stock structure of these fish stocks as discrete high seas stocks has been indicated in the FSA Review Conference Preparatory Report, at p. 20; Maguire *et al.*, The State of World Highly Migratory, Straddling and Other High Seas Fishery Resources and Associated Species, at p. 39. In addition to stocks currently regulated by NAFO, one stock of witch (a name of species) is also considered to be a discrete high seas stock: the Scientific Council concluded that the witch stock in Division 3M at the depth of less than 730 m do not appear to be linked with the witch stock in Division 2J+3KL. See Meeting Proceedings of the General Council and Fisheries Commission for 2002/2003, at p. 69.

362 Meeting Proceedings of the General Council and Fisheries Commission for 2002/2003, at pp. 69–70; Meeting Proceedings of the General Council and Fisheries Commission for 2003/2004, at pp. 89–90; Meeting Proceedings of the General Council and Fisheries Commission, September 2004-August 2005, at pp. 99 and 101–102; Fisheries Commission Report at the 28th Annual Meeting in 2006, at p. 7. For American plaice in Division 3M, the moratorium was continued. For the latest information, see Fisheries Commission Report at the 33rd Annual Meeting in 2011, at p. 7 and Annex 10.

363 Fisheries Commission Report at the 30th Meeting in 2008, at p. 7.

364 Fisheries Commission and STACTIC Report at the 31st Annual Meeting in 2009, at pp. 5–6 and Annex 6.

of the concerns expressed by the US, Iceland, Japan, Korea and Ukraine over the maintenance of the previous quota allocation.[365] In 2010, the TAC for cod in Division 3M was increased to 10000 tonnes with the same allocation scheme as the previous year.[366] At the same time, the bycatch requirements concerning reopened fishery were adopted.[367] In 2011, the TAC was slightly reduced to 9280 t, in accordance with the Scientific Council's advice.[368]

The redfish stock in Division 3M has been regulated through setting the TAC and its allocations to contracting parties. Between 2003 and 2006, while the Scientific Council's advice was set at the range of 3000–5000 tonnes, the annual TAC has continued to be set at 5000 tonnes.[369] In 2007, while the scientific advice was set at 5000 tonnes, the Fisheries Commission set the TAC at 8500 tonnes for 2008 and 2009.[370] The national quotas are the same as those for 2007.[371] In 2009, the Fisheries Commission set the TAC as 10000 tonnes without the allocation scheme as the previous year (i.e., "first come, first fish" in the words of the Danish delegation), but no new mesh size regulation was adopted.[372] In 2011, in accordance with the Scientific Council advice, the TAC was reduced to 6500 tonnes for 2012 and 2013.[373]

Shrimp fisheries in Division 3M have been managed by the imposition of restrictions on fishing efforts and their allocation. Until 2006, the Scientific Council advised 45000 or 48000 tonnes TAC, although it advised tentatively in 2007 that exploitation levels should not exceed the 2005–2006 levels. The restrictions on fishing days were maintained.[374] The straddling shrimp stock in

365 Ibid., at p. 6.

366 Fisheries Commission and STACTIC Report at the 32nd Annual Meeting in 2010, at p. 7 and Annex 6.

367 Ibid., at p. 7 and Annex 7.

368 Fisheries Commission Report at the 33rd Annual Meeting in 2011, at p. 7 and Annex 10.

369 Meeting Proceedings of the General Council and Fisheries Commission for 2002/2003, at pp. 69–70 and 109; Meeting Proceedings of the General Council and Fisheries Commission for 2003/2004, at pp. 89 and 125; Meeting Proceedings of the General Council and Fisheries Commission for 2005/2006, at p. 120.

370 Report of the Fisheries Commission, 29th Annual Meeting, 24–28 September 2007, Lisbon, Portugal, at pp. 5 and 7.

371 Ibid., at p. 7. In reality, the sum of national quotas exceeds the TAC; the footnote on the quota table specifies the procedures to be taken when the catch reaches 50% and 100% of the TAC.

372 Fisheries Commission and STACTIC Report at the 31st Annual Meeting in 2009, at p. 7. For this species, the Scientific Council in 2009 had recommended the TAC of 8500 tonnes (p. 4). The same TAC was maintained in 2010. Fisheries Commission and STACTIC Report at the 32nd Annual Meeting in 2010, at p. 7.

373 Fisheries Commission Report at the 33rd Annual Meeting in 2011, at p. 7 and Annex 10.

374 Meeting Proceedings of the General Council and Fisheries Commission for 2002/2003, at p. 70; Meeting Proceedings of the General Council and Fisheries Commission for 2003/2004, at pp. 88 and 90; Meeting Proceedings of the General Council and Fisheries Commission, September 2004-August 2005, at pp. 98 and 102; Meeting Proceedings of the General Council and Fisheries

the neighbouring Division 3L has been managed through the TAC and quota allocation.[375] In fact, although a majority of the contracting parties wished to maintain the current fishing effort limitation scheme in the absence of any new information on the stock in Division 3M, Iceland has continuously claimed that the current system is not effective in that the total catch exceeds what was advised by the Scientific Council and is thus contrary to the recommendation. In accordance with the Fisheries Commission mandate to review possible management systems and options for that stock,[376] an intersessional meeting was held in May 2008 and it decided to maintain the current management regime and not to increase the fishing effort or the TAC.[377] At the 2008 Annual Meeting, it was decided to continue the existing measure for this stock in 2009, but the Fisheries Commission requested the Standing Committee on International Control to review the relevant provisions of the CEM and to recommend any necessary improvements.[378]

Due to the concern over the drastic change in the scientific advice on shrimp in Division 3M, an intersessional meeting was held to discuss the management of shrimp in Division 3M. Divergent views were presented: while the EU presented the view (shared by Cuba, Denmark (in respect of the Faroe Islands and Greenland) and Russia) that it was difficult to support the moratorium recommendation, some contracting parties (including Canada, Iceland, Norway and the US) held the view that the Fisheries Commission should be mindful of the precautionary approach in its decision making. Through a voting procedure, the Fisheries Commission adopted an interim measure for 2010 of a 50% reduction in the fishing effort allocation subject to the condition that additional measures would be adopted for 2011 if the Scientific Council's advice indicates that the stock is still in the collapse zone.[379] In 2010, it was agreed that a fishery on this stock would not be permitted in 2011.[380]

Commission for 2005/2006, at p. 121; Fisheries Commission Report at the 28th Annual Meeting in 2006, at p. 7; Report of the Fisheries Commission, 29th Annual Meeting, 24–28 September 2007, Lisbon, Portugal, at pp. 5 and 7. See also Article 19 (Product Labelling Requirements) and Article 20 (Recording of Catch and Stowage).

375 See NAFO Conservation and Enforcement Measure, Annex I.A.

376 Fisheries Commission Report at the 29th Annual Meeting in 2007, at pp. 7–8

377 See press release from the Special Meeting of the NAFO Fisheries Commission, 30 April–7 May 2008, "NAFO takes action to protect the deep seas", available at <http://www.nafo.int/about/media/statement.html>.

378 Fisheries Commission Report at the 30th Meeting in 2008, at p. 7.

379 The Canadian proposal for a moratorium for 2010 and 2011 was rejected. Report of the Fisheries Commission Intersessional Meeting, 16 November 2009, London, at pp. 3–4.

380 Fisheries Commission and STACTIC Report at the 32nd Annual Meeting in 2010, at p. 7. See also Fisheries Commission Report at the 33rd Annual Meeting in 2011, at p. 6 and Annex 10.

In the course of the amendment to the NAFO Convention, the Scientific Council proposed to redefine the boundary of Division 3M to include a small rectangle, currently in 3L.[381] The proposal was not adopted in the final text of the amendment and the boundary of Division 3M remains unchanged. That amendment would have contributed to managing DHSFS on the basis of ecosystems since the current boundary does not include the south-western deeper part of the Flemish Cap where certain deep-water species live.

4.5.4 *Cooperation with Other Organizations*

The new Convention provides that the Commission shall seek to develop cooperative working relationships with other intergovernmental organizations which can contribute to NAFO's work and which have competence for ensuring the long-term conservation and sustainable use of living resources and their ecosystems.[382]

The Organization shall cooperate with other relevant RFMOs taking note of their conservation and management measures.[383] In recent times, as regards pelagic redfish which occur in the Norwegian EEZ as well as in the NEAFC Regulatory Area, catches were reported also in the NAFO areas. Cooperation in its management was initiated by these two RFMOs. NAFO has been allocated part of the TAC established by NEAFC, and the TAC has been allocated to NAFO contracting parties which are not members of NEAFC.[384]

It is also noted that the Performance Review Panel encouraged NAFO to continue developing cooperative relationships with other RFMO/As and international organizations, as appropriate, to achieve its objectives and facilitate its work.[385] It also encouraged NAFO to consider whether activities other than fishing (e.g., oil exploration, shipping and recreational activities) in the Convention Area may have an impact on the stocks and fisheries for which NAFO is responsible as well as biodiversity in the Regulatory Area.[386]

---

381 See Report of the Working Group on the Reform of NAFO, 12–15 September 2006, 17 September 2006, Lunenburg Co., Nova Scotia, Canada, NAFO/GC Doc. 06/3, at p. 4. See also Report of the General Council, 31st Annual Meeting, 21–25 September 2009, Bergen, Norway, at p. 5.

382 New NAFO Convention, Article XVII(b).

383 Ibid., Article XVII(c).

384 For the latest measure, see Annex I.A (Annual Quota Table) of NAFO Conservation and Enforcement Measures.

385 "General Council Decision regarding the establishment of NAFO General Council Working Group on the Future of NAFO on the development of Plans of Action necessary for the implementation of the Recommendations of the NAFO Performance Review Panel", Annex 16 of NAFO/GC Doc. 11/3, at p. 41.

386 Ibid., at p. 39.

4.6 Concluding Remarks

This chapter has examined the constitutive instruments and practice of five RFMOs that have competence to regulate straddling stocks (CCAMLR, the GFCM, SEAFO, NEAFC and NAFO). Focus has been placed on measures regulating fisheries for DHSFS and deep-water species as well as those aimed at protecting benthic ecosystems. Issues raised in this Chapter include whether these RFMOs have competence to manage the exploitation of DHSFS and sedentary species discrete to the Area; what principles apply to the management of fisheries for DHSFS and whether distinctions are made between DHSFS and straddling fish stocks under the constitutive instruments and practice of these RFMOs; whether deep-sea fisheries are addressed differently from pelagic fisheries in the constitutive instruments and practice of these RFMOs; whether and how conservation measures, especially area-based management to protect VMEs, are established by these RFMOs and coordinated with other international organizations; whether these RFMOs approach the issue of area closures differently between the outer continental shelf and the Area; whether the practice of these RFMOs has led, or is likely to lead, to an implicit or explicit amendment of the legal framework under the LOSC. This section attempts to synthesize the findings in the preceding sections and to draw conclusions.

The examination of the practice of the RFMOs concerned shows that many of the (probable) DHSFS managed by these RFMOs are deep-water species.[387] The regulated DHSFS species vary from one RFMO to another. For example, most of the species managed by SEAFO are deep-water species, while stocks regulated by NAFO are mostly those not included in deep-water species.

First of all, it has been confirmed that the constitutive instruments of the five RFMOs examined in this chapter may be interpreted to give them competence to regulate fisheries for both straddling fish stocks and DHSFS.

CCAMLR and the GFCM aim to conserve 'living marine resources', while NEAFC and NAFO limit the scope of fishery resources to fish, molluscs and crustaceans. SEAFO goes beyond the latter two organizations but falls short of conserving living marine resources as a whole: the SEAFO Convention defines fishery resources as fish, molluscs and crustaceans and 'other sedentary species'. Sedentary species beyond the outer limit of the continental shelf not classified as

[387] This observation accords with the observation made by the reports prepared for the FSA Review Conference by the FAO and the UN Division for Ocean Affairs and the Law of the Sea that most of the currently known high seas stocks, i.e., not highly migratory species and that occur exclusively in the high seas, are deep-water species. FSA Review Conference Preparatory Report, at p. 23; Maguire *et al.*, The State of World Highly Migratory, Straddling and Other High Seas Fishery Resources and Associated Species, at p. 49. For a list of species of potential discrete high seas stocks, see the Introduction.

fish as well as molluscs and crustaceans are included in fishery resources under the SEAFO Convention but not under the NEAFC or NAFO Convention.[388] The CCAMLR Convention does not explicitly refer to sedentary species, but, since the convention is applicable to all marine living resources, such organisms are within the scope of CCAMLR.[389] Benthic micro-organisms subject to bio-prospecting, rather than fishing, are not explicitly included in the mandate of any of the five RFMOs.[390]

As far as management units for deep-water species and DHSFS are concerned, NAFO and CCAMLR contrast starkly with the GFCM, SEAFO and NEAFC. CCAMLR and NAFO have adopted conservation and management measures on the basis of stocks. The conservation and management measures of the latter three RFMOs are mostly concerned with each species or with deep-sea fisheries in general, although NEAFC and SEAFO have recently adopted separate TACs for two management units in the Regulatory Area and the Convention Area, respectively. It remains to be seen whether stock-based management becomes prevalent when more scientific information on, *inter alia*, the stock biomass becomes available.

The adoption of TACs by NEAFC and SEAFO in 2007 means that the TAC is now used for deep-water species by four out of the five RFMOs concerned, although NEAFC still manages most of the deep-sea fisheries by virtue of effort regulation. Despite the hesitation shown in discussions within NEAFC and in the conclusion of the FAO Expert Consultation on deep-sea fisheries in 2006, it remains to be seen whether or not output controls become the norm in the management of deep-sea fisheries, when sufficient information is available.

The application of the precautionary approach is found in each and every RFMO investigated in this Chapter. None of the constitutive instruments of RFMOs predating the FSA explicitly provides for the precautionary approach, but three of them (i.e., NAFO, NEAFC and the GFCM) have amended their constitutive instruments to make reference to the application of the precautionary approach. SEAFO, which was established after 1995 to manage straddling stocks and DHSFS, has provisions for the application of the precautionary approach.

388 It is only the SEAFO Convention that explicitly uses the term 'sedentary species' in the context of the seabed and the subsoil thereof beyond the outer limit of the continental shelf.

389 Because the continental shelves of coastal states cover all benthic areas in the Mediterranean and the Black Sea, there is by definition no sedentary species beyond the outer limit of the continental shelf within the GFCM Convention Area. See also Molenaar, "Current Legal and Institutional Issues Relating to the Conservation and Management of High Seas Deep-Sea Fisheries", at p. 125.

390 This seems true even for CCAMLR. See ibid., at p. 131 note 121.

Three RFMOs (i.e., NAFO, NEAFC and CCAMLR) have clearly adopted the precautionary approach in practice in the context of deep-sea fisheries. With regard to the GFCM, despite critical comments on its application of the precautionary approach,[391] scientific advice on deep-sea fisheries and area closures seems to have been generally followed. On the other hand, critical remarks appear warranted for SEAFO.[392] Scientific advice is not fully reflected in its conservation measures.

Some of the provisions of the constitutive instruments explicitly refer to the FSA in the application of the precautionary approach. The application of the precautionary approach in practice is generally compatible with the provisions concerning the precautionary approach in the FSA and other relevant international instruments.[393]

The provisions for the precautionary approach in the constitutive instruments do not distinguish between straddling stocks and discrete high seas stocks.[394] Nevertheless, the vulnerability of deep-water species and VMEs has given rise to arguments in favour of more precautionary measures for deep-sea fisheries than those for other regulated fisheries. However, measures applying the precautionary approach are different from one another, and the degree to which the conservation measures reflect scientific advice varies among the RFMOs concerned. It is therefore not possible to conclusively argue that because of the characteristics of deep-water species such as the paucity of scientific information and data, stricter measures are adopted for deep-sea fisheries. It is still to be seen whether or not this sort of management approach, which was advocated in the 2006 FAO Expert Consultation, will be accepted in the actual practice of RFMOs. Stricter measures include: setting very precautionary TAC or fishing effort restrictions, much lower than the MSY level, and provisional closures.

All five RFMOs incorporate ecosystem considerations in various ways and to a varying degree. In response to the concerns of the international community, many regional *fisheries* management organizations and arrangements are now competent to protect marine ecosystems even though such measures are not taken with a view to maximizing the yield of fish. But their measures are still

391 Mooney-Seus and Rosenberg, Regional Fisheries Management Organizations (RFMOs): Progress in Adopting Precautionary Approach and Ecosystem-Based Management, at p. 44 (arguing that the GFCM has not really done much more than just state its intention to implement the precautionary approach).

392 For such a view, see ibid., at p. 144 (observing that the reluctance to institute provisional reference points analogous to those for similar or better known stocks or to establish interim measures is not in keeping with the precautionary approach). See also ibid., at p. 146.

393 For possible (minor) departures, see Section 4.4.1.1 (NEAFC) and Section 4.5.1 (NAFO) above.

394 As noted earlier, this is quite different in other issues such as the role of coastal states in decision-making and the compatibility of conservation measures.

part of the management of fisheries: no RFMO attempts to regulate activities not related to fisheries in order to protect marine ecosystems.

All of the five RFMOs examined in this Chapter have taken measures to protect benthic habitats from deep-sea fisheries. This is because of the characteristics of deep-sea fisheries, which are likely to impact on benthic communities and, thus, to affect VMEs such as seamounts, hydrothermal vents and coldwater corals. Some of the areas closed to bottom fisheries are located within the continental shelf of the coastal states, while others exist in the Area. There is no discernible difference between measures taken to protect ecosystem features within and beyond the outer limit of the continental shelf. But this does not necessarily imply that there is no tension between a regional organization and its coastal members with regard to the management of the latter's continental shelf, as vividly illustrated by the controversy over the establishment of MPAs in areas beyond national jurisdiction at the OSPAR Commission.

In many cases, the coastal members of the RFMOs have proposed that high seas areas should be closed (especially Canada in NAFO and the EC in the GFCM and NEAFC), regardless of whether or not the area concerned is located within the outer limit of the continental shelf.[395] Nothing in the record indicates that coastal members of the RFMOs have been given a special right to initiate proposals to address impacts on benthic ecosystems in the five RFMOs unless the area concerned partly covers the EEZ or the territorial sea, but it is reasonable to assume that the coastal states' proposals were at least partly motivated by concerns about the impacts of high seas fisheries on ecosystem features of their continental shelf.

So far, measures are focused on the protection of seamounts and, to a lesser degree, coldwater corals. Hydrothermal vents (e.g., the Rainbow Hydrothermal Vent in the NEAFC Regulatory Area) appear to have escaped the attention of RFMOs in spite of the relevant recommendations in the UNGA Resolutions. As far as the protocols for encounters with VMEs are concerned, NAFO, SEAFO and NEAFC define encounters by using the amount of the catches of corals and sponges while CCAMLR considers the VME indicator organism as any benthic organism listed in its VME Taxa Classification Guide.

The above observation leads to the next issue, namely the collaboration between RFMOs and regional environmental protection mechanisms. The RFMOs have retained independent authority to establish area-based conservation measures: formally, CCAMLR and NEAFC may control the designation of MPAs in their Convention areas;[396] the GFCM and the SPAMI List under the

395 SEAFO measures were considered on the basis of recommendations by the Scientific Committee, rather than proposals by members.

396 As noted in Section 4.1.4 above, CCAMLR's authority with regard to the approval of proposals for ASPAs and ASMAs goes beyond fisheries.

Mediterranean environmental regime have no formal relationship. But the regional environmental protection regimes may exert an indirect influence on their practice, *inter alia*, through establishing standards and criteria for the designation of such areas. So far, cooperation with regional environmental protection mechanisms in the establishment of protected areas by RFMOs has not yet fully materialized for two reasons. First, RFMOs have not established mechanisms to effectively take into account criteria or considerations of regional environmental protection mechanisms. Second, regional environmental protection mechanisms have mostly focused on the establishment of protected areas within areas under national jurisdiction. Ongoing developments at the global level, however, will undoubtedly accelerate the pace of establishing high seas MPAs. Collaboration (or conflicts) between RFMOs and regional environmental protection mechanisms that have the competence to establish high seas MPAs are likely to occur in the years to come.

Paragraph 83 of UNGA Resolution 61/105 with regard to bottom fisheries has been implemented by four of the RFMOs examined in this chapter. CCAMLR adopted conservation measures in 2007. NAFO, NEAFC and SEAFO followed by taking measures in 2008. While measures relating to impact assessment procedures taken by NAFO, NEAFC and SEAFO are concerned with all bottom fishing activities in new fishing areas or with bottom gear not previously used in the area concerned, CCAMLR Conservation Measure 22-06 applies to all bottom fishing activities commenced after 1 December 2008. Although the NAFO, NEAFC and SEAFO measures do not distinguish between commercial and scientific research fishing, certain scientific fishing activities are partly governed by the provisions of CCMALR Conservation Measure 24-01 in lieu of those of Conservation Measure 22-06. The measures taken by NAFO, NEAFC and SEAFO are mostly equivalent to each other but there are some differences. For example, due to its organizational structure, NEAFC receives advice from PECMAS that considers assessments made by ICES. NAFO and SEAFO receive advice and recommendations from their respective scientific bodies. NEAFC's dual process is more likely to result in departures from scientific advice. Apart from the closure of known seamounts and coral reefs, the GFCM has not implemented the relevant paragraphs of Resolution 61/105: in depths of less than 1000 m, the regulation of bottom fisheries is limited in the GFCM Agreement Area.

Chapter Five

Other Practice at the Regional Level

In contrast to the regions examined in Chapter 4, there are various regions where no RFMOs with competence to regulate fisheries for DHSFS are fully operational. This chapter examines the state practice in these regions addressing deep-sea fisheries and their impact on benthic marine ecosystems. The first section analyzes a regional fisheries management arrangement concluded for the Southern Indian Ocean, which only recently entered into force. The next two sections deal with regions where states have already concluded instruments establishing RFMO/As, focusing on both the interim measures aimed at addressing the impacts of bottom fisheries and the text of such instruments. The fourth section deals with other areas with a view to examining practice which may have implications for the conservation and management of discrete high seas stocks at the international level, exploring the possibility of establishing RFMO/As in these regions and analyzing factors which are potentially relevant to the establishment and structure of such RFMO/As.

5.1 *Southern Indian Ocean*

In the high seas area of the South-western part of the Indian Ocean, relatively large-scale fishing operations for orange roughy began in the 1990s by vessels from Australia, New Zealand and South Africa. In 2000, Namibian vessels, and probably vessels from Russia and China, were also involved in the fisheries.

Two parallel negotiations took place for the management of these fisheries.[1] On the one hand, in November 1999, Australia invited New Zealand and South Africa for preliminary talks on the negotiation of an RFMO to manage the fisheries. At the end of July 2000, Australia forwarded a draft text for an RFMO to the

[1] For an account of the negotiating history as well as the status of the fisheries, see Molenaar, "Management and Conservation of Orange Roughy", at pp. 109–115; Possible Options Regarding Issues Central to the Negotiations for the Establishment of a South West Indian Ocean Fisheries Commission, SAFR/DM/SWIO/04/4, at pp. 1–8; Mooney-Seus and Rosenberg, Regional Fisheries Management Organizations (RFMOs): Progress in Adopting Precautionary Approach and Ecosystem-Based Management, at p. 158.

two negotiating partners for their consideration and comment. Australia's intention, it is argued, was that, after an initial round of drafting, the draft would be lodged with the FAO to allow other states with a real interest to participate in the finalization of the draft.[2]

On the other hand, under FAO Council Resolution 1/116 in June 1999,[3] the FAO Council authorized the Director-General of the FAO to convene *ad hoc* meetings of the members of the former Committee for the Development and Management of Fisheries in the Southwest Indian Ocean (a subsidiary body of the Indian Ocean Fishery Commission); the meetings were aimed at completing the process of establishing a new fisheries commission for the area under Article XIV of the FAO Constitution and taking such interim action as may be required regarding the management of the fisheries resources of the area covered by the former Committee (i.e., FAO Statistical Area 51) pending the formal establishment of the new body. Two technical meetings attended by coastal states were held in 2000.[4] At the second technical meeting participants agreed on the establishment of an Article XIV body under the FAO. The focus of the discussions at the meeting remained on the common interest of the coastal states in the management of fisheries. But, it was recognized that deepwater fisheries were developing on the high seas in the southern Indian Ocean. In view of this, the technical meetings were followed by Intergovernmental Consultations with the broad participation of both coastal states and high seas fishing nations.[5] The above-mentioned negotiating track started by high seas fishing states was discontinued after the participants in that track took part in this process.

The first Intergovernmental consultation agreed on a new definition of the area of competence of the proposed Commission that would exclude areas under national jurisdiction, and the second Intergovernmental consultation reaffirmed this view.[6] However, the recognition of the special interests of developing states as well as the desirability of some form of consultative mechanism that would assist coastal states in developing their coastal fisheries and in confront-

[2] Molenaar, "Management and Conservation of Orange Roughy", at pp. 109–110.

[3] That resolution abolished the Indian Ocean Fishery Commission.

[4] Participants in these meetings were Comoros, France, Kenya, Madagascar, Mauritius, Mozambique, Seychelles, Somalia and Tanzania. See Molenaar, "Management and Conservation of Orange Roughy", at p. 110.

[5] In particular, Australia, the EC, Namibia, New Zealand and South Africa participated in this Consultation. See Report of the Intergovernmental Consultation on the Establishment of a Southwest Indian Ocean Fisheries Commission, Saint-Denis, Réunion, 6–9 February 2001, FAO Fisheries Report No. 647, SAFR/R647(Bi), at p. 1, para. 2.

[6] See ibid., at p. 5 (Article III); Report of the Second Intergovernmental Consultation on the Establishment of a Southwest Indian Ocean Fisheries Commission, Antananarivo, Madagascar, 25–28 September 2001, Fisheries Report No. 664, SAFR/R664 (Bi), at pp. 5–6 (Article 4).

ing common problems posed a challenge.[7] Doubts concerning the advisability of an Article XIV body were expressed due to the awareness that the present state of the resources and fishing activities does not warrant the establishment of heavy and expensive management machinery.

In January 2004, the third Intergovernmental Consultation decided to split the negotiation of the South West Indian Ocean Fisheries Commission (SWIOFC) into two tracks, one for coastal areas and the other for the high seas.[8] The track for coastal areas resulted in the establishment of the Statutes of the SWIOFC by FAO Council Resolution 1/127, as an advisory body under Article VI of the FAO Constitution.[9] Subsequently, the high seas track adopted SIOFA on 7 July 2006. The Agreement entered into force on 21 June 2012.[10]

The Agreement does not establish an RFMO, but fisheries are to be managed through Meetings of Parties.[11] This is one of the few fisheries management mechanisms established after the adoption of the FSA that may be qualified as an 'arrangement' in the sense of the FSA.

5.1.1 *The Southern Indian Ocean Fisheries Agreement (SIOFA)*

The Agreement applies to the area of the Southern Indian Ocean defined in Article 3, excluding waters under national jurisdiction.[12] Waters under national

---

7 See Report of the Intergovernmental Consultation on the Establishment of a Southwest Indian Ocean Fisheries Commission, Saint-Denis, Réunion, 6–9 February 2001, at pp. 10–11 (Article VI and the comments thereon).

8 Report of the Third Intergovernmental Consultation on the Establishment of a Southwest Indian Ocean Fisheries Commission, Nairobi, Kenya, 27–30 January 2004, Fisheries Report No. 742, SAFR/R742, at pp. 3–5, paras. 20–28.

9 Report of the Council of FAO, Hundred and Twenty-seventh Session, Rome, 22–27 November 2004, CL 127/REP, at para. 99.

10 See FAO "Regional Fishery Bodies Summary Descriptions: South Indian Ocean Fisheries Agreement (SIOFA)", available at <http://www.fao.org/fishery/rfb/siofa/en>.

11 During the negotiation, it was emphasized that the proposed Agreement should be "as light and inexpensive as possible". See, e.g., Report of the Third Intergovernmental Consultation on the Establishment of a Southwest Indian Ocean Fisheries Commission, Nairobi, Kenya, 27–30 January 2004, at p. 5, para. 25. As a matter of fact, it is claimed that the orange roughy fishery already collapsed before the management mechanism was established. See press release by the Deep Sea Conservation Coalition on 5 July 2006, "The Southern Indian Ocean needs real protection, not voluntary half measures", available at <http://www.savethehighseas.org/publicdocs/DSCC_PR_sealordresp_050706.pdf>. During the negotiations, participants were aware that, because of the longevity and aggregated distribution of the species targeted, there was an expressed urgent need for a management mechanism. See, e.g., Possible Options Regarding Issues Central to the Negotiations for the Establishment of a South West Indian Ocean Fisheries Commission, at p. 6.

12 SIOFA, Article 3(1).

jurisdiction bordering or enclosed by the high seas area are the area of competence for the SWIOFC.

Fishery resources in SIOFA mean fish, molluscs, crustaceans and other sedentary species within the area, but sedentary species subject to the jurisdiction of coastal states and highly migratory species are excluded from the application of the Agreement.[13] The formulation of fishery resources in SIOFA is the same as that contained in the SEAFO Convention.

A distinctive characteristic of the negotiation which led to the adoption of the Agreement is that participants were fully aware of the existence of DHSFS throughout the process. One would even argue that these fish stocks were the essential catalyst of the entire process of the negotiations on the management of high seas fisheries. In fact, Australia, New Zealand and South Africa proceeded from the presumption that the orange roughy stock in the region is a discrete high seas fish stock.[14]

The objectives of SIOFA are to ensure the long-term conservation and sustainable use of the fishery resources and to promote the sustainable development of fisheries in the Agreement's area of application.[15] Fishery resources are to be managed so that they are maintained at the levels capable of producing the MSY, and depleted stocks of fishery resources are to be restored to the said levels.[16]

Measures are to be adopted based on the best scientific evidence available to ensure the long-term conservation of fishery resources, taking into account the sustainable use of such resources and implementing an ecosystem approach to their management.[17] Due account shall be taken of the need to minimize any harmful impact that fishing activities may have on the marine environment.[18] In formulating the conservation and management measures necessary for ensuring the long-term sustainability of the fishery resources, the Meeting of the Parties shall take into account the need to protect marine biodiversity, based on the best scientific evidence available.[19]

At the same time, the precautionary approach shall be applied in accordance with the Code of Conduct and the FSA, whereby the absence of adequate scientific information shall not be used as a reason for postponing or failing to take

13 Ibid., Article 1(f).

14 Molenaar, "Management and Conservation of Orange Roughy", at p. 109.

15 SIOFA, Article 2.

16 Ibid., Article 4(d). Note the suggested reference to the maximum economic yield at a previous meeting. Report of the Fourth Intergovernmental Consultation on the Establishment of a Southwest Indian Ocean Fisheries Commission, Mahe, Seychelles, 13–16 July 2004, Fisheries Report No. 776, SAFR/R776, at Appendix F, Article 4(d).

17 SIOFA, Article 4(a).

18 Ibid., Article 4(e).

19 Ibid., Articles 4(f) and 6(d).

conservation and management measures.[20] Measures shall be taken to ensure that the level of fishing activity is commensurate with the sustainable use of fishery resources.[21] In addition, the Meeting of the Parties shall adopt generally recommended international minimum standards for the responsible conduct of fishing operations.[22]

Contracting parties, acting jointly under the Agreement, shall cooperate closely with other international fisheries and related organizations in matters of mutual interest, in particular with the SWIOFC and any other RFMO with competence over high seas waters adjacent to the Agreement's area of application.[23] In addition, in developing advice and recommendations, the Scientific Committee shall take into consideration the work of the SWIOFC as well as that of other relevant research organizations and RFMOs.[24]

5.1.2 *Practice*

Currently, contracting parties to this Agreement are Australia, the Cook Islands, the European Union, Mauritius and the Seychelles.[25]

The Conference on the Southern Indian Ocean Fisheries Agreement, in adopting SIOFA, also adopted, by unanimity, the Resolution on interim arrangements concerning the high seas in the Southern Indian Ocean and called on all interested states and REIOs to cooperate on a voluntary basis and in accordance with the Resolution towards the conservation and management of the fishery resources covered by the Agreement, while awaiting the entry into force of the Agreement.[26] The Resolution called upon all states, REIOs and fishing entities that have participated in the Intergovernmental Consultations or that have carried out or carry out fishing activities in the high seas in the Southern Indian Ocean to implement, among others, the measures outlined in the resolution on data collection originally adopted in the Seychelles in July 2004.[27] The

---

20 Ibid., Article 4(c).

21 Ibid., Article 4(b). The reference to the prevention or elimination of over-fishing and excess fishing capacity, which was included as one option in the previous draft discussed at the fourth Intergovernmental Consultation, was not adopted in the final text of the Agreement. For a comparison, see Fisheries Report No. 776, at Appendix F, Article 4(b).

22 SIOFA, Article 6(1)(e).

23 Ibid., Article 16.

24 Ibid., Article 7(1)(b).

25 FAO, "Regional Fishery Bodies Summary Descriptions: South Indian Ocean Fisheries Agreement (SIOFA)", available at <http://www.fao.org/fishery/rfb/siofa/en>.

26 Final Act of the Conference on the Southern Indian Ocean Fisheries Agreement, para. 15. The Final Act, the Resolution and other relevant instruments are available at FAO Legal Office, "Final act of the Conference on the Southern Indian Ocean Fisheries Agreement", <http://www.fao.org/fileadmin/user_upload/legal/docs/035t-e.pdf>.

27 Para. 1.

information recommended to be collected includes tow-by-tow information on target species, trawl type (bottom or mid-water), depth of gear and vessel latitude and longitude and net opening.[28] In response to a questionnaire circulated by the UN Secretary-General, both the EC and New Zealand indicated that they were committed to implementing the interim measures.[29]

At the side-event following the signing ceremony for SIOFA, four fishing companies, which are members of the Southern Indian Ocean Deepwater Fishers Association (SIODFA), announced voluntary closures to high-seas deepwater trawling.[30] The SIODFA, using the scientific knowledge gathered over a decade of activity in the Indian Ocean and in consultation with the FAO, has delimited 309,000 km^2 of the ocean floor in eleven separate benthic protected areas to be closed for fishing by vessels of its members.[31]

5.2 *South Pacific Ocean*

In the vast high seas areas of the South Pacific Ocean, the desire to resolve conflicts concerning fishery resources has led to the establishment of regional and sub-regional arrangements focusing on fisheries for certain species. More recently, the coastal states and high seas fishing nations successfully established a new RFMO covering the entire region.

The first sub-section looks at the (past) controversy between Australia and New Zealand over orange roughy fisheries in the Western South Pacific, illustrating the difficulties and complications associated with the management of fisheries for deepwater species on the high seas.[32] The experience of these countries

---

28 Annex 1 of the Resolution on data collection concerning the high seas in the Southern Indian Ocean.

29 A/62/260, at pp. 48–49, para. 155. In addition, the EU sought to adopt interim measures implementing UNGA Resolution 61/105 in this region. See Communication from the Commission to the Council and the European Parliament, Destructive fishing practices in the high seas and the protection of vulnerable deep sea ecosystems, COM(2007) 604 final, 17.10.2007, at p. 9.

30 For information, see FAO press release, "New agreement governing high-seas fishing in Indian Ocean", 12 July 2006, <http://www.fao.org/newsroom/en/news/2006/1000360/index.html>.

31 Southern Indian Ocean Deepwater Fishers' Association, press release, "Fishing companies announce world's first voluntary closures to high-seas deepwater trawling", 5 July 2006, available at http://www.scoop.co.nz/stories/BU0607/S00061.htm.

32 Another conflict in the background of the negotiations for the establishment of a new RFMO in the South Pacific is the conflict between Chile and distant water fishing nations, notably Spain, over swordfish and jack mackerel. On this issue, including proceedings initiated by the EC and Chile before WTO panel and the ITLOS Special Chamber, see M.C. Engler, Establishment and Implementation of a Conservation and Management Regime for High Seas Fisheries, with Focus on the Southeast Pacific and Chile: From Global Developments to Regional Challenges, Research Paper (UN – Nippon Foundation Fellow 2006–2007), at p. 105; Miles and Burke,

appears to have influenced subsequent negotiations for the establishment of an RFMO in this region. The next sub-section provides an overview of the negotiations with regard to the establishment of a South Pacific Regional Fisheries Management Organization (SPRFMO), followed by the analyses of the interim measures agreed during these negotiations and of the Convention on the Conservation and Management of High Seas Fishery Resources in the South Pacific Ocean (SPRFMO Convention) in the third and fourth sub-sections, respectively.

5.2.1 *Orange Roughy Fisheries in the South-West Pacific*

Orange roughy has been caught in the South Pacific by fishermen mainly from New Zealand, Australia and South Africa for more than 20 years; in the South-West Pacific, there are several orange roughy fisheries in the Australian and New Zealand EEZs and in the high seas bordered by these EEZs.[33] High seas

---

"Pressures on the United Nations Convention on the Law of the Sea of 1982 Arising From New Fisheries Conflicts: The Problem of Straddling Stocks", at pp. 347–348; M.A. Orellana, "The Swordfish Dispute between the EU and Chile at the ITLOS and the WTO", 71 *Nordic Journal of International Law* (2002), at pp. 55–81; M. Rau, "Comment: The Swordfish Case: Law of the Sea v. Trade", 62 *Zeitschrift für ausländisches öffentliches Recht und Völkerrecht* (2002), at pp. 37–42; Rayfuse, *Non-Flag State Enforcement in High Seas Fisheries*, at pp. 315–318; P.-T. Stoll and S. Vöneky, "The Swordfish Case: Law of the Sea v. Trade", 62 *Zeitschrift für ausländisches öffentliches Recht und Völkerrecht* (2002), at pp. 21–35; Orrego Vicuña, *The Changing International Law of High Seas Fisheries*, at p. 57 *et seq*. In this connection, see also discussions on the Chilean Presential Sea Doctrine: T.A. Clingan, Jr., "Mar Presencial (The Presential Sea): *Deja vu* all over again? A response to Francisco Orrego Vicuna", 24 *Ocean Development and International Law* (1993), at pp. 93–97; J.G. Dalton, "The Chilean Mar Presencial: A Harmless Concept or a Dangerous Precedent?" 8 *International Journal of Marine and Coastal Law* (1993), at pp. 397–418; Joyner and De Cola, "Chile's Presential Sea Proposal", at pp. 99–121; Kwiatkowska, "The High Seas Fisheries Regime: At a Point of No Return?" at pp. 340–341; F. Orrego Vicuña, "The 'Presential Sea': Defining Coastal States' Special Interests in High Seas Fisheries and Other Activities", 35 *German Yearbook of International Law* (1992), at pp. 264–292; F. Orrego Vicuña, "Toward an Effective Management of High Seas Fisheries and the Settlement of the Pending Issues of the Law of the Sea", 24 *Ocean Development and International Law* (1993), at pp. 81–92; Orrego Vicuña, *The Changing International Law of High Seas Fisheries*, at pp. 107–111; J.A. de Yturriaga, *The International Regime of Fisheries: From UNCLOS 1982 to the Presential Sea* (1997), at pp. 228–238.

33 The fishing history and other facts in this sub-section are based on the following literature: Lack *et al., Managing Risk and Uncertainty in Deep-sea Fisheries*; Molenaar, "Management and Conservation of Orange Roughy", at pp. 77–124; D. Staples, "Management of Shared Fish Stocks: Australian Case Studies", Papers Presented at the Norway-FAO Expert Consultation on the Management of Shared Fish Stocks – Bergen, Norway, 7–10 October 2002, reproduced in FAO Fisheries Report No. 695 Supplement, pp. 159–179, at pp. 160–163; J. Willing, "Arrangement between the Government of Australia and the Government of New Zealand for the Conservation and Management of Orange Roughy on the South Tasman Rise", Papers Presented at the Norway-FAO Expert Consultation on the Management of Shared Fish Stocks – Bergen, Norway, 7–10 October 2002, reproduced in FAO Fisheries Report No. 695 Supplement (2003), pp. 200–205.

areas where orange roughy fisheries have been reported include the Challenger Plateau, the Lord Howe Rise, the Louisville Ridge and the South Tasman Rise.[34] Some of the orange roughy stocks are considered to be DHSFS.

The Louisville Ridge is a seamount area in the high seas area east of the New Zealand EEZ in the South-West Pacific Ocean. An orange roughy fishery has been developed there since the 1990s. Populations of orange roughy in the area of the Louisville Ridge are considered to be a discrete high seas fish stock. Australian vessels have not been engaged in the fishery there since 1998/99, although New Zealand vessels have fished consistently since 1993/94. The fishery expanded rapidly, so that its commercial viability diminished and New Zealand efforts in the fishery have declined. Attempts by New Zealand to estimate pre-fished and current biomass for the fishery have been largely unsuccessful.[35]

Another area of orange roughy fisheries in the South-West Pacific is the South Tasman Rise. The South Tasman Rise is an area south of Tasmania, adjacent to the Australian EEZ. While modest amounts of orange roughy had been caught within the Australian EEZ, commercial fishing was primarily on the seamount in the high seas. Australian and New Zealand vessels were engaged in the fisheries there. In late 1997, fishing activities increased over a short period of time. Officials from Australia approached New Zealand raising concerns about the sustainability of the resource and the need to find a way to manage the fishery. The negotiations between the two governments led to the adoption of an arrangement in 1998 for the management of orange roughy on the South Tasman Rise that took effect as from 1 March 1998.[36] A precautionary TAC (2100 tonnes) for the high seas was agreed based on the 1997 catches, but Australian vessels landed 2052 tonnes in February 1998 before the Arrangement became effective. To obtain samples during the July-August spawning season, the two states agreed to set aside a further 150 tonnes for each party for scientific research. Following its expiration on 28 February 1999, the two governments agreed on a new arrangement in 2000.[37] The 1998 and 2000 Arrangements primarily focused

---

34 Lack *et al.*, *Managing Risk and Uncertainty in Deep-sea Fisheries*, at pp. 63–66.

35 See ibid., at pp. 43–44 and 47. For the population structure, see also p. 7 of the species profile of orange roughy developed by the South Pacific RFMO, as updated on 4 May 2007, available at <http://www.southpacificrfmo.org/assets/Species-Profiles-Final/orange%20roughy%20species%20profile%20040507.pdf>.

36 Arrangement between the Government of Australia and the Government of New Zealand for the Conservation and Management of Orange Roughy on the South Tasman Rise, Signed on 12 January 1998 and 18 February 1998, respectively.

37 Arrangement between the Government of Australia and the Government of New Zealand for the Conservation and Management of Orange Roughy on the South Tasman Rise, Signed for New Zealand on 17 February 2000 and for Australia on 25 February 2000. See also Willing, "Arrangement between the Government of Australia and the Government of New Zealand for

on three aspects: scientific research, conservation and management, and control of third party vessels.

At the early stages of the discussions, the debate focused on whether the stock was a straddling stock occurring both in the Australian EEZ and in the high seas adjacent thereto. Australia took the position that the resource is a single straddling stock, while New Zealand argued that two distinct stocks exist (one in the Australian Fishing Zone (AFZ) and the other in the high seas). Due to the disagreement between the two governments, there was a need to establish whether or not the orange roughy populations on the South Tasman Rise were a discrete high seas stock. Therefore, the 1998 Arrangement provided that the parties would carry out a scientific research programme from 1 March 1998 to 28 February 1999 "to provide clear information on the stock structure and relationship between orange roughy taken on the high seas and orange roughy occurring within the [AFZ]" and to "obtain information to enable a preliminary assessment of the status and productivity of the fishery".[38] The 1998 Arrangement provided that, during the period, the parties would prohibit fishing on the high seas except with the authorization to be given for the purposes of implementing the above-mentioned scientific research programme;[39] the parties would exchange all relevant scientific information relating to orange roughy on the South Tasman Rise including associated species and bycatch.[40] The parties would prepare a joint report on the outcomes of analyses of scientific information, with particular emphasis on the best available information in relation to stock structure, and consider new conservation and management measures in February 1999.[41] The 2000 Arrangement extended the prohibition from the orange roughy fishery to trawling and other demersal fishing targeting *all species* on the high seas area of the South Tasman Rise except with the authorization of

---

the Conservation and Management of Orange Roughy on the South Tasman Rise", at p. 204; Staples, "Management of Shared Fish Stocks: Australian Case Studies", at p. 161.

38 1998 South Tasman Rise Arrangement, para. 2. A 'precautionary' total catch limit of 2100 tonnes of orange roughy for this scientific research programme was shared between Australia and New Zealand in the proportion of verified catches made by vessels of each state in the high seas area of the South Tasman Rise during the period 1 January to 17 December 1997. 1998 South Tasman Rise Arrangement, paras. 3 and 8. Initially, Australia allegedly took a position in which it claimed *exclusive rights* to catch and manage the orange roughy fishery in the high seas area of the South Tasman Rise. Willing, "Arrangement between the Government of Australia and the Government of New Zealand for the Conservation and Management of Orange Roughy on the South Tasman Rise", at p. 201. But, this statement was denied by a former Australian official. Serdy, "Schrödinger's TAC", at p. 495 note 4. For the purpose of assessing the stock structure, several other provisions were made. 1998 South Tasman Rise Arrangement, paras. 5–7.

39 1998 South Tasman Rise Arrangement, para. 8.

40 Ibid., para. 9.

41 Ibid., para. 18.

the appropriate authorities.[42] The 2000 Arrangement provided that the parties would still carry out a scientific research programme for the purposes of "obtaining information to enable an assessment of the size of the stock(s) of orange roughy on the South Tasman Rise and of the sustainable yield" and "providing further information on the stock structure and relationship between orange roughy found in the high seas area of the South Tasman Rise and orange roughy found within the AFZ".[43]

Despite references to the FSA and some provisions implicitly confirming Australia's interest, these Arrangements appear to avoid a definite identification of the orange roughy stock as a straddling stock. In fact, both Arrangements contained a non-prejudice clause with regard to any future arrangements or agreements with respect to Tasman Sea fisheries generally and/or orange roughy in the high seas area.[44]

Scientific research conducted in accordance with these Arrangements has two major implications for fisheries in the Tasman Sea region beyond orange roughy fisheries on the South Tasman Rise. First, the 1998 Arrangement provided that if it was to turn out that the orange roughy stock was a straddling stock, the parties would establish a consistent approach towards conservation, management and allocation for all straddling fish stocks in the South Tasman Sea region, consistent with the FSA.[45] It could be inferred from this that if the stock was found to be a discrete high seas fish stock or, perhaps, if it was not conclusively determined either way, an approach to the management of the (probable) discrete high seas fish stock would be different from the approach to straddling stocks in other areas of the Tasman Sea region. Second, the methodologies developed through a scientific research programme for orange roughy on the South Tasman Rise,

---

42 2000 South Tasman Rise Arrangement, para. 2.

43 Ibid., para. 12. Any catch under the scientific programme would be taken within the TAC. Ibid., para. 13. The initial catch was set at 2400 tonnes for the 2001–2002 season and the TAC for subsequent seasons is subject to the variation made by decisions of the parties taking into account the outcomes of the scientific research and other circumstances and its quotas will remain in the same proportions. See ibid., paras. 3–6. The Australian and New Zealand fisheries ministers agreed in 2002 to management arrangements for the 2003–2004 to 2006–2007 seasons. They provide for the TAC to be lowered to 800 tonnes for the 2003–2004 season and then to be further reduced in each subsequent season unless fish in sufficient numbers are caught to trigger additional TAC. The website of the Australian Department of Agriculture, Fisheries and Forestry further states that "[t]he fishery is currently closed with a limited-entry 'international' fishery managed with New Zealand under an MOU in place". Information is available from the Australian Government, Department of Agriculture, Fisheries and Forestry, "New Zealand – Australia Fisheries Cooperation", <http://www.daff.gov.au/fisheries/international/cooperation/newzealand>.

44 See 1998 South Tasman Rise Arrangement, para. 22; 2000 South Tasman Rise Arrangement, para. 36. See also Molenaar, "Management and Conservation of Orange Roughy", at p. 88.

45 1998 South Tasman Rise Arrangement, para. 23.

including mutually acceptable criteria for determining whether or not a stock is a straddling stock, will assist in broader assessments of orange roughy stock characteristics in the Tasman Sea region.[46] In other words, it is likely that assessments of stock characteristics of newly discovered stocks in other areas of the region will be carried out in accordance with the methodologies developed in the research programmes for the South Tasman Rise.

As regards third parties, the 2000 Arrangement stipulates, in its Paragraph 31, that "[t]he Parties will jointly consider in terms of Article 11 of the [FSA] any request by a third country referred to in paragraph 30 to become a Party to this Arrangement".[47] Currently, Australia and New Zealand are the only parties to the Arrangement.[48] Since the Arrangement is essentially a bilateral endeavour to deal with high seas fisheries, the weakness of the 2000 Arrangement persists: in the future when the stock is rebuilt, the parties will again need to address third-party issues in order to ensure the effectiveness of the Arrangement unless an effective multilateral management scheme, whether legally-binding or not, is established with the involvement of all states having an interest in high seas fisheries.[49]

5.2.2 *International Consultations on the Establishment of the South Pacific Regional Fisheries Management Organization*

As illustrated above, fisheries management for the high seas, including the management of DHSFS, has been problematic in several parts of the South Pacific because of the lack of legitimacy, the lack of wide participation, or even the non-existence of a management system. The need to establish a broader fisheries management mechanism has been recognized.[50]

---

46 Ibid., para. 7; 2000 South Tasman Rise Arrangement, preambular para. 8.

47 However, Willing observes that since the 1998 Arrangement was between the two states and the issue of further membership was not anticipated, it had "many of the features usually found in a cooperative transboundary regime rather than a straddling or high seas regime". Willing, "Arrangement between the Government of Australia and the Government of New Zealand for the Conservation and Management of Orange Roughy on the South Tasman Rise", at p. 203.

48 FSA Review Conference Preparatory Report, at p. 53, para. 252.

49 Neither the 1998 Arrangement nor the 2000 Arrangement is considered to be a treaty. The texts of the arrangements are carefully drafted so as not to give an impression of creating obligations (e.g., the use of 'will' instead of 'shall'). Staples calls these arrangements 'MOU'. Staples, "Management of Shared Fish Stocks: Australian Case Studies", at pp. 161–162. See also Molenaar, "Management and Conservation of Orange Roughy", at p. 104 note 111; Molenaar, "Unregulated Deep-Sea Fisheries", at p. 237 note 70 and its accompanying text. Both states implemented the Arrangements in their domestic legislation. See Australia, Fisheries Management Regulations 1992, Regulation 4B(n); New Zealand, Fisheries (South Tasman Rise Orange Roughy Fishery) Regulations 2000 (SR 2000/11).

50 For example, see the media release by W. Truss, Australian Minister for Agriculture, Fisheries and Forestry, "Orange Roughy Agreement", 7 February 2000, available at <http://pandora.nla

In 2006, Australia, New Zealand and Chile, three coastal states in the South Pacific, started negotiations to establish a regional fisheries management organization or arrangement dealing with predominantly discrete high seas and straddling fish stocks in the entire South Pacific.[51] The negotiations were open to coastal states in the South Pacific and other interested parties, including Canada, China, the EU, the Republic of Korea, Japan, Russia, Ukraine and the US as well as the Chinese Taipei fishing entity as a special observer.

Participants in the negotiations were aware of the existence of DHSFS of both pelagic and deep-sea species.[52] A paper submitted by Australia, New Zealand and Chile devoted one section to "Specific Regimes", stating that it may be "worthwhile considering establishing more specific fisheries management regimes for particular species [i.e., species of discrete high seas stocks and straddling stocks] that take into account their unique characteristics". The paper continues:

> For discrete high seas [stock] species, a specific fisheries management regime could be developed for these species that takes into account their unique biology and habitat. The regime should be sufficiently flexible to respond to the discovery of new fish stocks. Although the [FSA] applies to straddling and highly migratory fish stocks, its principles could usefully be applied, as a minimum, to the management of discrete high seas stocks.

The need for a regime for discrete high seas stocks which is different from that for straddling stocks is further underlined where the paper states that for straddling stock species, "consideration should be given to establishing decision

---

.gov.au/pan/54888/20051219–0000/www.maff.gov.au/releases/00/012wt.html> ("Australia and New Zealand have also agreed to explore the possible development of a broad Tasman Sea Fisheries arrangement").

51 Information is available at the website of the South Pacific Regional Fisheries Management Organisation (formerly, "International Consultations on the Establishment of the South Pacific Regional Fisheries Management Organisation"), <http://www.southpacificrfmo.org/>. Hereinafter, the term 'non-highly migratory species' in this sub-section means discrete high seas stocks and straddling stocks to be managed by the RFMO. Note that all draft texts for the Convention as well as comments and proposals mentioned in this section are available at the above website.

52 An information paper submitted at an initial stage of the negotiation even stated that fisheries for species other than highly migratory species in the high seas areas of the South Pacific are "mainly discrete high seas stocks". Information Paper (SP/01/Inf3 rev1), at p. 1. Although this statement does not seem correct in view of the subsequent fishery research, there are still some DHSFS. For example, the Central Pacific stock of Chilean jack mackerel (*Trachurus murphyi*) has been suggested to be a discrete high seas fish stock whereas stocks of orange roughy on the Louisville Ridge and the Lord Howe Ridge appear to be DHSFS. The fish stock profiles also provide information on other high seas fish stocks which might potentially be regarded as discrete high seas stocks and subject to commercial fisheries. Information available at South Pacific Regional Fisheries Management Organisation, "SWG profiles", <http://www.southpacificrfmo.org/swg-profiles/>.

making processes (such as separate chambers) that allow for the adoption of effective conservation and management measures, while preserving the rights of coastal and fishing states under international law".[53] In other words, the paper illustrates distinct points for the management of fisheries for discrete high seas stocks. First, principles of the FSA could be used as a minimum, while *additional* principles taking into account their unique biology and habitat should be applied. Second, the regime should accommodate the possibility that new fish stocks will be discovered. Third, a consideration of coastal states' interests is less important, if not totally irrelevant, for discrete high seas stocks.

At the eighth meeting of the consultations in November 2009, the participants adopted the SPRFMO Convention.[54] At the third meeting, the negotiations succeeded in adopting interim measures, which were revised for pelagic fisheries at the eighth meeting in November 2009. At the same meeting, the Interim Measure for Deepwater Gillnets in the Convention Area was also agreed upon. After the adoption of the Convention, the three sessions of the Preparatory Conference to prepare for the first meeting of the Commission were held in July 2010, January 2011 and January–February 2012, respectively. At the second session of the Preparatory Conference, the participants agreed on the revised interim measures for *Trachurus* species, replacing the 2009 Revised Interim Measures and, as relevant, revising the 2007 Interim Measures. The third session of the Preparatory conference further revised the Interim Measures for *Trachurus* species, replacing the 2011 Interim Measures and, as relevant, revising the 2007 Interim Measures. As the Convention entered into force, the first Commission meeting is expected to take place in Auckland, New Zealand, from 28 January to 1 February 2013. The next sub-sections examine the interim measures and discuss the provisions of the Convention.

5.2.3 *Interim Measures*

The participants in the negotiations to establish the SPRFMO were aware of the need to control fisheries through interim measures pending the establishment of the RFMO and discussed interim measures from the outset of the process. They adopted interim measures at the third meeting in April–May 2007.

The interim measures are voluntary and are not legally binding.[55] They are to apply until the entry into force of the proposed Convention and the adoption of conservation and management measures; the participants are to review

53 Governance Principles and Their Application (SP/01/Inf6), at pp. 3–4.

54 Convention on the Conservation and Management of High Seas Fishery Resources in the South Pacific Ocean (SPRFMO Convention), Auckland, New Zealand, 14 November 2009. It entered into force on 24 August 2012.

55 Interim Measures adopted by Participants in Negotiations to Establish South Pacific Regional Fisheries Management Organisation, Reñaca, Chile, 30 April–4 May 2007, preambular para. 2.

these interim measures, as necessary, so that they may be revised at future meetings.[56]

The interim measures state that the participants are to take the interim measures, "taking into account an ecosystem approach to fisheries management and the precautionary approach", in order to achieve "the sustainable management of fish stocks and the protection of vulnerable marine ecosystems" in the high seas of the South Pacific (hereinafter 'the area').[57] While the northern boundary of the proposed RFMO was still under discussion at that time, the northern boundary of the area of application of the interim measures was defined as south of the Equator.

Both pelagic fisheries and bottom fisheries in the high seas of the South Pacific Ocean are addressed in the interim measures.[58] The interim measures stipulate various measures separately for pelagic fisheries and for bottom fisheries as well as a handful of measures which are common to both fisheries. As opposed to the Chair's draft presented to the second meeting, fisheries managed by existing (sub-regional) arrangements are not excluded from the scope of the interim measures.[59] The following paragraphs focus on the regulation of bottom fisheries by the interim measures, highlighting the difference between the measures for bottom fisheries and those for pelagic fisheries.

In respect of bottom fisheries, the participants agreed on the following, *inter alia*: freezing fishing efforts, identifying, mapping and closing VME sites, and assessing the impacts of individual bottom fishing activities.

---

56 See the section entitled "Period of Application and Review" of these Interim Measures.

57 SPRFMO Interim Measures, preambular para. 1. The chair's draft for interim measures stated various principles and measures applicable to both pelagic and bottom fisheries, including reference to the commitment to conservation and management in accordance with the LOSC and the FSA.

58 The interim measures, in paragraph 9 of the Pelagic Fisheries part, provide that these interim measures do not apply to squid fisheries in the area concerned. It is not clear from the text whether the Pelagic Fisheries part does not apply (and the Bottom Fisheries part *does* apply) to the squid fishery or whether the entire interim measures do not apply to the squid fishery at all. There is no definition of pelagic fisheries or bottom fisheries agreed by the participants in the negotiation. Note that the squid fishery in the South Atlantic was listed as one of the deep-sea fisheries in the Report of the 2006 FAO Expert Consultation on deep-sea fisheries in the high seas. See Report and Documentation of the Expert Consultation on Deep-sea Fisheries in the High Seas, Bangkok, Thailand, 21–23 November 2006, at p. 30 (Annex II).

59 The chair's draft, in preambular paragraph 1, stated that the proposed resolution would have been applicable to the high seas area of the South Pacific Ocean under discussion for fishery resources "not currently managed through a regional fisheries management organisation or arrangement". See also a paper entitled "Interim Arrangements", SP/01/Inf5, at para. 14 (raising the issue of, *inter alia*, the amalgamation of existing arrangements (e.g., the South Tasman Rise Arrangement) into the competence of the SPRFMO). See also SP/06/INF/18 (Cook Islands), Article 4(4)(b).

First, the participants are to limit the bottom fishing effort or catch to existing levels in terms of the number of fishing vessels and other parameters.[60] The criteria used here are obviously ambiguous.[61] Participants are not to expand bottom fishing activities into new regions where such fishing does not currently occur.[62] There is no exception clause for non-fishing states with a catch history unlike that in pelagic fisheries.[63] Before opening new regions or expanding fishing efforts or catch beyond existing levels in 2010 or later, the participants are to establish conservation and management measures to prevent SAIs from individual bottom fishing activities or to determine that such activities will not have adverse impacts.[64]

Second, participants agreed to cooperate to identify VMEs, on the basis of the best scientific information available, and to map sites of their location.[65] They are to require that their vessels cease bottom fishing activities within five miles of any site where evidence of VMEs is encountered in the course of fishing operations, and to report the encounter to the interim Secretariat so that appropriate measures can be adopted in respect of the relevant site.[66] In respect of areas where VMEs are known to occur or are likely to occur based on the best scientific information available, including the reporting of VME sites encountered

60 SPRFMO Interim Measures, Bottom Fisheries, para. 1.

61 As for pelagic fisheries, participants are to commit themselves to limit the total level of gross tonnage (GT) of their vessels fishing for pelagic stocks in 2008 and 2009 to the levels of total GT recorded in 2007. Ibid., Pelagic Fisheries, para. 1.

62 Ibid., Bottom Fisheries, para. 2. Notwithstanding this, participants are to undertake, as appropriate, scientific research activities for stock assessment purposes in identified parts of such regions and only in accordance with a research plan provided to the interim Secretariat in advance (para. 8). Flag states are required to submit the research plan in advance, but are not required to obtain approval from the meeting of the participants or the Scientific Working Group. This is the same in the case of pelagic fisheries. See SPRFMO Interim Measures, Pelagic Fisheries, para. 6.

63 Russia noted that it accepted these interim measures except paragraphs 1 and 2 of the bottom fisheries section. Report of the Third International Meeting on the Establishment of the Proposed South Pacific Regional Fisheries Management Organisation (SPRFMO 3rd Meeting Report), Reñaca, Chile, 30 April–4 May 2007, at para. 11. As of 2007, it had no bottom trawl vessel operating in this area. See press release by the Deep Sea Conservation Coalition, "Landmark Agreement First of Its Kind to Stop Destruction from High Seas Bottom Trawling", 4 May 2007, available at <http://www.savethehighseas.org/news/view.cfm?ID=155>. Compare the relevant paragraph with SPRFMO Interim Measures, Pelagic Fisheries, para. 2.

64 Ibid., Bottom Fisheries, para. 3. It is reasonable to assume that this does not require the prevention of adverse impacts in an absolute manner, given the context of this provision and other comparable provisions.

65 Ibid., para. 5. For the purposes of the interim measures, VMEs include seamounts, hydrothermal vents, cold water corals and sponge fields. SPRFMO Interim Measures, note 3 to Bottom Fisheries, para. 3.

66 SPRFMO Interim Measures, Bottom Fisheries, para. 7.

during bottom fishing activities as provided in paragraph 7, the participants are to close such areas to bottom fishing unless conservation and management measures have been established to prevent SAIs or it has been determined that such bottom fishing will not have SAIs.[67]

Third, participants agreed to assess, on the best scientific information available, whether individual bottom fishing activities would have SAIs on VMEs, and, if it is assessed that these activities would have SAIs, to ensure that they are managed to prevent such impacts, or they are not authorized to proceed.[68] In undertaking the assessments, participants are to take into account any international technical guidelines regarding standards, criteria or specifications for identifying VMEs and the impacts of fishing activities on such ecosystems that may have been developed.[69]

Procedures regarding the assessments are applied as follows. Participants are to submit to the interim Science Working Group their assessments, including the proposed management measures, and the Working Group will review the assessments and proposed measures and provide comments to the submitting participant.[70] Pending such comments, submitting participants may provisionally apply their proposed management measures. On the basis of their own assessments and comments by the Working Group, participants may authorize their vessels to undertake bottom fishing activities and require such vessels to implement conservation and management measures to prevent SAIs.[71]

Participants also agreed to request, individually or jointly, those states fishing for non-highly migratory fish species in the area but not participating in the negotiations to cooperate fully in the implementation of these interim measures and to consider participating in the negotiations.[72]

67 Ibid., para. 6.

68 Ibid., para. 11. It is not clear whether or not the assessment is required only under circumstances specified in paragraphs 3 and 6.

69 Ibid., para. 13.

70 The role of the interim Scientific Working Group is more limited for pelagic fisheries: paragraph 3 of the Pelagic Fisheries section of the interim measures provides that the Working Group will give advice to the meeting of the participants on the status of the pelagic stocks.

71 SPRFMO Interim Measures, Bottom Fisheries, para. 12. Paragraph 12(c) or any other paragraph relating to conservation and management measures does not specify who may definitely determine the conservation and management measures after comments on the assessment and proposed measures are received by flag states. However, in view of the commitment of participants to notify the interim Secretariat of the measures required under paragraph 12(c) and a list of vessels to which the measures relate, only flag states may ultimately determine the conservation and management measures applicable to their vessels. See para. 12(d). See also SPRFMO Interim Measures, Pelagic Fisheries, para. 4.

72 Ibid., Cooperation with other States.

5.2.4 *The Convention on the Conservation and Management of High Seas Fishery Resources in the South Pacific Ocean (SPRFMO Convention)*

At an initial stage, Australia, Chile and New Zealand had not decided whether or not to establish an RFMO. Their paper on Governance Principles and Their Application states that the legal framework "should establish an organisation that promotes the cooperation of coastal States and States with a real fishing interest in the region", but it also refers to the possibility to manage fisheries through Meetings of the Parties.[73] Although the chair's first draft was based on SIOFA, the great majority of participants in the second meeting made it clear that they were in a position to support, from the outset, a 'fully-fledged' organization with a commission and a permanent secretariat and enjoying separate legal personality. Therefore, the revised version of the chair's draft, prepared for the third meeting, reflected these views and provided for the structure of the proposed SPRFMO. The revised draft drew heavily on an EC non-paper presented prior to the second meeting, which in turn was based on the proposed revision of the NAFO Convention.[74]

The Convention provides that the Organization consists of a Commission, a Scientific Committee, a Compliance and Technical Committee, an Eastern Subregional Management Committee, a Western Sub-regional Management Committee, a Finance and Administration Committee, a Secretariat and any other subsidiary bodies to be established by the Commission.[75] Each Sub-regional Management Committee is to develop and make recommendations to the Commission on conservation and management measures and on participation in fishing for fishery resources for the parts of the Convention Area as described in Annex I.[76] Each Management Committee shall develop its recommendations on the basis of the advice and recommendations of the Scientific Committee.[77] As will be noted in Section 5.2.4.3 below, the issue of Sub-regional Management Committees was at the heart of the disagreements in the negotiations.

The Convention applies to waters of the Pacific Ocean beyond areas of national jurisdiction as defined in Article 5, whose boundary roughly corresponds to the following: (western boundary) the eastern boundary of SIOFA; (southern boundary) the northern boundary of the convention area of CCAMLR; (eastern boundary) the outer limit of the maritime jurisdictions of Chile, Peru, Ecuador and Colombia.[78] In the end, the northern boundary was determined as south of a line extending west along the 2° parallel of north latitude (excluding the maritime

73 Governance Principles and Their Application (SP/01/Inf6), at pp. 1 and 5.
74 SPRFMO, 3rd meeting, Chair's Cover Letter, at pp. 1–2.
75 SPRFMO Convention, Article 6(1)–(2).
76 Ibid., Article 12(1).
77 Ibid., Article 12(4).
78 Ibid., Article 5(1).

zones around the Galapagos Islands) to the intersection with the 150° meridian of west longitude, then due north along that longitude to its intersection with 10° parallel of north latitude, and then west along the 10° parallel of north latitude to its intersection with the outer limit of the Marshall Islands maritime zones. The Convention applies to high seas enclaves surrounded by the maritime zones of Pacific Island states in the waters bounded by the 10° parallel of north latitude and the 20° parallel of south latitude and by the 135° meridian of east longitude and the 150° meridian of west longitude.[79]

The objective of the Convention is, through the application of the precautionary approach and an ecosystem approach to fisheries management, to ensure the long-term conservation and sustainable use of fishery resources in the Convention Area and, in so doing, to safeguard the marine ecosystems in which those resources occur.[80] The proposed Convention appears to address fisheries only; it does not aim to regulate activities other than fisheries or ecosystem impacts caused by such other activities.

Fishery resources are defined in the Convention as all fish, including molluscs and crustaceans and other living marine resources as may be decided by the Commission, but excluding sedentary species subject to the national jurisdiction of coastal states pursuant to Article 77(4) of the LOSC, highly migratory species listed in Annex I of the LOSC, anadromous and catadromous species, and marine mammals, marine reptiles and seabirds.[81] The definition of fishery resources

---

79 Ibid., Article 5(2). One of the issues during the negotiations was the treatment of high seas areas adjacent to the EEZ of the Pacific Island states. Pacific Islands Forum states referred to an option to include areas of the high seas surrounded by their EEZs within the area of competence of the proposed SPRFMO in their declaration in 2006. Declaration on Deep-sea Bottom Trawling to Protect Biodiversity in the High Seas, Nadi, Fiji, 24–25 October 2006, para. 17(1). It is noted, in passing, that, in May 2008, the fisheries ministers of the eight parties to the Nauru Agreement (Federated States of Micronesia (FSM), Kiribati, Marshall Islands, Nauru, Palau, Papua New Guinea (PNG), Solomon Islands and Tuvalu) adopted conservation and management measures within the framework of the Nauru Agreement (Third Implementing Arrangement to the Nauru Agreement). It subjects the granting of fishing licences in their EEZs to any foreign fishing vessels to the prohibition of fishing in the two high seas pockets adjacent to the EEZs of the parties to the Nauru Agreement. Third Arrangement Implementing the Nauru Agreement Setting Forth Additional Terms and Conditions of Access to the Fisheries Zones of the Parties, Koror, Palau, 16 May 2008, Article I(3). The first high seas pocket is bound by the waters of the FSM, Indonesia, Palau and PNG; the second area is bound by the waters of the FSM, Fiji, Kiribati, Marshall Islands, Nauru, PNG, Solomon Islands and Tuvalu. The text of the Arrangement as amended on 11 September 2010 adds "any additional high seas areas" located within 10 degrees North and 20 degrees South latitude and 170 degrees East and 150 degrees West longitude. The measures are primarily intended to protect tuna resources in the EEZ of these states.

80 SPRFMO Convention, Article 2.

81 Ibid., Article 1(g).

implies two points which are relevant to this study. First, the Commission is competent to deal with the conservation and management of DHSFS. Second, while, unlike the SEAFO Convention and SIOFA, the category of sedentary species is not explicitly recognized in the context of the area beyond the outer limit of the continental shelf, the exclusion of such species on the continental shelf virtually means that the Commission may be competent to manage the exploitation of such species in the Area.

In the Convention, fishing is defined very broadly, including transshipment.[82] Particularly contentious was the treatment of fisheries for scientific research. The definition of fishing in the first revised draft included scientific research.[83] In that draft, conservation and management measures included measures to regulate, where necessary, the conduct of fishing for scientific purposes.[84]

Faced with clear opposition,[85] the relevant provisions of the Convention no longer explicitly include scientific research in the definition of fishing.[86] However, as the expression "for any purpose" in Article 1(1)(g)(ii) suggests, the Commission is still capable of dealing with fisheries for scientific research purposes. Besides, the Convention retains references to scientific research, *inter alia*, in the following two articles. First, it stipulates that the Commission shall promote the conduct of scientific research to improve knowledge of fishery resources and marine ecosystems in the Convention Area and of the same fishery resources in adjacent waters under national jurisdiction and, in collaboration with the Scientific Committee, establish procedures for the conduct of fishing for scientific purposes in the Convention Area.[87] Second, each flag state contracting party shall ensure that vessels flying its flag engaged in or intending to engage in research into fishery resources comply with any procedures established by the Commission for the conduct of scientific research in the Convention Area.[88] Read together, these provisions will prevent commercial fisheries from being conducted in fishing under the guise of scientific research. This is very significant in the context of fisheries for deep-sea species, which are vulnerable to fishing efforts and may be fished down in a short period.

---

82 Ibid., Article (1)(h). Also note that the definition of fishing vessel explicitly includes fish processing vessels. Ibid., Article (1)(i).

83 Revision 1 of the Draft South Pacific Ocean Regional Fisheries Management Agreement (Revised Draft SPRFMO Agreement), Article 1(f)(ii).

84 Ibid., Article 17(2)(g).

85 See, e.g., Revised U.S. Discussion Paper: Treatment of "Fishing Research Vessels".

86 See SPRFMO Convention, Article 1(1)(g).

87 Ibid., Article 8(d). Australia opposed the establishment of procedures for the conduct of fishing for scientific purposes. Australian drafting suggestions circulated, 11 September 2007.

88 SPRFMO Convention, Article 25(5).

5.2.4.1 *Precautionary Approach*

Decisions shall be based on the best scientific and technical information available.[89] In this connection, full and accurate data on fishing including information relating to *impacts on the marine ecosystems* in which fishery resources occur shall be collected, verified, reported and shared in a timely and appropriate manner.[90] Each member of the Commission shall collect, verify and report scientific, technical and statistical data pertaining to fishery resources and marine ecosystems in the Convention Area.[91] The Commission shall, taking full account of Annex I of the FSA, develop standards, rules and procedures for, *inter alia*, the collection, verification and timely reporting of all relevant data by contracting parties, and the compilation and management of accurate and complete data to facilitate effective stock assessment and to ensure that the provision of the best scientific advice is enabled.[92] This is intended, *inter alia*, to enhance the information base for the conservation and management of fishery resources, non-target and associated or dependent species and the protection of the marine ecosystems in which those resources occur.[93]

At the same time, the precautionary approach as described in the FSA and the Code of Conduct shall be applied.[94] In particular, the Convention singles out some aspects of the precautionary approach, including: the contracting parties shall be more cautious when information is uncertain, unreliable or inadequate, shall not use the absence of adequate scientific information as a reason for postponing or failing to take conservation and management measures, and shall take account of best international practices regarding the application of the precautionary approach, including Annex II of the FSA and the Code of Conduct.[95]

Fishing shall be commensurate with the sustainable use of fishery resources taking into account the impacts on non-target and associated or dependent species and the general obligation to protect and preserve the marine environment.[96] In determining a TAC or total allowable fishing effort for any fishery resource, the Commission shall take into account the factors specified in the Convention, including measures adopted by other intergovernmental organizations.[97]

---

89 Ibid., Article 3(1)(a)(5). The term 'scientific and technical information' began to appear as a consequence of the realization that the management of deep-sea fisheries needs to use not only scientific information but also technical information. See, e.g., FAO International Guidelines for the Management of Deep-sea Fisheries in the High Seas, para. 21(iv).

90 SPRFMO Convention, Article 3(1)(a)(iv).

91 Ibid., Article 24(1)(d).

92 Ibid., Article 23(1)(a)–(b).

93 Ibid., Article 23(1).

94 Ibid., Article 3(1)(b) and (2)(a).

95 Ibid., Article 3(2)(a)(i)–(iii).

96 Ibid., Article 3(1)(a)(ii). See also ibid., Article 20(1)(b)–(c).

97 Ibid., Article 20(3)(h).

Conservation and management measures may include the determination of reference points, including precautionary reference points as described in Annex II of the FSA, and the actions to be taken if those reference points are approached or exceeded.[98]

A fishery that has not been subject to fishing or has not been subject to fishing with a particular gear type or technique for ten years or more shall be opened as a fishery or opened to fishing with such gear type or technique only when the Commission has adopted cautious preliminary conservation and management measures in respect of that fishery and, as appropriate, non-target and associated or dependent species, and appropriate measures to protect the marine ecosystem in which that fishery occurs from adverse impacts of fishing activities.[99] Such preliminary conservation and management measures shall be consistent with the objective and the conservation and management principles and approaches of this Convention and shall ensure that the new fishery resource is developed on a precautionary and gradual basis until sufficient information is acquired to enable the Commission to adopt appropriately detailed conservation and management measures.[100] The Commission may adopt standard minimum conservation and management measures that are to apply in respect of some or all new fisheries prior to the commencement of fishing for such new fisheries.[101]

The term 'fishery' as opposed to 'fishery resource' or 'fishing' is not defined in Article 1 of the Convention. This makes it difficult to consider what is meant by 'fishery' in the context of 'new or exploratory fisheries'. In fact, there were discussions during the negotiations. The determination of existing and new fisheries seems to be understood, at least by one participant, to be based on species rather than stocks. Australian comments reject the formulation of the definition of 'target fishery' used in Revision 2 of the chair's text, stating as follows:

> [the] current definition based on a species-wide, convention-area wide definition is too generic to capture the reality of individual high seas stocks [...]. We would like previously unfished stocks of a species to be treated as new/exploratory, rather than as a target fishery simply because the same species has been fished elsewhere in the Convention Area. [...] [W]e will be seeking ways in which the definition can better respond to the needs of stocks management.[102]

Although the definition of 'target fishery' was dropped from the draft, the ambiguity concerning the management unit of fishery resources remains with regard to, *inter alia*, the control of new fisheries.[103]

---

98 Ibid., Article 20(2)(a)–(b).

99 Ibid., Article 22(1).

100 Ibid., Article 22(2).

101 Ibid., Article 22(3).

102 Specific Australian Comments on the Agreement Text (SP/03/Inf10), 20 June 2007, at para. 20.

103 See also the similar problem with regard to the compatibility requirement between coastal state measures and measures for the high seas in Section 5.2.4.2 below.

Where fishing presents a serious threat to the sustainability of fishery resources or the marine ecosystem in which these fishery resources occur or when a natural phenomenon or human-caused disaster has, or is likely to have, an SAI on the status of fishery resources, the Commission shall adopt emergency measures to ensure that fishing does not exacerbate such a threat or adverse impact.[104] Such measures shall be temporary and be based on the best scientific evidence available.[105] Such measures shall not be open to the objection procedure but may be the subject of dispute settlement procedures under the Convention.[106]

5.2.4.2 *Ecosystem Considerations*

The SPRFMO shall adopt an ecosystem approach to fisheries management.[107] Although the term 'ecosystem approach to fisheries management' is not defined, it shall be applied through "an integrated approach under which decisions in relation to the management of fisheries resources are considered in the context of the functioning of the wider marine ecosystems in which they occur to ensure the long-term conservation and sustainable use of those resources and in so doing, safeguard those marine ecosystems".[108]

The Convention stipulates that the conservation and management measures shall include measures, *inter alia*, to maintain or restore populations of non-target and associated or dependent species to above levels at which their reproduction may become seriously threatened and to protect the habitats and marine ecosystems in which fishery resources and non-target and associated or dependent species occur from the impacts of fishing.[109] Moreover, an allowance for discards and any other incidental mortality and the catch of non-target and associated or dependent species and impacts on the marine ecosystems in which the fishery resource occurs are among the factors to be taken into account in determining the TAC or total allowable effort.[110]

The conservation and management measures include those to protect the habitats and marine ecosystems from the impacts of fishing, including measures to prevent SAIs on VMEs and precautionary measures where it cannot adequately be determined whether VMEs are present or whether fishing would cause SAIs on VMEs.[111] The measures are to ensure the long-term sustainability of fishery resources and promote the objective of their responsible utilization.[112]

---

104 SPRFMO Convention, Article 20(5)(a).
105 Ibid., Article 20(5)(b).
106 Ibid.
107 Ibid., Article 3(1)(b).
108 Ibid., Article 3(2)(b).
109 Ibid., Article 20(1)(c)–(d).
110 Ibid., Article 20(3)(d)–(e).
111 Ibid., Article 20(1)(d).
112 Ibid., Article 20(1)(a).

Except for the reference to the need to preserve biodiversity in preambular paragraph 8, the Convention lacks any explicit reference to biodiversity. The reference to biodiversity in the marine environment in Revision 3[113] was replaced by the reference to marine ecosystems.[114] Although specific references to deep-sea ecosystems found in previous drafts[115] no longer exist in the current draft, this should not be considered to imply a loss of interest in the protection of deep-sea ecosystems: the insertion of the words "ecosystems which have long recovery times following disturbance" certainly covers deep-sea ecosystems and appears to be aimed at expanding the scope of protection.[116]

Cooperation and coordination among contracting parties shall be promoted to ensure that conservation and management measures adopted by the Commission and those applied in respect of the same fishery resources in areas under national jurisdiction are compatible.[117] Since the terms 'fishery resources' and 'straddling fishery resources', rather than 'fish stocks', are used in the Convention, their meaning is unclear. Although a reasonable interpretation would be to equate the terms 'fishery resources' and 'fish stocks', the Convention uses the term 'fish stocks' in the sense of a sub-category of 'fishery resources'.[118] Therefore, Article 4 could be interpreted to mean that measures for a particular stock shall take into account measures by coastal states for another stock of the same species under their national jurisdiction and shall ensure that the measures adopted by the Commission do not undermine the effectiveness of such coastal state measures.[119] As a result, theoretically, coastal state measures for a stock that is confined to its EEZ or straddles the high seas and its EEZ may influence measures for a discrete high seas fish stock.

5.2.4.3 *Participatory Rights*

As noted earlier, the Convention stipulates that organs of the SPRFMO include Sub-regional Management Committees, whose recommendations will be the basis of conservation and management measures and decisions referred to in Articles 20 and 21, respectively, to be adopted by the Commission.[120] The membership

---

113 Revision 3 of the Draft SPRFMO Convention, Article 3(1)(i).

114 SPRFMO Convention, Article 3(1)(a)(vii).

115 E.g., Revised Draft SPRFMO Agreement, Article 3(1)(h).

116 On the protection of deep-sea ecosystems, see also SPRFMO Convention, Article 20(1)(d).

117 Ibid., Article 3(1)(a)(vi). See also Article 4, which was inserted to articulate the requirement of compatibility. Note that initial drafts did not explicitly refer to the 'adjacent' high seas areas. See also Russia's Comments on Article 3(1)(g) in SP/03/Inf11.

118 SPRFMO Convention, Article 8(a)–(b).

119 As noted earlier, the term 'fishery resource' was understood, at least by one participant, to mean 'species' rather than 'stock'. See the comments on the Australian Comments in the context of new fisheries cited above. See also a proposal by Peru, SP/06/INF/08.

120 SPRFMO Convention, Article 12(7).

of each Sub-regional Committee includes those contracting parties which are situated adjacent to the part of the Convention Area for which that Committee is responsible, or whose vessels are currently fishing or have fished within the past two years in that area or are fishing for a specific fishery resource assigned to that Committee pursuant to Article 12(3), including in areas under national jurisdiction adjacent to the Convention Area.[121] In addition, any member of the Commission that gives notice to the Secretariat of an intention to fish within two years of the notice in that part of the Convention Area for which a Committee has responsibility shall become a member of that Committee.[122] Therefore, participation in fisheries management at the Sub-regional Committee level is limited to coastal states and (existing or future) distant water fishing states.

While several coastal states consider the creation of sub-regional management committees to be essential to the proposed RFMO,[123] a number of distant water fishing states have strongly opposed such committees.[124] While most of the proposals of distant water fishing states intended to establish one Committee dealing with all fisheries covered by the Convention, interestingly, Ukraine proposed to establish committees on the basis of biological characteristics such as one for "Fish of submerged ridges (beryx and other)" whose membership was not limited to coastal states and high seas fishing states.[125]

The competences of Sub-regional Committees (and thus the influence of coastal states in each sub-region) are substantially curtailed in the Convention,[126] while potential conflicts over the interpretation of the term 'basis' may give rise to uncertainty concerning the discretion of which the Commission may avail itself in adopting conservation and management measures and decisions on participation in fishing for fishery resources.

The Commission shall take into account the factors specified in Article 21(1) in decision-making concerning participation in fishing for any fishery resource. The special requirements of developing states shall be taken into account.[127] The Commission shall, when appropriate, review decisions regarding participation in

121 Ibid., Article 12(5)(a).

122 Ibid., Article 12(5)(b).

123 See, e.g., Permanent Commission for the South Pacific (CPPS) Resolution No. 17, November 2007, para. 2. But see SP/06/INF/08 (Peru), Articles 6.2(b) and 9.1.

124 The EU (EC amendments to the chair's text Revision 2, 14 September 2007), Japan (Japan's Proposed Amendments to Revision 2 of the Chair's Draft Text, 17 October 2007), the Republic of Korea (Korea's Proposals, October 2007; SP/05/INF/8), Russia (Proposals, 14 September 2007; SP/05/INF/11, Russian Federation proposal concerning main articles of the Convention text, SP/06/INF/16) and Ukraine (SP/05/INF/7). But see SP/06/INF/02 (indicating the EC's willingness to accept two subcommittees).

125 Comments on Article 11 in SP/05/INF/7.

126 Compare with, e.g., Revision 3 of the Draft SPRFMO Convention, Articles 11(5) and 14(6).

127 SPRFMO Convention, Articles 3(1)(a)(viii) and 19.

fishing for fishery resources, including allocation, taking into account the provisions of Article 21 and the interests of new contracting parties.[128] Again, it is possible that the TAC is set for species, not for stocks.[129]

5.2.4.4 *Cooperation with Other Organizations*

The Commission shall cooperate, as appropriate, with other RFMOs, the FAO, other UN specialized agencies and other relevant organizations on matters of mutual interest.[130] In particular, the Commission shall take account of the conservation and management measures or recommendations adopted by other RFMOs or other relevant intergovernmental organizations that have competency in relation to the Convention Area or to the adjacent areas (e.g., CCAMLR and SIOFA)[131] or in respect of particular living marine resources not covered by the Convention, including non-target and associated species (e.g., tuna RFMOs as well as the IWC) and that have objectives that are consistent with, and supportive of, the objectives of the Convention. The Commission shall also endeavour to ensure that its own decisions are compatible with, and supportive of, such conservation and management measures or recommendations.[132]

5.3 *North Pacific Ocean*

In the North Pacific, some of the high seas areas have been subject to extensive fisheries for straddling stocks, and coastal and distant water fishing states have attempted to establish regimes regulating these fisheries in two areas, i.e., the central Sea of Okhotsk[133] and the central part of the Bering Sea.[134]

---

128 Ibid., Article 21(6).

129 Cf. proposed Article 17(n)) in SP/03/Inf11 (Russia).

130 SPRFMO Convention, Article 31(1).

131 The term used here is 'regional fisheries management organisations', but it is reasonable to assume that this provision extends to 'arrangements'.

132 SPRFMO Convention, Article 31(2).

133 On fisheries in the central part of the Sea of Okhotsk (also referred to as the 'Peanut Hole') as well as coastal state actions, see Oude Elferink, "Fisheries in the Sea of Okhotsk High Seas Enclave", at pp. 1–18; A.G. Oude Elferink, "The Sea of Okhotsk Peanut Hole: De facto Extension of Coastal State Control", in O.S. Stokke (ed.), *Governing High Seas Fisheries: The Interplay of Global and Regional Regimes* (2001), at pp. 179–205.

134 For the fisheries in the high seas part of the Central Bering Sea (the so-called 'Donut Hole' or 'Doughnut Hole') and the CBS Convention, see D.A. Balton, "The Bering Sea Doughnut Hole Convention: Regional Solution, Global Implications", in O.S. Stokke (ed.), *Governing High Seas Fisheries: The Interplay of Global and Regional Regimes* (2001), at 143–177; W.V. Dunlap, "The Donut Hole Agreement", 10 *International Journal of Marine and Coastal Law* (1995), at pp. 114–126; Kaye, *International Fisheries Management*, at 305–353; Miovski, "Central Bering Sea Overfishing", at pp. 525–574. The objectives of the CBS Convention include: "to cooperate in the

In recent years, an initiative was launched to address the implementation of UNGA Resolutions on sustainable fisheries in relation to bottom fisheries on the high seas in this region.[135] Following UNGA Resolution 59/25, three states whose nationals engaged in bottom trawling in the North-West Pacific, i.e., Japan, the Republic of Korea and Russia, started an initiative to establish a mechanism for the management of such fisheries in the area.[136] Later, the three high seas fishing states and the United States (a coastal state) met to exchange information and discuss the issue in August 2006.[137] China also expressed its willingness to participate in the work to establish an RFMO in the North-West Pacific.[138] The participants at the second meeting agreed that any new participants with a real interest should be encouraged to respect the previous decisions of the meeting.[139] The participants have held ten meetings and adopted a convention establishing an RFMO. The participants also adopted an interim mechanism for the management of high seas bottom fisheries in this region.

This section considers this initiative, focusing on the interim mechanism and the newly adopted Convention on the Conservation and Management of High Seas Fisheries Resources in the North Pacific Ocean.

---

gathering and examining of factual information concerning pollock and other living resources in the Bering Sea"; and "to provide, if the Parties agree, a forum in which to consider the establishment of necessary conservation and management measures for living marine resources other than pollock in the Convention Area as may be required in the future". CBS Convention, Article II. While not legally binding, a statement in the Record of Discussions provides that fishing operations for species other than pollock, which could occur in the future, by vessels of any party, should only be conducted pursuant to specific authorization issued by that party. Cited in Balton, "The Bering Sea Doughnut Hole Convention: Regional Solution, Global Implications", at p. 157. In relation to marine living resources other than pollock, no major development was reported. For the latest COP, see Report of the 16th Annual Conference of the Parties to the Convention on the Conservation and Management of Pollock Resources in the Central Bering Sea, 22 September–4 November 2011 (Virtual Conference).

135 Targeted fish stocks affected by bottom fisheries on the high seas within the North Pacific Ocean include slender armourhead, alfonsino, warty oreo, deep-sea crabs, deep-sea sharks, channelled rockfish and rockfishes nei. A. Bensch *et al.*, Worldwide review of bottom fisheries in the high seas, FAO Fisheries and Aquaculture Technical Paper 522 (2008), at p. 86.

136 FSA Review Conference Report, at p. 13, para. 61. Japan and Russia are also coastal states of the high seas area in the region.

137 Fisheries Agency of Japan, Press release on the outcome of the first inter-governmental consultations on the management of bottom trawling in the high seas area of the North-West Pacific on 25 August 2006. All the press releases and other materials of the Fisheries Agency of Japan mentioned in this section are available at <http://www.jfa.maff.go.jp/>.

138 *ENB*, vol. 7, No. 62, at p. 3.

139 Record of the Second Inter-governmental Meeting on NWP Bottom fisheries, at p. 3. The record of the negotiations is available at the website of "Multilateral Meeting on Management of High Seas Fisheries in the North Pacific Ocean", <http://nwpbfo.nomaki.jp/>.

5.3.1 *The Interim Mechanism*

The Second Meeting adopted a document entitled "Establishment of new mechanisms for protection of vulnerable marine ecosystems and sustainable management of high seas bottom fisheries in the North Western Pacific Ocean". In the document, participating states agreed to take urgent action on an interim basis, including interim measures set out in paragraph 4, while working on a long-term agreement to achieve the identified objectives.[140] The interim mechanism is not limited to interim measures *per se*. It also provides a (provisional) framework for future negotiations on the establishment of a permanent mechanism. The interim mechanism consists, *inter alia*, of general provisions (i.e., scope, purpose and principles), the regulation of bottom fishing activities (i.e., interim measures, contingent action and control of bottom fishing vessels) and scientific activities (i.e., a scientific working group and the collection and sharing of scientific information).

The scope of coverage of the interim mechanism extends to high seas areas of the North Western Pacific Ocean, defined as those occurring within FAO statistical area No. 61. Within this statistical area, all areas and marine species are included except for those already covered by existing international fisheries management instruments and for high seas areas surrounded by the EEZ of a single country.[141] This provision excludes, among others, pollock fisheries in the western part of the so-called Donut Hole, which is covered by the CBS Convention, from the scope of the interim mechanism. By way of the latter saving clause, it is clear that the high seas enclave surrounded solely by the Russian EEZ is also excluded from the scope of this arrangement.

The management target of the interim mechanism is bottom fisheries conducted by vessels operating on the high seas.[142] The document states that the general purpose of the action to be taken by the participants is the sustainable management of fish stocks and the protection of VMEs in the high seas areas of

---

140 See Record of the Second Inter-governmental Meeting on NWP Bottom fisheries, at pp. 2–3. Measures specified in the document will be applied on a voluntary basis. Establishment of New Mechanisms for Protection of Vulnerable Marine Ecosystems and Sustainable Management of High Seas Bottom Fisheries in the North Western Pacific Ocean, reproduced in Record of the Second Inter-governmental Meeting on Management of High Seas Bottom Fisheries in the North Western Pacific, Busan, Republic of Korea, 31 January–2 February 2007, Attachment 6 (hereinafter 'North Western Pacific Ocean Interim Mechanism'), at para. 10. At para. 11, the document stipulates that paragraph 4 (interim measures) will become effective upon the adoption of the document, while the other provisions of the document will become applicable and operational no later than 31 December 2007.

141 North Western Pacific Ocean Interim Mechanism, at para. 1A.

142 Ibid., at para. 1B. It is recalled that the target of management was extended from bottom trawling to bottom fisheries in general after the first meeting. In this regard, the title of the meeting was changed after the first meeting.

the North Western Pacific Ocean.[143] Although the term 'vulnerable marine ecosystems in the high seas' is different from that used in UNGA resolution 61/105 ('vulnerable marine ecosystems in areas beyond national jurisdiction'), this does not necessarily mean that the drafters intended to imply something different from the UNGA Resolution.[144]

The principles of the interim mechanism set out in the document are as follows: the implementation of the interim mechanism will be based on the best scientific information available, be in accordance with existing international laws and agreements, establish appropriate and effective conservation and management measures, be in accordance with the precautionary approach, and incorporate an ecosystem approach to fisheries management.[145]

Interim measures are composed of several elements. First, the participants agreed to freeze bottom fisheries at the current level. Each country will limit its fishing efforts in bottom fisheries in the high seas part of the area covered by the interim mechanism in terms of the number of fishing vessels and other parameters which reflect the level of fishing effort, fishing capacity or potential impacts on marine ecosystems.[146] Furthermore, each country will not allow bottom fisheries to expand into areas of the North Western Pacific Ocean where no such fishing is currently occurring, in particular by limiting such bottom fisheries to seamounts located south of 45 degrees and provisionally prohibit them in other areas of coverage.[147] At the sixth meeting, the participants agreed on the procedures for the commencement of bottom fishing activities in new fishing areas or with bottom gear not previously used in the existing fishing areas ('exploratory fisheries'), including the prior submission of plans and impact assessments to the Scientific Working Group, their review by the Group, and its decision on whether the fishing activities had SAIs and a recommendation on the continuation of the fisheries concerned and possible additional management measures.[148]

Second, the participants will work to establish science-based standards and criteria for determining that any proposed fishing activity will not have an SAI on marine species or any VMEs.[149] Upon the adoption of such criteria, exceptions to the above restrictions may be provided in cases where increased fishing activity

143 Ibid., at para. 2.

144 Cf. ibid., at preambular para. 4.

145 Ibid., at para. 3. Compare them with the CBS Convention, which lacks any reference to the precautionary approach or the protection of the ecosystem of the Bering Sea. On these aspects of the CBS Convention, see Kaye, *International Fisheries Management*, at pp. 336–337.

146 North Western Pacific Ocean Interim Mechanism, at para. 4A.

147 Ibid., at para. 4B.

148 Record of the Sixth Inter-Governmental Meeting on Management of High Seas Bottom Fisheries in the North Pacific Ocean, Busan, Korea, 18–20 February, 2009, NPO7 Refi, at pp. 22–23 (Attachment 4, Annex 1).

149 North Western Pacific Ocean Interim Mechanism, at para. 4D.

or fisheries in new areas would not have an SAI on marine species or any VME.[150] Such determinations may be made by any flag state or pursuant to any subsequent arrangement, and will be made publicly available through agreed means.[151] The fifth meeting revised the Interim Mechanism by incorporating, *inter alia*, the science-based standards and criteria for the identification of VMEs and an assessment of SAIs on VMEs and marine species.[152] The Exploratory Fishery Protocol, which applies to fisheries allowed as exceptions under paragraph 4.C, was adopted at the sixth meeting.[153]

Third, in areas where, in the course of fishing operations, evidence of VMEs, such as cold-water corals or other associated species, is encountered, the participants will require their vessels to cease bottom fishing activities. The location and species will be reported to the interim secretariat. The latter will notify the other parties so that appropriate measures can be adopted in respect of the relevant site.[154] In the third meeting, on the basis of accumulated information regarding fishing activities in the North Western Pacific Ocean, the species whose encounter triggers the move-on rule was changed to cold-water corals only and the distance of at least five nautical miles was specified for relocation from the relevant site.[155] Proposals for an encounter protocol, including the threshold for triggering the protocol in terms of the quantity of indicator species, were discussed at the sixth meeting, but no consensus was reached and some participating states declared that they would apply their own standard based on the NAFO encounter protocol.[156]

In addition to interim measures, the document specifies contingent action. In areas where VMEs are known to occur or are likely to occur, based on the best scientific information available, bottom fisheries shall cease by 31 December 2008, unless conservation and management measures have been established to

---

150 Ibid., at para. 4C and 4E.

151 Ibid., at para. 4F.

152 See the Record of the Fifth Inter-Governmental Meeting on Management of High Seas Bottom Fisheries in the North western Pacific Ocean, Tokyo, Japan, 17–18 October, 2008, NPO6/Refi, at p. 2.

153 Record of the Sixth Inter-Governmental Meeting on Management of High Seas Bottom Fisheries in the North Pacific Ocean, Busan, Korea, 18–20 February, 2009, NPO7 Refi, at p. 2.

154 North Western Pacific Ocean Interim Mechanism, at para. 4G.

155 See ibid., para. 4.G, revised on 26 October 2007, reproduced in Record of the Third Intergovernmental Meeting on Management of High Seas Bottom Fisheries in the North Western Pacific, Honolulu, Hawaii, United States of America, 24–26 October 2007, NWPBT/04/Ref 3, Attachment 3. The document as revised in 2009, in that paragraph, indicates tentative agreement that the cold water corals include Alcyonacea, Antipatharia, Gorgonacea and Scleractinia. See the Record of the Sixth Inter-Governmental Meeting on Management of High Seas Bottom Fisheries in the North Pacific Ocean, at p. 18 (Attachment 4).

156 Record of the Sixth Inter-Governmental Meeting on Management of High Seas Bottom Fisheries in the North Pacific Ocean, at p. 2.

prevent SAIs on VMEs, consistent with the relevant provisions of UNGA Resolution 61/105 and such international standards as may be developed pursuant thereto.[157] At the sixth meeting, the US proposed the closure of specific areas, but Korea, Japan and Russia only agreed to the closure of a portion of the area originally proposed by the US and that of the C-H Seamounts. Due to the lack of consensus, the amendment of the Interim Measures was not adopted, but the three fishing countries showed their willingness to implement their proposed closure on the Koko Seamount and the C-H Seamounts.[158]

Countries will exercise full and effective control over each of their bottom fishing vessels operating in the coverage area, including by means of fishing licences, authorizations or permits, and will ensure the maintenance of a record of these vessels.[159] The collection and exchange of information as required in the FAO Compliance Agreement will be carried out, and each participant will ensure that all these vessels are equipped with an operational VMS no later than 31 December 2007 or earlier if so decided by the flag state.[160]

Following the adoption of UNGA Resolution 64/72 in 2009, participants in the eighth meeting expressed a willingness to consider interim measures for the Northeast Pacific Ocean at the ninth meeting. They agreed to apply certain preliminary measures, including the limitation of fishing effort in bottom fisheries on the high seas to the existing level and the collection and submission of scientific information from bottom fishing vessels operating in the area.[161] The draft interim measures for the protection of vulnerable marine ecosystems in the Northeast Pacific Ocean, which provide for, *inter alia*, actions to be taken by flag states and fishing entities and control of bottom fishing vessels, were adopted at the 10th meeting and the attendees also agreed that the Exploratory Fishery Protocol agreed for the Northwest Pacific interim measures would also apply to the Northeast Pacific interim measures.[162]

---

157 North Western Pacific Ocean Interim Mechanism, at para. 5. Interestingly, this paragraph, together with paragraph 7 on the collection of scientific information, uses 'shall', while other paragraphs use 'will' or 'should'.

158 Record of the Sixth Inter-Governmental Meeting on Management of High Seas Bottom Fisheries in the North Pacific Ocean, at pp. 2–3.

159 North Western Pacific Ocean Interim Mechanism, at para. 8A.

160 Ibid., at para. 8B and 8C.

161 Record of the 8th Multilateral Meeting on Management of High Seas Fisheries in the North Pacific Ocean, Jeju, Korea, 20–22 January 2010, at p. 2. Note that no commercial fishing activities have been taking place in this area in recent years. Multilateral Meeting on Management of High Seas Fisheries in the North Pacific Ocean, "Background" available at <http://nwpbfo.nomaki.jp/>.

162 Record of the 10th Multilateral Meeting on Management of High Seas Fisheries in the North Pacific Ocean, Vancouver, British Columbia, Canada, 27 February–4 March 2011, at p. 1. The attendees also agreed on a definition of VMEs for the purposes of the interim measures in the Northeast and Northwest Pacific; the interim measures for the Northeast Pacific Ocean

5.3.2 *The Convention on the Conservation and Management of High Seas Fisheries Resources in the North Pacific Ocean*

Discussions took place over the long-term mechanism since the second meeting in 2007. On the basis of the US proposal, the participants in the fifth consultations provisionally agreed to extend the scope of the long-term mechanism to areas north of the 20 degree north of the entire North Pacific, while the Interim Mechanism remains only applicable to bottom fisheries in the North-Western Pacific.[163] The meeting also agreed to extend the scope of species included in the long-term mechanism but noted that there was no need for managing fisheries other than bottom fisheries immediately.[164]

On 4 March 2011, the participants at the 10th session of the negotiations (Canada, China, Japan, the Republic of Korea, Russia and the US) finalized the English text of the Convention on the Conservation and Management of High Seas Fisheries Resources in the North Pacific Ocean.[165] The Convention was adopted at Tokyo on 24 February 2012 and opened for signature at Seoul on 1 April 2012.[166] It will enter into force 180 days from the date of receipt by the Depositary of the fourth instrument of ratification, acceptance, approval or accession.[167] Apart from States that participated in the Multilateral Meetings on the Management of High Seas Fisheries in the North Pacific Ocean[168] (i.e., Canada, China, Japan, the

in para. 2 state that, for the purpose of these interim measures, the term 'VMEs' is "to be interpreted and applied in a manner consistent with the International Guidelines on the Management of Deep Sea Fisheries on the High Seas adopted by the FAO on 29 August 2008" (Attachment 4).

163 As a consequence, Canada and Mexico were to be invited to the meeting in Busan, Korea in February 2009. See the Record of the Fifth Inter-Governmental Meeting on Management of High Seas Bottom Fisheries in the North western Pacific Ocean, at p. 2. Canada newly participated in the sixth meeting. Record of the Sixth Inter-Governmental Meeting on Management of High Seas Bottom Fisheries in the North Pacific Ocean, at p. 1. The participants at the seventh meeting requested that the Interim Secretariat send a letter to Mexico reminding that it has been invited to participate in this process, but it did not attend any meeting of this process.

164 Record of the Fifth Inter-Governmental Meeting on Management of High Seas Bottom Fisheries in the North western Pacific Ocean, at p. 2. From the sixth meeting on, the relevant part of the title of the meeting was changed from 'North Western Pacific' to 'North Pacific'.

165 Chinese Taipei also attended the meeting as an observer. The Faroe Islands also attended the eighth and ninth meetings as an observer.

166 See Article 23(1). The first preparatory meeting was held from 29 August to 2 September 2011 in Busan, the Republic of Korea. See Multilateral Meeting on Management of High Seas Fisheries in the North Pacific Ocean, "Background" available at <http://nwpbfo.nomaki.jp/>.

167 Article 25(1).

168 The participants in the sixth meeting agreed to change the title of the meeting to 'Multilateral Meeting on Management of High Seas Fisheries in the North Pacific Ocean'.

Republic of Korea, Russia and the US),[169] after the entry into force of the Convention, the contracting parties may, *by consensus*, invite other states or REIOs whose fishing vessels wish to conduct fishing activities for fisheries resources in the Convention Area and other coastal states of the Convention Area to accede to the Convention.[170]

The Convention takes into account the trend in fisheries management at the global level, in particular the UNGA. In fact, the Convention explicitly refers to the UNGA calls for the protection of VMEs from SAIs of destructive fishing practices and, interestingly, for recognizing that the general principles of the FSA should also apply to discrete fish stocks in the high seas.[171]

The geographic scope of the Convention is the high seas areas of the North Pacific Ocean, excluding the high seas areas of the Bering Sea and other high seas areas that are surrounded by the EEZ of a single state. The southern boundary of the Convention Area is defined in Article 4(1) as follows: at 20 north latitude between the eastern limit of the EEZ of the Northern Mariana Islands and 180 east longitude, at 10 north latitude between 180 east longitude and 140 west longitude, and at 20 north latitude between 140 west longitude and the western limit of the Mexican EEZ. By virtue of Article 4(1), high seas areas of the Sea of Okhotsk and north, west and east of the seaward limit of the waters surrounding Okinotorishima subject to Japan's jurisdiction are excluded from the Convention Area.[172] It is noted that there are several high seas areas left between the Convention Areas of this Convention and the South Pacific RFMO Convention.

The Convention establishes the North Pacific Fisheries Commission.[173] It shall, *inter alia*, adopt conservation and management measures to ensure the long-term sustainability of the fisheries resources within the Convention Area, including the establishment of the TAC and total allowable efforts, and conser-

169 At the ninth meeting, the representative of the Faroe Islands expressed the interest of his government in participating in future meetings as a full participant and offered to present information relating to the real interest of the Faroe Islands in the relevant fisheries at the beginning of the 10th meeting. Record of the 9th Multilateral Meeting on the Management of High Seas Fisheries in the North Pacific Ocean, at p. 3. However, the Faroe Islands did not attend the 10th meeting and there was consensus that no change would be made to the status of the Faroe Islands in the negotiations. Record of the 10th Multilateral Meeting on the Management of High Seas Fisheries in the North Pacific Ocean, at pp. 1–2.

170 Article 24(2).

171 Preambular para. 3.

172 The area north of Okinotorishima is surrounded by the EEZ of Japan. Even for those who argue for the application of Article 121(3) of the LOSC to Okinotorishima, this area and the areas west and east of Okinotorishima would be outside the application of this Convention, in view of the formulation of this article that the southern boundary line extends from the seaward limit of waters under the jurisdiction of the US around the Commonwealth of the Northern Mariana Islands at twenty degrees north latitude proceeding 'East'.

173 Article 5(1).

vation and management measures for species belonging to the same ecosystem or dependent upon or associated with the target stocks, where necessary, based on the best scientific information available and the advice of the Scientific Committee.[174] It shall also adopt conservation and management measures to prevent SAIs on VMEs in the Convention Area, including measures for conducting and reviewing impact assessments, measures to address unexpected encounters with VMEs in the course of normal bottom fishing activities and, as appropriate, measures that specify locations in which fishing activities shall not occur.[175] It shall determine the nature and extent of participation in existing and new fisheries; for new fisheries in the Convention Area, it shall also establish the terms and conditions; decisions concerning new fisheries are taken by consensus.[176]

For the purposes of the Convention, the words 'fisheries resources' are defined as "all fish, molluscs, crustaceans and other marine species caught by fishing vessels within the Convention Area", excluding (1) sedentary species subject to the sovereign rights of coastal states and indicator species of VMEs as listed in or adopted pursuant to Article 13(5) of the Convention, (2) catadromous species, (3) marine mammals, marine reptiles and seabirds, and (4) other marine species already covered by pre-existing international fisheries management instruments.[177] Fishing activities are defined broadly, including engaging in any activity that can reasonably be expected to result in locating, catching, taking or harvesting of fisheries resources for any purpose and the processing at sea and the transhipping at sea or in port, and operations at sea in direct support of or in preparation for any activity described in Article 1(i)(i)–(iii).[178]

The Convention aims to ensure the long-term conservation and sustainable use of the fisheries resources in the Convention Area while protecting the marine ecosystems of the North Pacific Ocean in which these resources occur.[179]

Article 3 provides for general principles, which echoes many of the similar provisions found in the SPRFMO Convention. One provision clearly reflecting

---

174 Article 7(1)(a)–(c).

175 Article 7(1)(e).

176 Article 7(1)(f)–(g). Decision-making on substance, except for those explicitly requiring consensus, is made by a three-quarters majority. On the implementation of Commission decisions, especially on the objection procedure, see Article 9.

177 Article 1(h). Such instruments would include those establishing the WCPFC, the Inter-American Tropical Tuna Commission (IATTC) and the North Pacific Anadromous Fish Commission. Thus, all highly migratory species except sauries in the Western and Central Pacific, tuna and tuna-like species in the Eastern Pacific and seven salmonid species in the North Pacific are excluded. Species covered in the Convention include deep-sea species such as alfonsino and armourhead as well as flying squid and sauries.

178 Article 1(i).

179 Article 2.

the main concern of the present Convention, namely, impacts of bottom fisheries, reads as follows:

> ensuring that any expansion of fishing effort, development of new or exploratory fisheries, or change in the gear used for existing fisheries, does not proceed without prior assessment of the impacts of those fishing activities on the long-term sustainability of fisheries resources and a determination that those activities would not have [SAIs] on [VMEs], or ensuring that those activities are managed to prevent those impacts or are not authorized to proceed [...].[180]

Thus, this provision expands the requirement of a prior impact assessment to all fisheries. Although experimental or scientific fishing activities are not mentioned here, it should be noted that the Commission shall establish the terms and conditions for any experimental, scientific and exploratory fishing activities in the Convention Area.[181]

The Convention extensively addresses bottom fisheries on the high seas.[182] In addition to measures already mentioned above, the Commission shall adopt and amend from time to time a list of indicator species for VMEs for which directed fishing shall be prohibited.[183] It will be assisted by the Scientific Committee, which shall, *inter alia*, (1) develop a process to identify VMEs and areas or features where VMEs are known or likely to occur and the location of bottom fisheries in relation to these areas or features, (2) identify and advise the Commission on additional indicator species for VMEs for which directed fishing shall be prohibited, and (3) establish science-based standards and criteria to determine if bottom fishing activities are likely to produce SAIs on VMEs or marine species in a given area based on international standards and make recommendation for measures to avoid such impacts.[184]

The duties of flag states extend to a wide range of matters and the Convention includes some provisions specifically dealing with bottom fisheries in this regard. Flag states shall prohibit vessels entitled to fly their flag for engaging in directed fishing on four scientific orders of cold-water corals (Alcyonacea, Antipatharia, Gorgonacea and Scleractinia) as well as any other indicator species for VMEs identified by the Scientific Committee and adopted by the Commission.[185] Fishing vessels engaged in bottom fishing in the Convention Area shall have 100% observer coverage.[186]

180 Article 3(h).

181 Article 7(3)(d).

182 Cf. preambular para. 6.

183 Article 7(3)(e).

184 Article 10(4)(e)–(g).

185 Article 13(5).

186 Article 13(6). The level of observer coverage for fishing vessels engaged in other types of fishing activities in the Convention Area is to be decided by the Commission.

The Commission shall cooperate with other organizations on matters of mutual interest, especially with those RFMOs or arrangements with responsibility for fisheries in marine areas near or adjacent to the Convention Area, and it shall take into account the conservation and management measures or recommendations adopted by RFMOs and arrangements and other relevant intergovernmental organizations that have the competence in relation to areas adjacent to the Convention Area or in respect of fisheries resources not covered by the Convention, species belonging to the same ecosystem or dependent upon or associated with the target stocks and that have objectives that are consistent with and supportive of the objective of the Convention.[187]

5.4 *Other Areas*

There are several more areas with no operational RFMO/A competent to deal with DHSFS and/or deep-sea fisheries. These include: areas of the Arctic, the Western and Eastern Central Atlantic, the South-West Atlantic and areas of the Pacific.[188] The rest of this section examines the practice of states and international organizations in these regions with a view to considering the possibility of establishing RFMO/As and, for the South-West Atlantic, the regulations currently in place for vessels engaged in fishing on the high seas.

5.4.1 *Arctic Ocean*

The maritime area north of the Arctic Circle (66°30' north) is bordered by five coastal states, namely, Canada, Denmark (Greenland), Norway, Russia and the United States (hereinafter the 'Arctic Ocean coastal states').[189] A notable characteristic of the central Arctic Ocean is that a large part of the seabed and ocean floor of the high seas area may be considered to be outer continental shelves.[190]

187 Article 21(1)–(2).

188 In addition, there are high seas areas north of the Convention Area of SIOFA (i.e., FAO Statistical Area 04) but, according to a European Commission staff working paper, these areas are not significant in terms of hosting high seas fisheries. Commission Staff Working Document, Accompanying document to the Communication from the Commission to the Council and the European Parliament, Destructive fishing practices in the high seas and the protection of vulnerable deep sea ecosystems, SEC(2007) 1314, 17 October 2007, at p. 3.

189 While a tiny part of the Icelandic coastline is within the Arctic Circle, it is not normally characterized as an Arctic Ocean coastal state. See D.M. Johnston, "The Future of the Arctic Ocean: Competing Domains of International Public Policy", 17 *Ocean Yearbook* (2003), at p. 596 note 3. Finland, Iceland and Sweden have Arctic territories and are thus considered to be Arctic states.

190 See the map of the International Boundary Research Unit, "Maritime jurisdiction and boundaries in the Arctic region", available at <https://www.dur.ac.uk/ibru/resources/arctic/>. On

The existence of vast continental shelves in this area would lead to a potential RFMO/A with competence to regulate almost only non-sedentary species since sedentary species on the continental shelves fall under the sovereign rights of coastal states. It is still to be seen to what extent such an RFMO/A may regulate deep-sea fisheries given the ambiguity in relationships between the jurisdiction of coastal states over the exploitation of sedentary species and the competence of RFMO/As to regulate fisheries employing fishing gear which may affect marine ecosystems, including benthic communities.[191]

At present, almost all of the high seas area of the central part of the Arctic Ocean is covered by ice for most of the year, and the possibility of major fishing taking place in the high seas area is limited.[192] Nevertheless, climate change caused by global warming might alter the situation.[193] It is possible that some marine capture fisheries will take place on the high seas part since climate

---

the implementation of Article 76 in the Arctic, see, generally, A.G. Oude Elferink, "The Continental Shelf in the Polar Regions: Cold War or Black-letter Law?", XL *Netherlands Yearbook of International Law* (2009), at pp. 139–160. Note that the areas of the Arctic Ocean beyond 200 nautical miles from the baselines are considered high seas in spite of the presence of ice. See R.R. Churchill, "Claims to Maritime Zones in the Arctic: Law of the Sea Normality or Polar Peculiarity?" in A.G. Oude Elferink and D.R. Rothwell (eds.), *The Law of the Sea and Polar Maritime Delimitation and Jurisdiction* (2001), at p. 123. Furthermore, note that state practice indicates that the existing legal framework is being, and should be, used for the Arctic as well. For example, the five coastal states bordering on the Arctic Ocean stated in the 2008 Ilulissat Declaration issued at the Arctic Ocean Conference on 27–29 May 2008 that the existing legal framework, including the law of the sea, "provides a solid foundation for responsible management [...] of this Ocean through national implementation and application of relevant provisions" and they "see no need to develop a new comprehensive international legal regime to govern the Arctic Ocean". In the Declaration, the reference to the use of the existing framework for the Arctic Ocean is preceded by an observation that "[b]y virtue of their sovereignty, sovereign rights and jurisdiction in large areas of the Arctic Ocean the five coastal states are in a unique position to address these possibilities and challenges". In the fisheries context, Chair's Summary of the Arctic Ocean Foreign Ministers' Meeting held in Canada in 2010 stipulates that "Arctic Ocean coastal states have a unique interest and role to play in current and future efforts for the conservation and management of fish stocks in this region". "Chair's Summary", Arctic Ocean Foreign Ministers' Meeting, Chelsea, Quebec, 29 March 2010, available at <http://www.international.gc.ca/polar-polaire/arctic-meeting_reunion-arctique-2010_summary_sommaire.aspx?lang=en>.

191 Cf. R. Rayfuse, "Melting Moments: The Future of Polar Oceans Governance in a Warming World", 16 *Review of European Community & International Environmental Law* (2007), at p. 209 (referring to a patchwork of multiple small, discrete pockets of the Area completely surrounded by extended continental shelf as 'managerial wastelands').

192 See Johnston, "The Future of the Arctic Ocean", at pp. 616–617; Molenaar, "Current Legal and Institutional Issues Relating to the Conservation and Management of High Seas Deep-Sea Fisheries", at p. 128.

193 For an account of the potential effects on animal species in general, see ACIA, *Impacts of a Warming Arctic: Arctic Climate Impact Assessment* (2004), at pp. 58–76.

change will result in the reduction of ice and fish population shifts.[194] In particular, areas north of the Atlantic and the Central and Eastern Canadian Arctic and West Greenland are likely to see an increase in fisheries.[195]

The Atlantic part of the central Arctic Ocean, the Norwegian Sea and the Barents Sea are already covered by NEAFC. Under Article 1 of the new NEAFC Convention, high seas parts of the Arctic between 44° west longitude and 51° east longitude are covered by the NEAFC Convention.[196]

Except for the Atlantic side of the Arctic Ocean, there exists no RFMO/A dealing with straddling fish stocks or DHSFS for the high seas part of the area north of the Arctic Circle. Therefore, the following paragraphs examine the current and likely future situations concerning the management of high seas fisheries in this region.

In 1996, the Arctic Council was established to provide a high-level forum to discuss issues relating to the Arctic.[197] It aims to provide "a means for promoting cooperation, coordination and interaction among the Arctic States [...] in particular issues of sustainable development and environmental protection in the Arctic".[198] Any organ of the Arctic Council has not so far undertaken the role of high seas fisheries management. Nevertheless, it should be recalled that the Arctic Council has gradually been involved in policy-making as witnessed by the adoption of the Arctic Search and Rescue Agreement in 2011.[199] It is expected that it will become more active than before in policy-making, including the development of a legally-binding instrument.

---

194 For an account of the potential effects on fisheries in particular, see ibid., at pp. 62–64. See also R. Huebert, "The Law of the Sea and the Arctic: An Unfulfilled Legacy", 18 *Ocean Yearbook* (2004), at p. 200.

195 ACIA, *Impacts of a Warming Arctic: Arctic Climate Impact Assessment*, at pp. 18–19.

196 See also Molenaar, "Current Legal and Institutional Issues Relating to the Conservation and Management of High Seas Deep-Sea Fisheries", at p. 128 note 107 and its accompanying text (arguing that the Convention Area extends to the North Pole and the proposed amendments to the Convention do not change that). The maps of NEAFC documents, which are made for illustrative purposes only, are inconsistent as far as the inclusion of the Arctic Ocean in the Regulatory Area is concerned. One phrase in the NEAFC Convention might qualify the competence of NEAFC with regard to a slender portion in the Arctic Ocean. The second delineation of the scope ("within that part of the Atlantic Ocean north of 59° north latitude and between 44° west longitude and 42° west longitude") does not refer to the Arctic Ocean as opposed to the former delineation which refers to both the Atlantic and Arctic Oceans. Thus, it may be argued that the Arctic part of the area between 44° west longitude and 42° west longitude (i.e., north of Greenland) is not included within the scope of the NEAFC Convention.

197 Declaration on the Establishment of the Arctic Council, Ottawa, 19 September 1996.

198 Ibid., Article 1(a).

199 Agreement on Cooperation on Aeronautical and Maritime Search and Rescue in the Arctic, Nuuk, Greenland, 12 May 2011.

One interesting initiative which is relevant to the issue of fisheries management is the ongoing Arctic Ocean Review (AOR) project within the framework of the Protection of the Arctic Marine Environment Working Group (PAME). Following the first phase from 2009 to 2011, the second phase of the AOR project started in 2011 and the final report is expected to be presented to the 2013 Ministerial Meeting. The key objectives of the AOR project include analyzing "global and regional protection measures for environmental, economic and social/cultural outcomes" and developing "a report addressing protection measures for the Arctic marine and coastal environment, including possible impediments to achieving a regionally integrated, sustainable development approach to the Arctic Ocean in light of current and emerging trends".[200] Phase II of the AOR aims, *inter alia*, to "[a]nalyze potential opportunities in global and regional instruments and measures to achieving environmental, economic and socio-cultural outcomes" and to "[o]utline options to address potential opportunities to strengthen the conservation and sustainable use of the Arctic marine environment".[201] Like the recommendations of the Arctic Marine Shipping Assessment report, which are currently contemplated in various bodies of the Arctic Council, the outcome of the AOR project is expected to give guidance to the future policy options for Arctic states on matters within and beyond the Arctic Council.

The United States Senate launched an initiative by passing a resolution to recommend the executive to initiate negotiations for the international management of Arctic fisheries.[202] Among other things, it stipulated that the US should take the necessary steps with other *Arctic nations* to negotiate an agreement or agreements for managing migratory, transboundary and straddling fish stocks in the Arctic Ocean and to establish a new international fisheries management organization or organizations for the region;[203] that the above-mentioned agreement or agreements to be negotiated should conform to the requirements of the FSA;[204] and that until the agreement or agreements come into force and measures consistent with the FSA are in effect, the US should support international efforts

---

200 Arctic Council, Senior Arctic Official (SAO) Report to Ministers, Tromsø, Norway, April 2009, at p. 59.

201 "Project Plan for AOR Phase II", in PAME, "PAME Work Plan 2011–2013", at p. 26. See also the "Scope and Approach" section at pp. 26–27 and the "Main components, Timeline and Major Milestones" section at p. 27 ("Develop new mechanisms as appropriate").

202 The 2007 Senate Joint Resolution 17, directing the United States to initiate international discussions and take necessary steps with other nations to negotiate an agreement for managing migratory and transboundary fish stocks in the Arctic Ocean, having been passed in the House in May 2008, was singed by President Bush into law on 4 June 2008.

203 2007 Senate Joint Resolution 17, para. 1.

204 Ibid., para. 2.

to halt the expansion of commercial fishing activities in the high seas of the Arctic Ocean.[205]

The US Presidential Directive states that the US supports the application in the Arctic region of the general principles of international fisheries management outlined in the FSA and similar instruments and that US endorses the protection of VMEs in the Arctic from destructive fishing practices. It commands the relevant departments and agencies to (1) continue to identify ways to conserve, protect and sustainably manage Arctic species and, for species whose range includes areas both within and beyond US jurisdiction, the US shall continue to collaborate with other governments to ensure effective conservation and management and (2) seek to develop ways to address changing and expanding commercial fisheries in the Arctic, including through the consideration of international agreements or organizations to govern future Arctic fisheries.[206]

The initiative to negotiate a new RFMO/A has received mixed reactions. On the one hand, when the Senior Arctic Official of the United States brought the issue to the attention of the Senior Arctic Officials in the Arctic Council in November 2007, the participants at the meeting expressed strong support for building on and considering this issue within the context of existing mechanisms.[207] On the other hand, in recent years, support for a new RFMO/A has been expressed by certain Arctic Ocean coastal states and the EU.[208]

Apart from an RFMO/A, it should be noted that there was a call by scientists for a moratorium on commercial fisheries in the Central Arctic Ocean, which was initiated by the Pew Environment.[209] The call was addressed to the leaders of the five Arctic Ocean coastal states.

If the process of negotiations for an Arctic RFMO (or arrangement) starts in multilateral settings, the geographical scope of the proposed regime will be a critical issue: high seas only or also the EEZ?; the entire Arctic Ocean or a part thereof (e.g., separate negotiations for the Atlantic part and the part north of Alaska)? And, will the negotiations be limited to Arctic states only or open to all interested states? The choice of the geographical scope would determine the relevant coastal and high seas fishing states and, eventually, influence the substance of the proposed regime. If the proposed management regime only intends to regulate high seas fishing on the Atlantic side, one may argue that

205 Ibid., para. 4

206 National Security Presidential Directive and Homeland Security Presidential Directive, 9 January 2009, NSPD-66 and HSPD-25, Section III.H.

207 Meeting of Senior Arctic Officials, Final Report, 28–29 November 2007, Narvik, Norway, at p. 12.

208 E.J. Molenaar, "Arctic Fisheries and International Law: Gaps and Options to Address Them", 1/2012 *Carbon and Climate Law Review* (2012), at p. 73.

209 Pew Charitable Trusts, "Leaders Urged to Impose Moratorium on Industrial Fishing in the Arctic", <http://www.pewtrusts.org/news_room_detail.aspx?id=85899382339>.

it is sufficient to activate NEAFC to perform its management function in that area. On the other hand, if one attempts to fill a coverage gap by addressing any potential fisheries, a new RFMO/A needs to be negotiated with the geographical coverage of either the entire high seas part of the central Arctic Ocean or that part excluding the NEAFC Regulatory Area in the Arctic.[210]

5.4.2 *Central Atlantic Ocean*

In the Central Atlantic, there is no regional fishery body with a management mandate with respect to straddling stocks or DHSFS. On the other hand, there are several regional fishery bodies with an advisory mandate, including the Fishery Committee for the Eastern Central Atlantic (CECAF) and the Western Central Atlantic Fishery Commission (WECAFC). These two bodies have for many years discussed the potential to transform them into RFMOs.

CECAF has discussed possibilities for upgrading it to a management body under Article XIV of the FAO Constitution, but so far has decided not to do so.[211] During this period, the Committee reviewed the Terms of Reference of its Statutes in 2002 and endorsed the revision.[212] The new version in its preamble provides that the purpose of the Committee shall be to promote the sustainable utilization of the living marine resources of the area by the proper management and development of the fisheries and fishing operations, and to this end its functions include "to promote, encourage and coordinate research in the area related

---

210 In the first option, the proposed mechanism needs to coordinate fisheries management in the overlapping area.

211 For the history of discussions within CECAF, see Report of the Fifteenth Session of the Fishery Committee for the Eastern Central Atlantic, Abuja, Nigeria, 1–3 November 2000, FAO Fisheries Report No. 642, RAFI/R642, at pp. 4–5, paras. 27–34; Report of the Technical Consultation on the Future of the Fishery Committee for the Eastern Central Atlantic (CECAF), Lagos, Nigeria, 27–30 November 2001, FAO Fisheries Report No. 687, RAFI/R687, at p. 6, paras. 32–33; Report of the Sixteenth Session of the Fishery Committee for the Eastern Central Atlantic, Santa Cruz de Tenerife, Spain, 22–24 October 2002, FAO Fisheries Report No. 693, RAFI/R693, at pp. 6–7, paras. 37–38 and 40–41; Report of the Seventeenth Session of the Fishery Committee for the Eastern Central Atlantic, Dakar, Senegal, 24–27 May 2004, FAO Fisheries Report No. 754, RAFI/R754, at p. 7, para. 44; Report of the Eighteenth Session of the Fishery Committee for the Eastern Central Atlantic, Douala, Cameroon, 3–5 October 2006, FAO Fisheries Report No. 828, RAFI/R828, at p. 8, para. 53. In the course of the discussions, the Committee noted that an option to establish a new body competent for the high seas and to continue with CECAF competence limited to areas under national jurisdiction could constitute a general orientation of work for long-term arrangements, although it considered that that option would not be an option for immediate implementation. FAO Fisheries Report No. 693, at pp. 6–7, para. 37.

212 FAO Fisheries Report No. 693, at p. 7, para. 42. The terms of reference were approved by the 124th session of the FAO Council in June 2003.

to the living resources [...] and to draw up programmes required for this purpose and to organize such research as may appear necessary".[213]

In 2004, the Committee adopted a resolution calling on all members of CECAF and non-members fishing in the high seas in the CECAF region for species other than tuna to provide reports on their catches to CECAF.[214] The resolution refers to ensuring the long-term conservation and sustainable use of all living resources and safeguarding the environment and marine ecosystems in which the resources occur, and implementing the precautionary approach in the management of fishery resources; it recognizes, in particular, the fragility of the ecosystems of seamounts on which exploratory fisheries have taken place; it calls on members of the Committee fishing for species other than tuna in the high seas to report on their fishing activities and on the state of the fishery stocks on which their activities are based; it also invites non-members to submit the same information on a voluntary basis.

Interestingly, an independent performance review of CECAF, which was conducted during 2011, recommended, among other things, that CECAF should restrict its area of competence to the EEZ of its members.[215] Having examined discussions in the process for the establishment of the SWIOFC and SIOFA, it states that:

> The present statutes of CECAF, as an Article VI, FAO body, do not allow it to have enforcement (regulatory) powers, which are essential for the management of high-seas stocks. By restricting its mandate to the EEZ of member States, CECAF not only would have its actual focus formally institutionalized, but would also open the way for another RFMO, with an appropriate institutional and legal framework, to be created in the area. One possible alternative would be for SIOFO [*sic*] to expand its area of competence to the north.[216]

This recommendation was discussed during the Twentieth Session of CECAF in March 2012. The meeting "considered problematic the possible restriction of the Convention Area to the EEZ of coastal member countries, requiring further

213 For the new version, see Appendix E of the Report of the 124th FAO Council. For the old version, see Options for Future Arrangements for Cooperation in Fisheries Management in the Area of Competence of the Fishery Committee for the Eastern Central Atlantic, CECAF/XVI/2002/7, at p. 22. A need for an update of the Statutes in order to incorporate modern concepts of fisheries management such as the ecosystem approach and the precautionary approach has been suggested by the CECAF Performance Review and subsequently agreed on by the 20th session of CECAF in 2012. "CECAF Performance Review", CECAF/XX/2012/5, at pp. 19–21; "Summary of Recommendations of Regional Fishery Bodies Established within the FAO Framework", COFI/2012/SBD.4, at p. 7.

214 FAO Fisheries Report No. 754, at p. 7, paras. 44–45 and p. 45 (Appendix F).

215 See CECAF/XX/2012/5, at pp. 9–10.

216 Ibid. The word 'SIOFO' may be a typo intended to be 'SEAFO'.

analysis and discussion".[217] Therefore, it is not clear whether negotiations for an RFMO with the competence to regulate high seas fisheries in this region will be initiated.[218]

In the Western Central Atlantic, the geographical area of competence of WECAFC is partly covered by the Caribbean Regional Fisheries Mechanism (CRFM) within the context of the Caribbean Community (CARICOM). On the one hand, WECAFC is composed of Members and Associate Members of the FAO that are coastal states in that region or non-coastal states whose vessels engage in fishing in the Commission's area of competence.[219] The area of competence of WECAFC covers the entire Western Central Atlantic, as provided in its Statutes and the Commission covers all living marine resources in the area.[220] On the other hand, the CRFM is primarily concerned with EEZ and transboundary aquatic resources in the Caribbean region. The objectives of the Mechanism are, *inter alia*, the efficient management and sustainable development of marine and other aquatic resources within the jurisdictions of Member States and the promotion and establishment of co-operative arrangements among interested states for the efficient management of shared, straddling or highly migratory marine and other aquatic resources.[221] The representatives of the CRFM Secretariat attended meetings of WECAFC and the contribution of the CRFM to the process of strengthening WECAFC was acknowledged by the UNGA.[222]

The poor state of many resources in the Western Central Atlantic led WECAFC to discuss its transformation to an FAO Article XIV body (an FAO body with a management mandate), but WECAFC decided to remain an advisory body.[223]

---

217 COFI/2012/SBD.4, at p. 7.

218 As noted in Chapter 2, the European Commission proposed a Regulation for the management of deep-sea fisheries in the North-east Atlantic in July 2012, which will also apply to international waters in the CECAF areas 34.1.1, 34.1.2 and 34.2 in addition to the Regulatory Area of the NEAFC. Proposal for a Regulation of the European Parliament and of the Council establishing specific conditions to fishing for deep-sea stocks in the North-East Atlantic and provisions for fishing in international waters of the North-East Atlantic and repealing Regulation (EC) No 2347/2002, 19 July 2012, COM(2012) 371 final, 2012/0179 (COD), Article 2.

219 Revised Statutes of the Western Central Atlantic Fishery Commission, Rome, 20–25 November 2006, Adopted at the Hundred and Thirty-first Session of the Council of the FAO, Article 5.

220 Ibid., Articles 3–4.

221 Agreement establishing the Caribbean Regional Fisheries Mechanism, Belize City, Belize, 4 February 2002, Article 4. See also the information available at the website of the CRFM, "About CRFM" <http://www.caricom-fisheries.com/main/about_us.asp>.

222 UNGA Resolution 59/25, para. 55.

223 For the history of the discussions, see Restructuring of WECAFC: Options and Implications, WECAFC/IX/99/5, at para. 33; Report of the Ninth Session of the Western Central Atlantic Fishery Commission and of the Sixth Session of the Committee for the Development and Management of Fisheries in the Lesser Antilles, Castries, Saint Lucia, 27–30 September 1999, FAO Fisheries Report No. 612, SLAC/R612, at p. 8, paras. 46–47; Report of the Tenth Session of the

Instead, the Commission adopted a "recommendation on strengthening WECAFC as an organization" and recommended the establishment of an Intersessional Working Group in 2003.[224] In response to the work of the Working Group, members of the Commission have cooperated to strengthen it through the revision of its statutes. The revised Statutes were subsequently adopted by the Council of the FAO in November 2006. The general objective of the Commission includes promoting effective conservation, management and the development of the living marine resources of the area of competence of the Commission, in accordance with the FAO Code of Conduct.[225] To this effect, the Statutes, in their general principles, provide that the Commission shall have due regard for and promote the application of the provisions of the FAO Code of Conduct and its related instruments, including the precautionary approach and the ecosystem approach to fisheries management.[226]

Its revised Statutes now explicitly state that the Commission covers all living marine resources.[227] While no reference is made to DHSFS as opposed to transboundary and straddling stocks (whose conservation, management and development under Members' respective national jurisdictions is explicitly mentioned), this does not necessarily imply that discrete high seas stocks are excluded from the mandate of the Commission.

The Commission is to contribute, among other things, to improved governance through institutional arrangements that encourage cooperation amongst Members.[228] The Commission may, as appropriate, organize or undertake research related to the living marine resources in the Commission's area of competence, including on the interactions between fisheries and the ecosystem, and design

Western Central Atlantic Fishery Commission and of the Seventh Session of the Committee for the Development and Management of Fisheries in the Lesser Antilles, Bridgetwon, Barbados, 24–27 October 2001, FAO Fisheries Report No. 660, SLAC/R660, at p. 11, para. 70; Report of the Eleventh Session of the Western Central Atlantic Fishery Commission and of the Eighth Session of the Committee for the Development and Management of Fisheries in the Lesser Antilles, St George's, Grenada, 21–24 October 2003, FAO Fisheries Report No. 725, SLAC/R725, at p. 12, paras. 64–67; Report of the First Meeting of the Intersessional Working Group on the Strengthening of WECAFC, Punt Cana, Dominican Republic, 12–14 July 2005, WECAFC/XII/05/11, at pp. 2–3, paras. 9–12; Report of the Twelfth Session of the Western Central Atlantic Fishery Commission and of the Ninth Session of the Committee for the Development and Management of Fisheries in the Lesser Antilles, Port of Spain, Trinidad and Tobago, 25–28 October 2005, FAO Fisheries Report No. 788, at p. 14, para. 75.

224 Report of the Eleventh Session of the Western Central Atlantic Fishery Commission and of the Eighth Session of the Committee for the Development and Management of Fisheries in the Lesser Antilles, St George's, Grenada, 21–24 October 2003, at pp. 12 and 70–71 (Appendix D).

225 WECAFC Revised Statutes, Article 1.

226 Ibid., Article 2(a).

227 Ibid., Article 4.

228 Ibid., Article 6(a).

programmes required for this purpose and undertake the collection, exchange and dissemination of statistical, biological, environmental and socio-economic data and other marine fishery information as well as its analysis or study.[229] The Commission is to provide the necessary support and advice to enable Members to ensure that fishery management decisions are based on the best scientific evidence available, and to provide advice on management measures to Member Governments and competent fisheries organizations.[230] The Commission is to promote and encourage the utilization of, *inter alia*, the most appropriate gear and fishing techniques in accordance with the FAO Code of Conduct.[231]

WECAFC, at its Fourteenth session, held discussions over the role of WECAFC in the management of high seas fisheries. In preparation for the meeting, a report was produced.[232] It recommended priorities to be considered for the implementation of the FAO International Guidelines on Dee-sea Fisheries such as:

> a. identify the role(s) WECAFC could play, if any, to assist and facilitate implementation of the UNGA Resolutions and the FAO Guidelines, [...]
> b. as a next step, and in response to calls by the UNGA and the Guidelines, consider transforming WECAFC to an Article XIV body with a management mandate to govern deep-sea fishing in the WECAFC Area;
> c. for Members with vessels that engage in deep-sea fishing in the WECAFC area, ensure implementation of the UNGA Resolutions and FAO Guidelines, and in particular the management and conservation components of the Guidelines; [...][233]

The Commission was invited to "[e]laborate on the role of WECAFC in high seas fisheries management and deep sea fisheries in the high seas in particular".[234] The draft resolution, which the Commission was invited to review and finalize, read as follows:

> AGREE TO ESTABLISH a WECAFC subsidiary body on the management of deep sea fisheries [in the high seas], to guide the management of such fisheries by WECAFC members in such a manner as to [...] promote responsible fisheries that provide economic opportunities while ensuring the conservation of marine living resources and the protection of marine biodiversity and to facilitate the implementation of the FAO International Guidelines for the Management of Deep-sea Fisheries in the High Seas.[235]

229 Ibid., Article 6(e)–(f).
230 Ibid., Article 6(g)–(h).
231 Ibid., Article 6(k).
232 "Support to the implementation of international fisheries instruments", WECAFC/XIV/2012/6, February 2012.
233 Ibid., at p. 49.
234 Ibid., at p. 58.
235 "Draft resolution of the members of the Western Central Atlantic Fishery Commission on strengthening the implementation of international fisheries instruments", WECAFC/XIV/2012/7, February 2012 (the first set of square brackets in original).

The Commission issued a resolution on strengthening the implementation of international fisheries instruments in the Caribbean region.[236] Furthermore, it has been noted that "[i]n the coming years a decision will need to be taken on maintaining WECAFC as a FAO Article VI body, restructuring it to an Article XIV body, or abolishment, as the commission was not able to carry out its functions properly in the last decade".[237]

If CECAF and WECAFC seek to establish a management body or bodies for the high seas in the future, factors to be taken into consideration include: the area of coverage (i.e., one for the entire Central Atlantic or one for each region (east and west)); the form (i.e., RFMO or arrangement); the relationship with the existing bodies; interaction with areas under national jurisdiction (i.e., whether only high seas or both the high seas and areas under national jurisdiction); and the relationship with the FAO (i.e., Article XIV body or autonomous body).

One factor in this consideration is the high seas area of the Caribbean Sea. It was recognized in WECAFC that the Caribbean Sea can be considered as a semi-enclosed sea and that countries in the region "have a common interest in protecting the ecosystem, conserving its resources and managing its use".[238] Moreover, the overlap of the areas of competence between WECAFC and the CRFM may lead to a division of work between them.

With regard to the institutional design of the new RFMO/A, if the states concerned decide to exclude areas under national jurisdiction from the scope of the new RFMO/A(s), the relatively low amount of existing fisheries for straddling fish stocks in the high seas area of the Central Atlantic would be one of the factors in the consideration of balancing the possible financial burdens and potential interests in establishing an organization exclusively for the high seas. As Sydnes commented in the context of SEAFO, although the lack of vested interests might facilitate the process, "[i]t is doubtful that States will [...] commit resources to [an] RFMO managing "paper fish", without any benefits in return".[239]

5.4.3 *South-West Atlantic Ocean*

In the high seas area of the South-West Atlantic, EU (Spanish), Korean, Chinese and Taiwanese vessels have been engaged in high seas deep-sea fisheries.[240] Currently, there is neither an RFMO/A nor an advisory body competent to deal with

236 COFI/2012/SBD.4, at p. 14.

237 Ibid.

238 FAO Fisheries Report No. 612, at p. 4, para. 22.

239 Sydnes, "New Regional Fisheries Management Regimes: Establishing the South East Atlantic Fisheries Organisation", at p. 361.

240 Based on Commission Staff Working Document, Accompanying document to the Communication from the Commission to the Council and the European Parliament, Destructive fishing practices in the high seas and the protection of vulnerable deep sea ecosystems, at p. 8.

straddling fish stocks or DHSFS in the region. Until 1997, there was a regional fisheries advisory body under Article VI of the FAO Constitution. However, the Regional Fisheries Advisory Commission for the Southwest Atlantic (CARPAS), which last met in 1974, was abolished by FAO Conference Resolution 13/97. Membership of CARPAS was limited to FAO Members which belong to the American Continent and have coasts bordering on the Western Atlantic Ocean south of the equator. In fact, the only members were Argentina, Brazil and Uruguay.[241] At the time of abolishment, the area of competence of CARPAS partly overlapped with those of CCAMLR and WECAFC.[242]

The Falklands/Malvinas conflict between Argentina and the UK is a factor that needs to be taken into account in any attempt to establish an RFMO/A in the high seas area of this region.[243] The UK and Argentina agreed to cooperate on fisheries management in areas surrounding the islands: they established the Falklands Outer Conservation Zone (FOCZ) on 28 November 1990, starting from the outer limit of the 150-mile Falkland Islands Interim Conservation and Management Zone (FICZ) and extending up to 200 miles from the baseline of the islands in the north, east and south of the islands. In the FOCZ, the two states were to cooperate through the establishment of the South Atlantic Fisheries Commission (SAFC) and the temporary total prohibition of commercial fishing by vessels of any flag in the maritime area defined in the Annex to the Joint Statement for conservation purposes.[244] There had been 27 meetings of the SAFC and 23 meetings of its scientific sub-committee by mid-2005; but, since late 2005, a number of meetings of the SAFC have been suspended.[245]

Although the Argentina-UK Joint Statement of July 1999 gave an impetus to the negotiation of a multilateral fisheries arrangement to extend conserva-

241 S.H. Marashi, The Role of FAO Regional Fishery Bodies in the Conservation and Management of Fisheries, Fisheries Circular No. 916, FIPL/C916, at section IV(6).

242 Molenaar, "Addressing Regulatory Gaps in High Seas Fisheries", at p. 541.

243 The above-mentioned Working Paper, at p. 8, notes that two out of three main harvesting areas are located in international waters bordering the Argentine EEZ. For an overview of the conflict in relation to fisheries, see G.A. Bisbal, "Fisheries Management on the Patagonian Shelf: A Decade After the 1982 Falklands/Malvinas Conflict", 17 *Marine Policy* (1993), at pp. 213–229.

244 Joint Statement on the Conservation of Fisheries between the Government of the Argentine Republic and the Government of the United Kingdom of Great Britain and Northern Ireland, London and Buenos Aires, 28 November 1990, para. 2. In the same paragraph, the two Governments further agreed to review the Joint Statement annually.

245 Information is available at the website of the Falkland Islands Government, "International Relations", <http://www.falklands.gov.fk/international-relations.php>. It appears that the UK and Argentine governments have maintained contacts with regard to fisheries management in this area. Merco Press, "'Old fishing law needed updating', says Falkland's Fisheries Director" (19 March 2007), available at <http://en.mercopress.com/2007/03/19/old-fishing-law-needed-updating-says-falklands-fisheries-director>.

tion measures to the high seas, there has been little progress on this issue. Any attempt to establish an RFMO with the participation of all coastal and fishing states may encounter a revived territorial dispute since Argentina would oppose the involvement of the UK as a coastal state. At the sixth Informal Consultations of States Parties to the FSA in 2007, the Republic of Korea expressed an interest in the development of a new RFMO for the South-West Atlantic.[246] To date, nevertheless, no formal negotiations have been initiated to this end.[247]

The EU (Spanish) vessels are still engaged in high seas fishing in the high seas areas of the South-West Atlantic.[248] As the establishment of a new RFMO/A is unlikely in the near future, it is important to know what regulations flag states have put in place for vessels flying their flag. In this respect, the EU enacted a Regulation aimed at the regulation of bottom fisheries in high seas areas where there is no RFMO/A or interim measure in place. The Regulation is primarily intended to address fisheries in high seas areas of the South-West Atlantic.[249] While it is intended to implement UNGA Resolution 61/105, some of the provisions of the Regulation may be seen as going beyond what was agreed in Resolution 61/105.

In 2007, the European Commission put forward a proposal for a Council Regulation on the protection of VMEs in the high seas from the adverse impacts of bottom fishing gear, among others, to implement UNGA Resolution 61/105.[250] The resultant Regulation applies to Community fishing vessels carrying out

---

246 *ENB*, vol. 7, No. 62, at p. 3.

247 The European Union regulation, discussed below, implies that any multilateral management mechanism involving both coastal and high seas fishing states is not forthcoming. Also note that one European Commission document refers to the requirement of a specific licence to fish in the high seas of the region as part of the licences delivered by the Falkland Islands Government under which Spanish bottom trawling vessels operate within the FICZ and in the high seas. Commission Staff Working Document, Accompanying document to the Communication from the Commission to the Council and the European Parliament, Destructive fishing practices in the high seas and the protection of vulnerable deep sea ecosystems, at p. 8. To the contrary, see Schiffman, *Marine Conservation Agreements: The Law and Policy of Reservations and Vetoes*, at p. 19 (commenting that a new body is under negotiation for the South-West Atlantic).

248 See European Association of Fish Producers Organisations, "Observations by the EAPO Regarding the Proposed Regulation by the Council for the Protection of Vulnerable Marine Ecosystems on the High Seas Due to the Adverse Effects of the Use of Bottom Catch Methods (COM 2007) 605 Final" (28 January 2008), available at http://eapo.com/index2.php?page=publicaties&lang=eng.

249 See, e.g., Communication from the Commission to the Council and the European Parliament, Destructive fishing practices in the high seas and the protection of vulnerable deep sea ecosystems, at p. 10.

250 EU, Proposal for a Council Regulation on the protection of vulnerable marine ecosystems in the high seas from the adverse impacts of bottom fishing gears, 18 June 2007, COM(2007) 330 final, 2007/0114 (CNS).

fishing activities with bottom gear in the high seas where there is neither an RFMO/A nor interim measures to protect VMEs from the destructive impacts resulting from the use of bottom gear.[251] Therefore, it does not apply to bottom fisheries in areas covered by RFMOs (the North Atlantic, the South-East Atlantic, the Mediterranean and the Southern Ocean) or by multilaterally-agreed interim measures (the South Pacific and the North-West Pacific). First of all, in order to conduct the fishing activities concerned, Community fishing vessels must have a special fishing permit.[252]

Second, the Regulation intends to make bottom fisheries conditional upon prior environmental impact assessments, as other marine resource exploitation activities.[253] It does so by requiring the submission of detailed fishing plans specifying, *inter alia*, the intended location of the activities, the depth of the gear being deployed and the configuration of the bathymetric profile of the seabed in the intended fishing grounds[254] and by subjecting the issuance of the special fishing permit to an assessment on the potential impacts of the proposed fishing activities on VMEs.[255] Where the assessment indicates SAIs to VMEs, the competent authorities shall specify the assessed risks and allow applicants to amend the fishing plan to avoid them. In the absence of such amendments, they shall refrain from issuing the permit.[256] The competent authorities of Member States shall rely on the best scientific information available concerning the location of VMEs, including scientific data on the basis of which the likelihood of the occurrence of such ecosystems can be estimated.[257] The competent authorities shall apply precautionary criteria and, in case of doubt, they shall consider that the likely adverse impacts resulting from the scientific advice provided are significant.[258] The main effect of these clauses is that the applicant needs to persuade the competent authorities that the proposed fishing activities are very unlikely to result in SAIs on (identified or estimated) VMEs.

It is recalled that the collection of information on VMEs, including their identification, can be pursued both by conducting independent scientific research

---

251 EU, Council Regulation (EC) No. 734/2008 of 15 July 2008 on the protection of vulnerable marine ecosystems in the high seas from the adverse impacts of bottom fishing gears, Article 1.

252 Ibid., Article 3(1).

253 Communication from the Commission to the Council and the European Parliament, Destructive fishing practices in the high seas and the protection of vulnerable deep sea ecosystems, at p. 6.

254 EU, Council Regulation (EC) No. 734/2008, Article 4(1).

255 Ibid., Article 4(2). See also ibid., Article 4(4). The Regulation defines, for the purpose of the regulation, "vulnerable marine ecosystem" in Article 2(b).

256 Ibid., Article 4(6).

257 Ibid., Article 4(3).

258 Ibid., Article 4(5).

and by collecting data from individual fisheries.[259] Whereas the Regulation does not attempt to encourage improved scientific research in itself,[260] it seeks to achieve that individual fishing activities contribute to data collection with regard to scientific observers.[261]

Third, in the case of unforeseen encounters with VMEs, the Regulation allows the resumption of operations only when the vessel concerned has reached an alternative site at a minimum distance of five miles from the site of the encounter[262] and, if a new encounter takes place in the alternative site, the vessel shall keep relocating until a site where no VMEs are found is reached.[263]

Fourth, on the basis of the best scientific information available on the occurrence or on the likelihood of the occurrence of VMEs, Member States shall identify areas to be closed to bottom fishing.[264] The Commission shall submit, where appropriate, proposals to the Council for the adoption of Community measures to protect these areas on the basis of an examination of the information contained in the report submitted by Member States in accordance with Article 12 or it shall consider such measures on its own initiative.[265]

In addition, the Commission proposal for the Regulation provided for ancillary requirements that were not included in the relevant UNGA Resolution. Among others, the use of bottom gear at depths beyond 1000 m would have been prohibited under the draft Regulation.[266] The same provision was already in place in Council Regulation 1967/2006 concerning Mediterranean fisheries and GFCM Recommendation 2005/1. The Commission considered that the expansion of the same obligation into all oceans would be appropriate. However, this provision is not retained in the adopted Regulation.

---

259 Communication from the Commission to the Council and the European Parliament, Destructive fishing practices in the high seas and the protection of vulnerable deep sea ecosystems, at p. 6.

260 Note that the European Union has already significantly contributed to the improvement of scientific research in deep-sea ecosystems. Ibid., at p. 4 note 2 and its accompanying text.

261 EU, Council Regulation (EC) No. 734/2008, Article 11(1)–(2).

262 Ibid., Article 7(1).

263 Ibid., Article 7(2).

264 Ibid., Article 8(1).

265 Ibid., Article 8(2).

266 EU, Proposal for a Council Regulation on the protection of vulnerable marine ecosystems in the high seas from the adverse impacts of bottom fishing gears, Article 6. This provision was opposed by the European Parliament, which questioned the adequacy of this particular depth. See Committee on Fisheries, Report on the proposal for a Council regulation on the protection of vulnerable marine ecosystems in the high seas from the adverse impacts of bottom fishing gears (COM(2007)0605 – C6-0453/2007–2007/0224(CNS) A6-0183/2008, 14.5.2008, at p. 11 (Amendment 14).

The status of the implementation of the Regulation was reported by the European Commission in November 2010.[267] Noting that Spain restricted the area where bottom fishing activities could be carried out by its vessels to two specified areas in the South West Atlantic where such activities had been carried out for at least two years, the report notes that an 'ex-ante' evaluation of the potential impact of bottom fisheries was not implemented by Spain and a more stringent application of the precautionary approach should have been implemented.[268]

5.4.4 *Certain Areas in the Pacific Ocean*

As the RFMOs to be established for the South Pacific and the North Pacific that have the competence to manage straddling fish stocks in the high seas and DHSFS do not cover the entire Pacific, there will remain a regulatory gap in the coverage of RFMO/As in several areas of the Pacific Ocean. Although, currently, there is no initiative to establish an additional RFMO/A in the Pacific Ocean, two issues relating to the existing regional organizations in this region should briefly be mentioned here.

With respect to bottom trawling, in 2006 the Pacific Islands Forum called upon its members:

> to take actions consistent with international law to prevent destructive fishing practices on seamounts in the Western Tropical Pacific Islands Area (WTPIA) and to prevent destructive fishing practices in other areas of high seas in the WTPIA until an appropriate environmental impact assessment has been carried out, effective conservation and management measures are implemented to protect affected ecosystems, and effective monitoring, control, surveillance and enforcement measures are in place to measure to ensure [*sic*] that the measures are properly implemented and adhered to.[269]

This commitment falls short of a call for a moratorium on bottom trawling in the high seas areas. The subsequent paragraphs refer to a moratorium by calling upon the members to *advocate* (rather than to take actions) for an interim prohibition on destructive fishing practices, including bottom trawling in areas

---

267 Report from the Commission to the European Parliament and the Council on the implementation of Council Regulation (EC) No 734/2008 on the protection of vulnerable marine ecosystems in the high seas from the adverse impacts of bottom fishing gears, 10 November 2010, COM(2010) 651 final. The report contains proposed amendments to the Regulation concerning, *inter alia*, the expansion of the scope of the Regulation, introduction of capacity or effort limitation, and lower thresholds for VME encounter and larger move-on distances.

268 Ibid., at p. 6.

269 Declaration on Deep-sea Bottom Trawling to Protect Biodiversity in the High Seas, para. 13. In 2007, the Pacific Islands Forum reaffirmed the 2006 Declaration and welcomed UNGA Resolution 61/105. See Vava'u Declaration on Pacific Fisheries Resources, Nuku'alofa, Tonga, 16–17 October 2007.

beyond national jurisdiction until appropriate conservation and management measures are in place.[270]

The WCPFC indicated in its reply to a questionnaire by the DOALOS that although it had no direct mandate to regulate bottom fisheries, some issues could be addressed through the general power given to the WCPFC and its members to protect marine biodiversity in the Convention Area and to protect dependent and associated species.[271] Furthermore, the WCPFC has the potential competence to regulate non-tuna fisheries. The Convention defines 'highly migratory fish stocks' as "all fish stocks of the species listed in Annex 1 of the [LOSC] occurring in the Convention Area, *and such other species of fish as the Commission may determine*" (emphasis added).[272] If the Commission adds another species in this regard, it needs to consider how to avoid overlaps between the conservation and management measures of the WCPFC and those of other RFMOs in the Pacific.

5.5 *Concluding Remarks*

This chapter has examined state practice in regions not dealt with in Chapter 4. The first three sections dealt with regions where a regional fisheries management arrangement or the constitutive instrument of an RFMO has been concluded; the fourth section dealt with other regions, where no formal negotiations to establish an RFMO/A are ongoing.

First of all, SIOFA was negotiated and concluded in order to address deep-sea fisheries, keeping in mind that many of the target stocks in the region may be DHSFS. SIOFA is furthermore different from the mechanisms examined in Chapter 4: it is to manage fisheries through meetings between contracting parties, rather than establishing an RFMO. Notwithstanding these characteristics, the substantive, as opposed to institutional, provisions of SIOFA are not dissimilar to the RFMOs examined in Chapter 4. This characteristic might suggest that future negotiations for establishing RFMO/As in other regions need not be specially tailored to deep-sea fisheries even if the proposed mechanism aims to address deep-sea fisheries: all-encompassing mechanisms filling regulatory or institutional gaps in fisheries management in each region would be a better course of action. Interim measures, on the other hand, need to address specific

270 Pacific Islands Forum Declaration on Deep-sea Bottom Trawling to Protect Biodiversity in the High Seas, paras. 14–15.

271 A/62/260, at p. 29, para. 92.

272 WCPFC Convention, Article 1(f). If the regulatory scope of the WCPFC is extended to DHSFS, this appears to be contrary to the intention of the drafters of the WCPFC Convention.

threats caused by particular fisheries. A particular characteristic in the SIOFA negotiations was the lack of a cautious attitude before the Agreement was concluded. No multilateral interim measures limiting existing deep-sea fisheries were agreed and the fishery collapsed before the Agreement was adopted.

As regards the South Pacific, it is noted that the existing fisheries management mechanisms were problematic. In one area, problems have been caused by limitations on the effectiveness of the existing management mechanism (e.g., the 2000 South Tasman Rise Arrangement). In another area, no management mechanism exists (e.g., the Louisville Ridge). In still another area, the existing management mechanism lacks legitimacy (e.g., the Galapagos Agreement). In fact, the South Tasman Rise Arrangement is a bilateral arrangement and primarily limited to orange roughy fisheries. This is why coastal states in the region recognize the need for a new, effective and comprehensive mechanism for fisheries management and they launched such an initiative in a transparent way inclusive of other coastal states and distant water fishing nations. The content of the existing sub-regional management mechanisms in the South Pacific had a limited influence on the substance of the SPRFMO Convention text; instead, developments in other regions (e.g., SIOFA and the new NAFO Convention) and at the global level (e.g., UNGA Resolution 61/105) have contributed more to the negotiations.

In the North Pacific, the momentum behind the new initiative in this region was obviously different from the one in the South Pacific. The negotiations have been attempting to address bottom fisheries (for species other than pollock), rather than reforming or replacing the existing regimes. The major states engaged in bottom fisheries in this region are largely limited to the coastal states of the region; the negotiations have been conducted among a limited number of states concerned in a less transparent manner, compared to the negotiations for the SPRFMO.

Both negotiations succeeded in adopting interim measures to address certain high seas fisheries pending the establishment of RFMO/As. Both of them attempt to control bottom fisheries in accordance with UNGA Resolution 61/105, in particular paragraph 83. However, they are different from one another and from the measures adopted within the existing RFMOs examined in Chapter 4.

Both of the interim measures refer to the incorporation of internationally developed criteria in other fora, including the FAO International Guidelines on deep-sea fisheries, and consequently it is likely that these two processes would employ the same criteria. Nevertheless, it is interesting to note that, in the context of the VME encounter protocol, the North Pacific process solely focused on cold-water corals. Indeed, it could be argued that, as threats to VMEs are not the same, different VMEs require different conservation and management measures. For example, while seamounts are susceptible to fisheries, threats to hydrothermal vents mainly come from bioprospecting and the exploitation of mineral resources of the seabed.[273]

The interim measures both in the North Pacific and in the South Pacific attempt to control bottom fisheries by freezing fishing efforts at the current levels and limiting fishing areas to those currently subject to bottom fisheries. However, since these measures lack a precise definition of the current level of fishing effort, it is questionable to what extent the interim measures will be effective in this respect.

Now, the two processes attempt to control both pelagic and bottom fisheries. In the North-West Pacific, the negotiations started with the aim to establish a management mechanism only for bottom fisheries. Whereas it is not clear from the text of the North-West Pacific interim measures if the CBS Convention area is excluded from the scope of these interim measures, the CBS Convention may not live up to the standards required in modern approaches to fisheries such as an ecosystem approach. Nor does it appear that the parties to the CBS Convention are ready to address bottom fisheries through meetings of the parties. The high seas part of the Sea of Okhotsk continues to be governed by an informal regime taking into account Russian interests in fisheries in the sub-region.

Both of these new RFMO constitutive instruments refer to global fisheries instruments such as the LOSC and the FSA, take modern approaches to fisheries such as the precautionary and ecosystem approaches, and address bottom fisheries, including exploratory fisheries and fisheries for scientific purposes, on the high seas. These negotiation tracks, which were remarkably different in terms of inclusiveness and transparency, also produced certain notable differences.

In other regions, there is no formal ongoing negotiation to establish an RFMO/A with competence to regulate fisheries for straddling stocks or discrete high seas stocks. The circumstances in these regions are very different from one another. Therefore, future development depends on the peculiar factors of each region.

In the Arctic Ocean, part of the high seas is not covered by any existing RFMO/A. The existing cooperative mechanisms for the protection of the Arctic environment have hardly addressed the management of high seas fisheries. As global climate change continues, it is not unlikely that commercial fisheries will

[273] J.-P. Lévy, "La Première Décennie de l'Autorité Internationale des Fonds Marins", 109 *Revue Générale de Droit International Public* (2005), at pp. 117–118. In addition, note a conclusion from the 2000 InterRidge Workshop referred to by Glowka: "participants concluded that the management of all hydrothermal vent sites was not only unnecessary but unrealistic. Instead efforts should focus on the most visited sites whether within or beyond the limits of national jurisdiction". Glowka, "Putting Marine Scientific Research on a Sustainable Footing at Hydrothermal Vents", at p. 305.

take place in the Arctic Ocean where ice melts in the summer. The practice of Arctic states shows that they are willing to build on the existing legal framework, including the LOSC. The US Senate has initiated a step to negotiate the establishment of a new RFMO in the central Arctic high seas area in conformity with the FSA. Initiatives for the establishment of one or more RFMO/As in the region need to take account of the existing mechanisms and various kinds of interests, including the geographic coverage of the proposed RFMO. The outcome of the ongoing AOR project is another factor that might impact upon the future of the international regime for the management of high seas fisheries in the Arctic Ocean.

In the Central Atlantic, repeated calls have been made for upgrading fisheries advisory bodies to fisheries management bodies. These initiatives resulted in the revision of their mandates reflecting modern approaches to fisheries management, rather than structural changes. Notwithstanding this, the recent calls to fill regulatory gaps in high seas fisheries certainly strengthened arguments for such a reform, and this momentum may bring about the establishment of one or more RFMO/As dealing with, *inter alia*, deep-sea fisheries in high seas areas of the region. But, before any initiative starts, general agreement on pivotal issues such as the scope of the proposed RFMO/A is essential.

In the South-West Atlantic, there is no regional fishery body for the high seas. Any initiative to involve coastal states in the establishment of a new RFMO which is competent to regulate high seas fisheries is likely to result in a deadlock due to the recurring question of sovereignty over the Falkland (Malvinas) Islands, in view of the current strained relations between Argentina and the UK as witnessed by the suspended operation of the SAFC. Given these circumstances, a feasible course of action to be taken by fishing states is to hold negotiations only between some high seas fishing states themselves with a view to agreeing on an arrangement for their fishing vessels, or to individually adopt regulations for vessels flying their flag. The new EU Regulation shows that the latter option will continue, at least for the time being.

In certain areas of the Pacific Ocean, regulatory gaps were left as a consequence of the determination of the area coverage of the SPRFMO and the North Pacific Fisheries Commission. For these gap areas, negotiations for RFMO/As are not forthcoming while the activities of some regional organizations such as the WCPFC are potentially relevant.

Chapter Six

Conclusions

As stated in the introductory chapter, the objectives of this book are to articulate the general principles of high seas fisheries and their implications for new challenges in the conservation and management of the marine living resources of the high seas, with a view to considering what further action should be taken in addressing the regulatory gaps in high seas fisheries. This study has attempted to answer these research questions from the perspective of international law by looking at treaties and other international instruments, national legislation, state practice at the global, regional and national levels as well as the literature on international law. The first two chapters identified the general principles of high seas fisheries under the current international legal framework and their potential implications for addressing the new challenges in high seas fisheries. Chapter 1 traced the development of the general principles up to the adoption of the LOSC in 1982. Chapter 2 examined the provisions of the LOSC in the light of the practice of states and international organizations after the adoption of the LOSC. The next three chapters examined the practice of states and international organizations in the field of new challenges in high seas fisheries, namely, DHSFS, deep-sea fisheries and the protection of vulnerable marine ecosystems. Chapter 3 examined the relevant practice in international institutions at the global level. Chapter 4 examined the practice of the five operational RFMOs relevant to the subject of the present book (i.e., CCAMLR, the GFCM, SEAFO, NEAFC and NAFO). Chapter 5 investigated state practice in areas where there is no operational RFMO, focusing on the regional fisheries management agreement for the South Indian Ocean as well as RFMOs in the South Pacific and the North Pacific and situations in other areas.

The final chapter draws some conclusions from the findings of the preceding chapters. The first two sections recapitulate and synthesize the content of Chapters 1–5 in the light of the above-mentioned research questions and attempt to articulate suggested actions that the international community may take in addressing new challenges in high seas fisheries. The last section is devoted to some brief general comments.

6.1 *General Principles Applicable to High Seas Fisheries and Their Implications*

Chapters 1 and 2 have examined what principles are applicable to high seas fisheries. As a point of departure, it was noted that the LOSC provides for an overall framework in which all activities in the oceans, including high seas fisheries, shall be conducted. Chapter 1 traced the development of the general principles of high seas fisheries up to the adoption of the LOSC with a view to understanding the content of the general principles. Chapter 2 examined the general principles of high seas fisheries in the light of the drafting history of the LOSC, the case law, subsequent agreements and the practice of contracting parties to the LOSC and other states. On the basis of these two chapters, the present study concludes that there are three general principles applicable to high seas fisheries, namely: (1) freedom of fishing on the high seas; (2) cooperation between states; and (3) the conservation of marine living resources, ecosystems and biological diversity.

First, states may exercise the restricted freedom of fishing on the high seas. This freedom is contrasted with the absolute freedom prevalent in the period up to the Second World War. Before the Second World War, the need to regulate high seas fisheries was not generally recognized except in a handful of special regimes. The freedom of fishing was therefore the single dominant principle of high seas fisheries. The technological developments and aspirations for more control over resources led to a change in the legal regime of high seas fisheries from the 1950s onwards. Although most of the detailed rules of the regime of high seas fisheries discussed in the ILC as well as at the 1958 Geneva Conference concerned fisheries in areas adjacent to the territorial sea and few regimes were in place with regard to fisheries conducted in more distant areas, these developments had considerable impacts on the regime of high seas fisheries in general. Notably, this is reflected in the HSC and the HSFC: the former clarified that the freedom of the high seas is subject to reasonable regard for the interests of other states in their exercise of high seas freedoms and the latter formulated the freedom of high seas fishing as a 'right'. In the judgments in the *Fisheries Jurisdiction* cases, the ICJ explicitly denied the concept of absolute freedom and considered the due regard requirement to be a rule of general international law. Under the LOSC the freedom of fishing, which remains applicable to areas beyond the 200-mile EEZ, is formulated as a right and is circumscribed by the following considerations.

Freedoms of the high seas are subject to restraints of a general nature. As the formulation as a 'right' suggests, an abusive exercise of freedom of fishing on the high seas is prohibited. Furthermore, just as other freedoms of the high seas, the freedom of high seas fishing needs to be exercised with due regard to the exercise of high seas freedoms by other states and rights with respect to activities in the Area. The latter aspect stipulated in Article 87(2) of the LOSC is a novel one: under Articles 87(2) and 147 of the LOSC, freedom of high seas fishing and activities in the Area need to be reconciled. This might be implemented by restricting mining activities in order to protect existing, important fishing activities on a certain scale targeting bottom-dwelling species in the same area, if mining activities render continuing bottom fisheries virtually impossible in a wide area.

The scope of freedom of fishing is determined by the scope of the term 'fishing'. In this regard, related activities such as transshipment, processing and refuelling are often considered to be part of fishing, or fishing-related activities, in the case law of international tribunals as well as in recent international fishery-related instruments. Fishing for scientific purposes, on the other hand, is not necessarily governed by the regime of fishing. Fisheries for scientific purposes are often subject to a regime which is different from that for commercial fisheries, although this does not imply that such fisheries are governed by the regime for MSR. In practical terms, however, a distinction between fisheries for scientific purposes and MSR is difficult to make.[1]

Under Article 116(a)–(c) of the LOSC, freedom of fishing is subject, *inter alia*, to the rights, interests and duties of coastal states as well as to the rules of conservation and management in Part VII section 2 of the LOSC. Arguably, the words 'subject to' could be regarded as depriving flag states of the right to fish on the high seas if their vessels are in serious and repeated breach of the rules for the conservation and management of marine living resources. With regard to the special position of coastal states, the sovereign rights of the coastal state for the purpose of exploring the continental shelf and exploiting its resources allow the coastal state to regulate fisheries in the water column of the high seas superjacent to its continental shelf. The jurisdiction of the coastal state in this matter is restrained by Article 78(2) of the LOSC: it needs to be exercised in a manner which does not infringe or result in unjustifiable interference with the rights and freedoms of other states as provided for in the LOSC. For instance, a blanket ban on fishing for bottom-dwelling species on the high seas above the entire continental shelf of a particular state appears to be incompatible with the LOSC.[2] As coastal states also have the right to regulate and authorize MSR on the continental shelf, distant water fishing states shall provide relevant information if MSR activities are a requisite for high seas fisheries on the outer continental shelf.

The right to fish on the high seas is formulated as the right of states concerning their nationals. Although the drafters of the LOSC appear to have intended

1 See Section 2.1.2 and Chapters 4 and 5 above.

2 See Section 2.1.3 above.

that the term 'nationals' denotes vessels flying the flag of the state concerned, the term is used in a number of recent international fishery-related instruments to mean natural and juridical persons rather than vessels. However, this should not be taken to imply that the essential aspect of freedom of fishing has been modified in state practice. These changes appear to be aimed at improving control of fishing activities by urging non-flag states to exercise jurisdiction over (beneficial owners as well as) operators and crew of the vessel engaged in IUU fishing on the high seas.[3]

Second, cooperation between states is another pillar of the regime for high seas fisheries. States engaged in high seas fisheries are required to cooperate with each other with a view to taking conservation measures. The duty to cooperate assumed by high seas fishing states does not by itself command a particular course of action to be taken by the states concerned. But some particular forms of cooperation are required in a specific context: negotiations in good faith; the establishment of RFMO/As as appropriate; the exchange of data on catches and efforts and other information. In particular, the duty to participate in the existing RFMO/As developed in the years after the LOSC was adopted. Under the present legal regime, high seas fishing states and RFMO/As have reciprocal duties: states fishing on the high seas are to participate in the regulatory regime of the competent RFMO/A either as members/participants or as cooperating non-members; the RFMO/A, for its part, is to accept high seas fishing states as members or participants.[4] New entrants are also required to cooperate through RFMO/As in order for their nationals to engage in high seas fisheries regulated by RFMO/As. The major difference between new entrants and non-member states whose vessels are already engaged in the high seas fishery concerned is that the former are not given the 'right' to participate in the work of the RFMO/As as members or participants; the existing members of or participants in the RFMO/A have discretion to decide whether or not to accept new entrants.[5]

Recently, the duty of states to take measures in their ports for the conservation and management of fishery resources has been contained in some treaties; some other international fishery-related instruments also stipulate actions that should be taken by port states and/or market states. A precursor of this approach already appeared in Article 8 of the HSFC; the participants in the Geneva Conference discussed the role of market states in a regulatory regime for high seas fisheries in connection with Draft Article 56 of the 1956 ILC draft articles.[6]

In addition, coastal states are entitled to participate in RFMO/As regardless of the involvement of their vessels in the high seas fishery concerned. With regard

---

[3] See Section 2.1.5 above.

[4] See Section 2.2.2 above.

[5] See ibid.

[6] See Section 1.2.2 above.

to straddling and highly migratory fish stocks, coastal states are required to give effect to the duty to cooperate by actually participating. In the case of DHSFS, there is no equivalent link between high seas fisheries and coastal states and therefore no obligation for the coastal states in the region to join the RFMO/A concerned. In reality, however, coastal states are likely to participate in the regulatory regime simply because, in the practice of RFMOs, both straddling fish stocks and DHSFS are managed in a single organization and all members take part in decision-making concerning the management of any fishery conducted in the region. On the other hand, the entitlement of non-coastal states to participation without any intention to engage in fishing or research is not widely recognized in treaties or state practice.[7]

Third, states are required to take measures for their nationals as may be necessary for the conservation of the marine living resources of the high seas. The duty of conservation is reinforced by the provisions concerning the protection and preservation of the marine environment. In other words, international environmental law has been integrated as an essential part of the international law of high seas fisheries. After the adoption of the LOSC, the duty to take conservation measures has been developed: in particular, the concept of conservation was elaborated in the light of the contemporary development of international environmental law.

In the codification process during the 1950s, the concept of conservation was formulated in respect of the management of fishery resources in high seas areas adjacent to the territorial sea: the maximization of sustainable yields became the primary objective, to be pursued on the basis of scientific information.[8] The LOSC establishes a framework for the determination of conservation measures – rather than detailed rules for this purpose – specifying the factors to be taken into account: the maintenance or restoration of harvested populations to levels capable of producing the MSY, qualified by environmental and economic factors, taking into account, among other things, generally recommended international minimum standards.[9]

These standards and criteria for conservation and management measures have been developed, taking into account the development of international environmental law. Four interconnected aspects in the development of this issue are discernible.[10] First, a major shift is found in the objective(s) of fisheries management: while it still contains the utilization aspect, emphasis is placed on the sustainability aspect as seen in phrases such as 'sustainable fisheries' and 'long-term conservation and sustainable use'. One may even imagine that the utilization

7 See Section 2.1.2.2 above.

8 See Section 1.2.3 above.

9 See Section 2.3.1.1 above.

10 See Section 2.3.2 above.

aspect will be outweighed by environmental concerns with regard to high seas fisheries, as in the regime for the exploitation of marine mammals.[11] Second, the consideration of ecosystem impacts has been strengthened and now fisheries management needs to adopt an ecosystem approach. Attempts to define the term 'ecosystem approach' have not been successful, but its main elements are threefold: consideration of species interactions such as the conservation of associated or dependent species and bycatch reductions, habitat protection and the compatibility of conservation measures in spatial terms. Third, in addition to the conservation of each ecosystem component, biodiversity as such is a factor which is relevant to the concept of conservation. Biodiversity includes the diversity of ecosystems, between species and within species. Fourth, the precautionary approach needs to be adopted. The application of the precautionary approach to fisheries management means that states are obliged to address the risk of serious or irreversible harm to the target species as well as to the marine ecosystem in which they occur; uncertainty surrounding fisheries management may not be an excuse for postponing action; and, as actions to be taken depend on a concrete circumstance, possible precautionary measures range from enhanced scientific research, to setting precautionary reference points, to imposing a moratorium on fishing.

While the general principles of high seas fisheries provide a useful framework to address high seas fisheries problems, there are ambiguities in the scope and implications of the general principles for high seas fisheries. For instance, the application of the precautionary approach to high seas fisheries does not generally specify concrete measures to be taken. Similarly, in a situation where the existing knowledge of the population distribution does not conclusively indicate whether the population is composed of a single straddling stock or two or more stocks (e.g., a discrete high seas fish stock and an EEZ stock), the requirement of compatibility may indicate the need for addressing the fishery as a single stock in an integrated manner, while the consideration of biodiversity may suggest considering the population concerned as two stocks and dealing with them separately.[12] With regard to newly developed fishing methods such as bottom fisheries on the high seas in particular, the following two issues should be considered for the purpose of gauging the extent to which the general principles of high seas fisheries impact on contemporary challenges in high seas fisheries.

First, it is not clear from the text of the LOSC how the general principles need to be implemented with regard to fisheries for DHSFS. This is due to the following factors: (1) the main concerns of the negotiators at UNCLOS III did not relate to the issue of the conservation and management of DHSFS; (2) the FSA

[11] See Section 2.3.2.1 above.

[12] See discussions on the management of fisheries for orange roughy in Section 5.2.1 above.

does not apply to DHSFS because the scope of the FSA is limited to straddling fish stocks and highly migratory fish stocks; (3) although modern approaches to the management of high seas fisheries reflected in other international fisheries-related instruments are formulated in an all-inclusive manner, the primary focus of these instruments is on the conservation and management of straddling fish stocks and highly migratory fish stocks; (4) ostensibly, the special interests of coastal states recognized for straddling fish stocks are irrelevant to DHSFS.[13]

Second, it is not clear whether or not the general principles reflected, *inter alia*, in Part VII of the LOSC are applicable to organisms belonging to sedentary species that occur beyond the outer limit of the continental shelf. This is because: (1) the concept of sedentary species in international law was developed for the regime of the continental shelf; (2) as the ongoing debates over the legal regime governing bioprospecting for genetic resources indicate, it is controversial whether the principles of Part XI of the LOSC apply to living resources of the Area.[14]

Chapters 3–5 examined how the ambiguities about the implications of the general principles have been addressed in state practice at the global, regional and national levels in the context of DHSFS, deep-sea fisheries and the protection of vulnerable marine ecosystems.

6.2 *Practice Addressing New Challenges in High Seas Fisheries*

States and international organizations have addressed new challenges in high seas fisheries at the global, regional and national levels. Chapter 3 of this study analyzed the practice of global international institutions, in particular the UNGA and the FAO. Chapter 4 examined the constitutive instruments and practice of five existing RFMOs: CCAMLR, the GFCM, SEAFO, NEAFC and NAFO. In Chapter 5, the present study examined state practice in other areas. As indicated in Chapters 4 and 5, the constitutive instruments and/or practice of the five RFMOs, SIOFA and the SPRFMO Convention as well as the Convention on the Conservation and Management of High Seas Fisheries Resources in the North Pacific Ocean have all adopted modern approaches to high seas fisheries, including the precautionary approach and an ecosystem approach. While the CCAMLR Convention does not explicitly use the term 'precautionary approach', the practice of CCAMLR has been the model for other RFMO/As in applying it. The other three existing RFMOs predating the FSA have amended their constitutive instruments

[13] See Section 2.1.4 above.

[14] See ibid.

since 1995, although the revision of the GFCM Agreement was rather limited except for institutional restructuring.[15]

In the South-West Atlantic, the Central Atlantic, part of the Arctic, some parts of the Pacific and the Northern Indian Ocean, no formal negotiations to establish an RFMO/A have been initiated at the multilateral level although, for the Arctic Ocean or the Central Atlantic, multilateral negotiations might start to establish an RFMO/A. On the other hand, the 'Peanut Hole' in the Sea of Okhotsk and the 'Donut Hole' in the Bering Sea will remain outside the scope of the envisaged North Pacific Fisheries Commission. The geographic coverage of new RFMOs for the North Pacific and for the South Pacific will leave certain areas between the southern boundary of the North Pacific Fisheries Commission and the northern boundary of the SPRFMO without RFMO/As.

6.2.1 *Discrete High Seas Fish Stocks*

In a number of international fora at the global level, including the UNGA, participants recommended recognizing that some principles of the FSA should be made applicable to DHSFS. The UNGA further called on states to address fisheries for DHSFS by adopting necessary measures aimed at ensuring long-term conservation and sustainable use, applying the precautionary approach and an ecosystem approach, and collecting and reporting data and other information. Consensus now exists that some principles of the FSA should be applied to the conservation and management of DHSFS. State practice at the global level is insufficient to conclude which principles should be applied under general international law and, more importantly, how they should be applied in the management of DHSFS.[16]

With regard to management bodies for DHSFS, new cooperative mechanisms need not be established separately from the existing and proposed RFMO/As dealing with straddling fish stocks. In fact, the Review Conference on the FSA found that RFMO/As with the competence to regulate straddling fish stocks also have the necessary competence with regard to DHSFS.[17] It is not clear which states are entitled or required to participate in the regulatory mechanism for DHSFS.[18]

---

15 The revision of the NEAFC Convention was not significant in terms of the number of amended provisions but, in terms of substance, major changes were made such as its objective, general principles and dispute settlement procedures.

16 See Section 3.1 above.

17 See ibid.

18 Note that UNGA Resolutions call on 'States' in respect of bottom fisheries, as opposed to 'relevant Coastal States and States fishing on the high seas' in respect of straddling fish stocks and highly migratory fish stocks, to cooperate to establish RFMO/As. Given the fact that most of

At the regional level, there are five operational RFMOs that have competence to regulate DHSFS. In the constitutive instrument of these RFMOs, there is no distinction between straddling fish stocks and DHSFS. This is also true for SIOFA. This finding is important, given the fact that the participants in the negotiations for SIOFA have been fully aware of the existence of DHSFS within the proposed regulatory area of these RFMO/As and the main target of the regulatory mechanism was potential DHSFS. Similarly, the amendments to the NEAFC and NAFO Conventions do not show any distinction between DHSFS and straddling fish stocks, even though the NAFO members were aware that certain stocks of its Regulatory Area are DHSFS and NEAFC members knew that some stocks of deep-sea species in its Regulatory Area may be DHSFS. Interestingly, the Convention on the Conservation and Management of High Seas Fisheries Resources in the North Pacific Ocean explicitly refers to discrete fish stocks in the high seas in its preamble. It remains to be seen how it will be implemented, in particular whether or not it will be implemented in the same manner for straddling fish stocks and DHSFS.

In NAFO, CCAMLR and NEAFC, fishing opportunities have been allocated to members with regard to (potential) DHSFS. Allocation keys differ from one stock to another and sometimes more fishing opportunities are allocated to coastal states for DHSFS in comparison with straddling fish stocks. Thus, it cannot be said that coastal state interests are not necessarily relevant to the allocation of catches taken in fisheries for DHSFS.

In short, DHSFS are under the competence of relevant RFMOs and have been managed without notable differences from straddling fish stocks. The only potential distinction lies in the irrelevance of the consideration of coastal state interests, but in some cases coastal state interests appear to have been taken into account in the wording of the constitutive instruments of the relevant RFMOs and/or in the actual management of DHSFS.

Chapter 3 briefly mentioned that two approaches could be taken towards identifying principles of the FSA which are applicable to DHSFS.[19] First, states could select a group of principles to be applied to DHSFS. Second, and conversely, states could exclude some principles in the FSA as being inappropriate because of their inherent links to straddling and highly migratory fish stocks. Typical of this approach would be to remove the compatibility requirement from the list of applicable principles. It is submitted, however, that the difference between these two approaches is a matter of emphasis, rather than substance.

It is submitted that attempts to further elaborate the applicable principles at the global level will not by themselves result in better conservation of DHSFS for

---

the currently known DHSFS are deep-water species, the same ambiguities exist in respect of DHSFS.

19 See Section 3.4 above.

the following two reasons. First, negotiations would not be easy and might result in watering down to the very common denominator (e.g., the precautionary and ecosystem approaches as well as timely, complete and accurate data collection and sharing). Second, disagreement on which provisions of the FSA constitute a 'principle' might hamper such negotiations.

An inductive approach to find principles commonly applied in practice at the regional level would be meaningful as the practice of RFMO/As generally shows no distinction between DHSFS and straddling fish stocks. In this way, we can see which principles are considered to be useful, at least by members of RFMO/As, for the conservation of DHSFS.

In addition to the precautionary and ecosystem approaches and timely, complete and accurate data collection and sharing, it can be concluded that most of the general principles enumerated in Article 5 of the FSA are applicable in practice, including: ensuring the long-term sustainability of fish stocks; science-based management; the promotion and conducting of scientific research; and effective monitoring, control and surveillance.

Apart from Article 5 (General Principles), some other provisions found in the FSA could be recognized as suitable principles for DHSFS, including: compatibility between conservation measures (of neighbouring RFMO/As), the entitlement of high seas fishing states to participate in the work of RFMO/As, the prohibition of fishing without becoming a member of or a participant in RFMO/As competent to regulate high seas fisheries or without cooperating with them, and the denial of fishing without authorization from flag states. Some of the other provisions of the FSA, including Annexes on data collection and on the application of the precautionary approach, are also useful tools of reference for fisheries management.[20]

If states wish to conclude a global legally binding instrument governing DHSFS, it may take the form of an Implementation Agreement to the LOSC, the FSA or the CBD; an Annex to the FSA; or a Protocol to the FSA. Filling regulatory gaps by extending the detailed co-operative regime in the FSA in respect of DHSFS is a feasible option and some have already suggested doing so. For example, it

20 Article 1(1) of the FSA is useful in considering the meaning of some terms under international law, but in some cases questions may arise. For example, the exclusion of sedentary species from the scope of the FSA leaves room for different interpretations and it is an interesting issue when considered in the context of general international law. If the term 'sedentary species' in the FSA is considered to include organisms belonging to sedentary species in areas beyond the outer limit of the continental shelf, one might wonder whether or not a distinction between organisms belonging to sedentary species and other species also exists under general international law. If the term is intended to simply mean sedentary species on the continental shelf, no problem occurs and it is concluded that the same regime applies, in the area beyond the outer limit of the continental shelf, to organisms belonging to sedentary species and those not belonging to sedentary species. See also Section 2.1.4 above.

has been argued that the adoption of a protocol to the FSA that would achieve the proposed extension without reopening the FSA's provisions to negotiations seems an effective avenue for progress.[21] An annex under Article 48 of the FSA is another option: if the annex is adopted by consensus at a meeting of states parties, the revision can be effective from the date of its adoption.[22] In this way, problems associated with the lack of or a delay in the ratification of a new legally binding instrument may be avoided. Alternatively, states might wish to address the issue of DHSFS by producing a non-legally binding instrument such as technical guidelines and/or through governance by COFI or another organization (or organ). Still another option is to include the issue of the conservation and sustainable use of DHSFS in a future implementation agreement to the LOSC concerning biodiversity. This process, however, is expected to take a long time before such an agreement is finally concluded and enters into force. The potential lack of adherence to the agreement by high seas fishing states may be an obstacle. In addition, it is also possible to leave the issue to management by RFMO/As without any more guidance at the global level. For the time being, it is acceptable to refrain from producing a legally binding instrument at the global level. If the management of DHSFS in certain regions or by certain flag states becomes particularly problematic in the future, the adoption of an annex to the FSA might become a likely option. But, given the potential disagreement on the content of such a global legally binding instrument, the annex will remain a general declaration stating the obligation to apply the provisions of the FSA to DHSFS *mutatis mutandis*. The choice of principles to be applied will be left to RFMO/As.

6.2.2 *Deep-Sea Fisheries*

The UNGA and the FAO, among other bodies, have addressed the management of deep-sea fisheries at the global level. It seems that there is now a common understanding among states that the UNGA is the main forum to address the governance of deep-sea fisheries at the global level and the FAO supports it by dealing with the technical aspect of the same topic.[23]

The successive UNGA Resolutions assigned a principal role to regulate deep-sea fisheries on the high seas to the regional cooperative mechanism and, only where there are no RFMO/As or interim measures, the flag state has the primary

---

21 European Commission Background Paper No. 3, at p. 6. Note that the paper also states that "most of [discrete high seas stocks] consist of deep species whose fragility justifies particular care and rigour in their management". European Commission Background Paper No. 3, at p. 6. This could be interpreted to imply the need for a regime which is different from that for straddling and highly migratory fish stocks.

22 FSA, Article 48(2).

23 See Section 3.2 above.

role of determining conservation and management measures by itself. The Resolutions called for (1) the adoption of appropriate conservation and management measures in accordance with, among others, the guidance in UNGA Resolution 61/105, (2) the expansion of the substantive competence of the existing RFMO/As and (3) the establishment of a new RFMO/A in areas where there is no competent RFMO/A. The Resolutions do not require different types of management bodies for deep-sea fisheries and other fisheries on the high seas. Nevertheless, as pointed out in the sub-section on DHSFS above, the scope of states that are called on to participate in regional cooperative mechanisms is left ambiguous in the relevant UNGA Resolutions as far as deep-sea fisheries are concerned. While the recommendation to participate in the establishment of an RFMO/A for straddling and highly migratory fish stocks fisheries is addressed to the relevant coastal states and high seas fishing states, the corresponding recommendation for deep-sea fisheries is addressed to 'States' in general.

Standards and criteria employed for determining conservation and management measures for deep-sea fisheries are different from those used for pelagic fisheries. At the global level, there is wide agreement on the need for stricter measures for deep-sea fisheries on the high seas applying the precautionary and ecosystem approaches, as illustrated by the UNGA Resolutions and the FAO International Guidelines on the Management of Deep-sea Fisheries on the High Seas. Both the UNGA Resolutions and the FAO International Guidelines adopt the requirement of prior impact assessments; fisheries are prohibited unless it is shown that they do not have an SAI on VMEs or until appropriate conservation and management measures have been taken. The Guidelines have further advanced this requirement by stipulating that if it is not determined whether the proposed activity causes an SAI on VMEs, that activity can only be conducted with precautionary measures.

At the regional level, five existing operational RFMOs have the competence to address fisheries for deep-sea species and have adopted measures. SIOFA, the SPRFMO and the North Pacific Fisheries Commission are also granted the competence to manage deep-sea fisheries.

The five existing RFMOs have adopted measures to address bottom fisheries on the high seas. CCAMLR, NAFO, NEAFC and SEAFO have adopted data collection schemes specifically tailored to deep-sea fisheries. CCAMLR, NAFO and NEAFC have adopted fishing effort regulations for all fisheries or for some stocks whereas the GFCM has adopted a resolution on this issue. CCAMLR, NAFO, NEAFC and SEAFO have adopted catch limitations on some stocks or species. They (in particular, CCAMLR) have also controlled new fisheries. All five RFMOs have adopted area closures to protect benthic ecosystems from bottom fisheries. Similarly, these RFMOs have imposed various gear restrictions, including the prohibition of fishing with gear such as gillnets and trammel nets in all parts of their regulatory areas.

Most of the existing RFMOs and negotiations for the establishment of RFMOs in the South Pacific and the North Pacific have addressed the implementation of UNGA Resolution 61/105. These measures have incorporated important parts of the UNGA recommendations. They include prior impact assessments, the prohibition of fishing until it is proven that they do not involve SAIs and closures of VMEs to bottom fisheries. The GFCM has yet to address the implementation of UNGA Resolution 61/105, including additional closures of seamounts and cold-water coral reefs as suggested by its scientific body. Some of the measures of these RFMOs have imposed additional requirements, including the freezing and/or reduction of fishing efforts.

Interestingly, many of the existing measures stipulate that they will take into account standards and criteria at the global level, including the FAO International Guidelines. This will facilitate the coordination of measures among different RFMO/As. Whether or not this will actually happen is, however, still to be seen. So far, as far as encounter protocols are concerned, strict standards employed by CCAMLR have not been adopted in other RFMOs.

In addition to national legislation addressing high seas fisheries in general, the legislation of the flag state (or the regional economic integration organization) to control the operations of its vessels engaged in bottom fisheries in areas where there is no RFMO/A or multilaterally-agreed interim measures has particularly important impacts on fisheries management in such areas.

This type of legislation was adopted, *inter alia,* by the European Union. A number of Community vessels are engaged in deep-sea fisheries in high seas areas where there are neither competent RFMO/As nor interim measures. The relevant Regulation seeks to control these fishing activities through the requirement of fishing permits and prior environmental impact assessments. In addition, the Regulation also imposes requirements on member states to adopt more stringent measures than required under the relevant UNGA Resolutions.

For the time being, deep-sea fisheries on the high seas will be managed by RFMO/As and flag states in accordance with the guidance provided by the UNGA and the FAO International Guidelines. While the 2006 FAO Expert Consultation on deep-sea fisheries in the high seas referred to an International Plan of Action to follow the Technical Guidelines, COFI did not refer to any follow-up action for the International Guidelines in this regard. As the adopted document takes the form of 'International' (rather than 'Technical') Guidelines and states have agreed on the content of the Guidelines, it is doubtful whether the production of another non-legally binding instrument is useful. It is desirable to start negotiations for a legally binding instrument if and only if the International Guidelines have not been properly implemented by the RFMO/As or individual flag states. Otherwise, given the divergence in the biological characteristics of target species as well as the difference in fishing patterns in each region, it is desirable to leave the management of deep-sea fisheries to the discretion of RFMO/As based on the UNGA Resolutions and the FAO International Guidelines. In areas where no RFMO/A is likely to be established, practical support to developing states, including capacity-building, from international organizations such as the FAO or its subsidiary body will be useful.

6.2.3 *High Seas MPAs and Other Area-Based Management Tools*

Area-based management tools for regulating activities in the high seas have been used in many regions. In fisheries, states have supported the use of closed areas for the sustainability of fishery resources and for the protection of the marine ecosystem.

Disagreements emerged after states started discussing the establishment of integrated, multi-purpose MPAs. There are controversies over the compatibility of such MPAs with international law, including the provisions of the LOSC. Moreover, there are conflicts over which entity should have a leading role in designing and operating such MPAs. At the global level, there seems to be a general agreement on the leading role of the UNGA in policy discussions. Other organizations have refrained from discussing future policies on this issue as a lead agency. However, this does not mean that states agree that the UNGA has a leading role in designing and managing the MPAs.[24]

In view of the lack of consensus on how to address the overall framework for MPAs in areas beyond national jurisdiction, including the possible development of an implementation agreement under the LOSC, one feasible and desirable way forward is to initiate or enhance cooperation and coordination between RFMO/As and other regional mechanisms, together with other relevant stakeholders, in the establishment and management of area-based management tools.

In fact, consultations between RFMOs and regional environmental protection mechanisms have yielded some cooperation with regard to MPAs, in particular, in the North-East Atlantic and the Southern Ocean. NEAFC has held a dialogue with the OSPAR Commission for several years and concluded an MoU. The OSPAR Commission, together with NEAFC and other stakeholders, has been discussing the management of MPAs in areas beyond national jurisdiction. Since the OSPAR Commission does not have the competence to adopt measures for fisheries, the management of high seas MPAs in the North-East Atlantic needs to proceed in accordance with the competence of NEAFC to manage high seas fisheries and to close areas to fisheries. In the Antarctic Treaty System, the establishment of an ASPA or an ASMA in the marine area needs to be preceded by an approval from CCAMLR. CCAMLR has in fact approved some ASPAs and ASMAs.

---

24 See Section 3.3 above.

6.3 *Final Remarks*

The analysis of this study has shown that the LOSC still serves as an overall framework to meet the new challenges in high seas fisheries examined in this study. Within the constraints set by the general principles of high seas fisheries, states have been taking action to address problems caused by the new challenges at the global, regional and national levels. Action at these levels has complemented, rather than contradicted, each other.

The issues of DHSFS, deep-sea fisheries and the protection of VMEs remain a subject of discussion in the international community for the upcoming years. It should still be observed whether the urgency required for these issues is being matched by actions at the global, regional and national levels. In addressing the issues, the international community chose not to resort to radical actions such as a blanket moratorium on deep-sea fisheries on the high seas. However, uncertainty remains concerning the effectiveness of measures that have been already taken.

To promote compliance with the rules of high seas fisheries, states in various capacities need to cooperate with each other. The role of flag states and, to some extent, coastal states in promoting the conservation and management of high seas fisheries resources has been discussed in detail in this book. The role of port states and market states is also important. In this respect, an interesting development has taken place in the national legislation of the United States.

The United States enacted the Magnuson-Stevens Reauthorization Act of 2006.[25] It provides for various actions against IUU fishing, including sanctions such as the denial of port privilege, the prohibition on imports of fish and fish products and possible additional economic sanctions.[26] The Act stipulates that, among other fishing activities, "fishing activity that has an adverse impact on seamounts, hydrothermal vents, and cold water corals located beyond national jurisdiction, for which there are no applicable conservation or management measures or in areas with no applicable international fishery management organization or agreement" shall be included in the definition of IUU fishing.[27]

---

25 United States, Magnuson-Stevens Fishery Conservation and Management Reauthorization Act of 2006, Pub. L. 109–479, title IV, sections 402–403; 16 U.S. Code 1801(a) and 1826h–1826j.

26 See 16 U.S. Code 1826a(a) and (b)(3)–(4) and 1826j(d)(3)(A).

27 16 U.S. Code 1826j(e)(3)(C). This is included in the Federal Regulations defining IUU fishing. United States, Electronic Code of Federal Regulations (e-CFR), Title 50, Section 300.201(3). Note that, in the presidential memorandum of 2 October 2006, the term 'destructive fishing practices' was used as "practices that destroy the long-term natural productivity of fish stocks or habitats such as seamounts, corals, and sponge fields for short term gain". United States, Memorandum for the Secretary of State and the Secretary of Commerce on Promoting Sustainable Fisheries and Ending Destructive Fishing Practices, 2 October 2006.

Even though the Act does not seem to intend to unilaterally manage high seas fisheries by third state vessels, the process of identifying nations whose vessels are engaged in IUU fishing, initiating consultations with such nations and, unless appropriate corrective action has been taken or negative certification has been issued, imposing sanctions against such nations will influence the behaviour of other states in their capacity as a flag state. No nation has been identified under this prong of the definition of IUU fishing in 2009 and 2011 biennial reports submitted to Congress.[28] It remains to be seen how this Act will be implemented with regard to fishing activities impacting VMEs located in areas beyond national jurisdiction, especially how to avoid the criticism of arbitrary unilateral action.

The management of high seas fisheries is not pursued in isolation from other international processes which deal with different human activities. There is a clear trend in taking coherent approaches to issues concerning the management of areas beyond national jurisdiction. For example, the ISA Mining Code has provisions on serious harmful effects on VMEs quite similar to those on SAIs on VMEs in the context of high seas bottom fisheries in UNGA resolutions and RFMO measures.[29] In view of this convergence, it should carefully be monitored how high seas fisheries issues will be dealt with in the broader context of the conservation and management of marine biodiversity in areas beyond national jurisdiction.

---

28 U.S. Department of Commerce, Implementation of Title IV of the Magnuson-Stevens Fishery Conservation and Management Reauthorization Act of 2006, Report to Congress Pursuant to Section 403(a) of the Magnuson-Stevens Fishery Conservation and Management Reauthorization Act of 2006, January 2009, at pp. 96–97; U.S. Department of Commerce, Implementation of Title IV of the Magnuson-Stevens Fishery Conservation and Management Reauthorization Act of 2006, Report to Congress Pursuant to Section 403(a) of the Magnuson-Stevens Fishery Conservation and Management Reauthorization Act of 2006, January 2011, at pp. 91–92.

29 Regulations on Prospecting and Exploration for Polymetallic Sulphides in the Area, adopted at the 16th session of the International Seabed Authority (16 April–7 May 2010), Regulation 33(4) ("The Commission shall develop and implement procedures for determining, on the basis of the best available scientific and technical information, including information provided pursuant to regulation 20, whether proposed exploration activities in the Area would have serious harmful effects on vulnerable marine ecosystems, in particular hydrothermal vents, and ensure that, if it is determined that certain proposed exploration activities would have serious harmful effects on vulnerable marine ecosystems, those activities are managed to prevent such effects or not authorized to proceed.") This is absent in the Regulations on Prospecting and Exploration for Polymetallic Nodules in the Area, which were adopted in 2000. The Regulations on Prospecting and Exploration for Cobalt-rich Ferromanganese Crusts in the Area, adopted at the 18th session of the International Seabed Authority (9–27 July 2012), also refer to serious harmful effects on VMEs in Regulation 33(4). Interestingly, it explicitly refers to seamounts and cold water corals, as opposed to hydrothermal vents, which are mentioned in the Sulphides Regulations.

This study has exemplified that the international community may have confidence in the framework set out in the LOSC in dealing with issues of the law of the sea that may arise in the coming decades. A 'Constitution for the Oceans' serves to provide an order for the oceans which is stable and flexible. What the international community needs is not a revolution concerning the existing order, but a resolution to make every effort to address new challenges ahead.

Bibliography

ACIA, *Impacts of a Warming Arctic: Arctic Climate Impact Assessment* (Cambridge: Cambridge University Press, 2004).

Allen, Craig H., "Protecting the Oceanic Gardens of Eden: International Law Issues in Deep-Sea Vent Resource Conservation and Management", 13 *Georgetown International Environmental Law Review* (2001), pp. 563–660.

Allott, Philip, "Power Sharing in the Law of the Sea", 77 *American Journal of International Law* (1983), pp. 1–30.

Anderson, David, "The Regulation of Fishing and Related Activities in Exclusive Economic Zones", in Erik Franckx and Philippe Gautier (eds.), *The Exclusive Economic Zone and the United Nations Convention on the Law of the Sea, 1982–2000: A Preliminary Assessment of State Practice* (Bruxelles: Bruylant, 2003), pp. 31–49.

Anderson, David H., "The Straddling Stocks Agreement of 1995: An Initial Assessment", 45 *International and Comparative Law Quarterly* (1996), pp. 463–475.

Armas Pfirter, Frida M., "The Management of Seabed Living Resources in 'the Area' under UNCLOS", 11 *Revisita Electrónica de Estudios Internacionales* (2006), 29 p.

Attard, David Joseph, *The Exclusive Economic Zone in International Law* (Oxford: Clarendon Press, 1987).

Aust, Anthony, *Modern Treaty Law and Practice* (Cambridge: Cambridge University Press, 2000).

Balton, D.A., "Strengthening the Law of the Sea: The New Agreement on Straddling Fish Stocks and Highly Migratory Fish Stocks", 27 *Ocean Development and International Law* (1996), pp. 125–151.

——, "The Compliance Agreement", in Ellen Hey (ed.), *Developments in International Fisheries Law* (The Hague: Kluwer Law International, 1999), pp. 31–53.

——, "The Bering Sea Doughnut Hole Convention: Regional Solution, Global Implications", in Olav Schram Stokke (ed.), *Governing High Seas Fisheries: The Interplay of Global and Regional Regimes* (New York: Oxford University Press, 2001), pp. 143–177.

Balton, David A. and Dorothy C. Zbicz, "Managing Deep-Sea Fisheries: Some Threshold Questions", 19 *International Journal of Marine and Coastal Law* (2004), pp. 247–258.

Bangert, Kaare, "The Effective Enforcement of High Seas Fishing Regimes: The Case of the Convention for the Regulation of the Policing of the North Sea Fisheries of 6 May 1882", in Guy S. Goodwin-Gill and Stefan Talmon (eds.), *The Reality of International Law: Essays in Honour of Ian Brownlie* (Oxford: Clarendon Press, 1999), pp. 1–20.

Baslar, Kemal, *The Concept of the Common Heritage of Mankind in International Law* (The Hague: Martinus Nijhoff Publishers, 1998).

Beer-Gabel, Josette and Véronique Lestang, *Les Commissions de Pêche et Leur Droit* (Brussels: Bruylant, 2003).

Bensch, Alexis *et al.*, Worldwide review of bottom fisheries in the high seas, FAO Fisheries and Aquaculture Technical Paper 522 (2008).

Bernhardt, Rudolf, "Custom and Treaty in the Law of the Sea", 205 *Recueil des Cours* (1987-V), pp. 251–330.

Birnie, Patricia W. and Alan E. Boyle, *International Law and the Environment* (New York: Oxford University Press, 2nd edition, 2002).

Birnie, Patricia W., Alan. E. Boyle and Catherine Redgwell, *International Law and the Environment* (New York: Oxford University Press, 3rd edition, 2009).

Bisbal, Gustavo A., "Fisheries Management on the Patagonian Shelf: A Decade After the 1982 Falklands/ Malvinas Conflict", 17 *Marine Policy* (1993), pp. 213–229.

Bloom, Evan T., "Establishment of the Arctic Council", 93 *American Journal of International Law* (1999), pp. 712–722.

Bowman, Michael and Catherine Redgwell, "Introduction", in Michael Bowman and Catherine Redgwell (eds.), *International Law and the Conservation of Biological Diversity* (London: Kluwer Law International, 1996), pp. 1–4.

Boyle, Alan and David Freestone (eds.), *International Law and Sustainable Development: Past Achievements and Future Challenges* (New York: Oxford University Press, 1999).

Brierly, James Leslie, *The Law of Nations: An Introduction to the International Law of Peace* (Oxford: Clarendon Press, 6th edition, 1963).

Brown, E.D., *The International Law of the Sea*, vol. I (Aldershot: Dartmouth, 1994).

Buck, Eugene H., U.N. Convention on the Law of the Sea: Living Resources Provisions, CRS Report for Congress, Updated January 7, 2004.

Burke, William T., "U.S. Fishery Management and the New Law of the Sea", 76 *American Journal of International Law* (1982), pp. 24–55.

——, *The New International Law of Fisheries: UNCLOS 1982 and Beyond* (Oxford: Clarendon Press, 1994).

Caddy, J.F. and R.C. Griffiths, Living Marine Resources and Their Sustainable Development: Some Environmental and Institutional Perspectives, Fisheries Technical Paper No. 353 (1995).

Canino, Mike, "Summary of Genetic Stock Identification Studies in the Bering Sea", Central Bering Sea Pollock Workshop on Allowable Harvest Level and Stock Identification, Seattle, Washington, USA, 6–9 June 2005.

Carroz, J.E., "Institutional Aspects of Fishery Management under the New Regime of the Oceans", 21 *San Diego Law Review* (1984), pp. 513–540.

Churchill, Robin R., "The Barents Sea Loophole Agreement: A "Coastal State" Solution to a Straddling Stock Problem", 14 *International Journal of Marine and Coastal Law* (1999), pp. 467–483.

——, "Claims to Maritime Zones in the Arctic: Law of the Sea Normality or Polar Peculiarity?" in Alex G. Oude Elferink and Donald R. Rothwell (eds.), *The Law of the Sea and Polar Maritime Delimitation and Jurisdiction* (The Hague: Martinus Nijhoff Publishers, 2001), pp. 105–124.

Churchill, R.R. and A.V. Lowe, *The Law of the Sea* (Manchester: Manchester University Press, 3rd edition, 1999).

Clingan, Thomas A., Jr, "An Overview of Second Committee Negotiations in the Law of the Sea Conference", 63 *Oregon Law Review* (1984), pp. 53–72.

——, "Dispute Settlement among Non-parties to the LOS Convention with Respect to the Outer Limits of the Continental Shelf", in Thomas A. Clingan, Jr (ed.), *The Law of the Sea: What Lies Ahead?* (Honolulu: The Law of the Sea Institute, 1988), pp. 495–500.

——, "Mar Presencial (The Presential Sea): *Deja vu* all over again? A response to Francisco Orrego Vicuna", 24 *Ocean Development and International Law* (1993), pp. 93–97.

Corell, Hans, "Reflections on the Possibilities and Limitations of a Binding Legal Regime for the Arctic", Seventh Conference of Parliamentarians of the Arctic Region, Kiruna, Sweden, 3 August 2006.

Daggett, A.P., "The Regulation of Maritime Fisheries by Treaty", 28 *American Journal of International Law* (1934), pp. 693–717.

Dalton, Jane Gilliland, "The Chilean Mar Presencial: A Harmless Concept or a Dangerous Precedent?" 8 *International Journal of Marine and Coastal Law* (1993), pp. 397–418.

Davies, Peter G.G. and Catherine Redgwell, "The International Legal Regulation of Straddling Fish Stocks", 67 *British Year Book of International Law* (1996), pp. 199–274.

de La Fayette, Louise, "The OSPAR Convention Comes into Force: Continuity and Progress", 14 *International Journal of Marine and Coastal Law* (1999), pp. 247–297.

de Yturriaga, José A., *The International Regime of Fisheries: From UNCLOS 1982 to the Presential Sea* (The Hague: Martinus Nijhoff Publishers, 1997).

Division for Ocean Affairs and the Law of the Sea, *The Regime for High-Seas Fisheries: Status and Prospects* (New York: United Nations, 1992).

Division for Ocean Affairs and the Law of the Sea and International Seabed Authority, *Marine Mineral Resources: Scientific Advances and Economic Perspectives* (2004).

Dunlap, William V., "The Donut Hole Agreement", 10 *International Journal of Marine and Coastal Law* (1995), pp. 114–126.

Edeson, William, "Soft and Hard Law Aspects of Fisheries Issues: Some Recent Global and Regional Approaches", in Myron H. Nordquist, John Norton Moore and Said Mahmoudi (eds.), *The Stockholm Declaration and Law of the Marine Environment* (The Hague: Martinus Nijhoff Publishers, 2003), pp. 165–182.

Edeson, William, David Freestone and Elly Gudmundsdottir, *Legislating for Sustainable Fisheries: A Guide to Implementing the 1993 FAO Compliance Agreement and 1995 UN Fish Stocks Agreement* (Washington: World Bank, 2001).

Engler, M. Cecilia, Establishment and Implementation of a Conservation and Management Regime for High Seas Fisheries, with Focus on the Southeast Pacific and Chile: From Global Developments to Regional Challenges, Research Paper (UN – Nippon Foundation Fellow 2006–2007).

FAO Fisheries and Aquaculture Department, The State of World Fisheries and Aquaculture 2006 (2007).

——, The State of World Fisheries and Aquaculture 2008 (2009).

——, The State of World Fisheries and Aquaculture 2010 (2010).

——, The State of World Fisheries and Aquaculture 2012 (2012).

FAO Fisheries Department, World Review of Highly Migratory Species and Straddling Stocks, FAO Fisheries Technical Paper No. 337 (1994).

——, The State of World Fisheries and Aquaculture 2004 (2004).

Fleischer, Carl August, "The New Régime of Maritime Fisheries", 209 *Recueil des Cours* (1988-II), pp. 95–222.

——, "Fisheries and Biological Resources", in René-Jean Dupuy and Daniel Vignes (eds.), *A Handbook on the New Law of the Sea* (Dordrecht: Martinus Nijhoff Publishers, 1991), pp. 989–1126.

Franckx, Erik, *Maritime Claims in the Arctic: Canadian and Russian Perspectives* (Dordrecht: Nijhoff, 1993).

Freestone, David, "International Fisheries Law Since Rio: The Continued Rise of the Precautionary Principle", in Alan Boyle and David Freestone (eds.), *International Law and Sustainable Development: Past Achievements and Future Challenges* (New York: Oxford University Press, 1999), pp. 135–164.

——, "A Decade of the Law of the Sea Convention: Is It a Success?" 39 *George Washington International Law Review* (2007), pp. 499–542.

Freiwald, André, Jan Helge Fosså, Anthony Grehan, Tony Koslow and J. Murray Roberts, *Cold-water Coral Reefs: Out of Sight - No Longer Out of Mind* (Cambridge, UK: UNEP-WCMC, 2004).

Fulton, Thomas Wemyss, *The Sovereignty of the Sea: An Historical Account of the Claims of England to the Dominion of the British Seas, and of the Evolution of the Territorial Waters: With Special Reference to the Rights of Fishing and the Naval Salute* (Edinburgh: Blackwood, 1911).

Garcia-Amador, F.-V. *The Exploitation and Conservation of the Resources of the Sea: A Study of Contemporary International Law* (Leyden: A.W. Sythoff, Second and enlarged edition, 1963).

Garcia, S.M., The Precautionary Approach to Fisheries and its Implications for Fishery Research, Technology and Management: An Updated Review", in Precautionary Approach to Fisheries, FAO Fisheries Technical Paper No. 350–2 (1996).

Garcia, S.M. and J. Majkowski, "State of High Seas Resources", in Tadao Kuribayashi and Edward L. Miles (eds.), *The Law of the Sea in the 1990s: A Framework for Further International Cooperation* (Honolulu: The Law of the Sea Institute, 1992), pp. 175–236.

Garner, Bryan A. (ed.), *Black's Law Dictionary* (St. Paul, Minnesota: West, Second pocket edition, 2003).

Gavouneli, Maria, *Functional Jurisdiction in the Law of the Sea* (Leiden: Martinus Nijhoff Publishers, 2007).

Gianni, Matthew, *High Seas Bottom Trawl Fisheries and their Impacts on the Biodiversity of Vulnerable Deep-Sea Ecosystems: Options for International Action* (Gland, Switzerland: IUCN, 2004).

Gidel, Gilbert, *Le Droit International Public de la Mer: Le Temps de Paix*, vol. I (Chatreauroux: Mellottée, 1932).

Gjerde, Kristina M., "High Seas Fisheries Management under the Convention on the Law of the Sea", in David Freestone, Richard Barnes and David M. Ong (eds.), *The Law of the Sea: Progress and Prospects* (New York: Oxford University Press, 2006), pp. 281–307.

Gjerde, Kristina M. and David Ong, "Protection of Particularly Sensitive Sea Areas under International Marine Environmental Law: Report of the International Meeting of Legal Experts on Particularly Sensitive Sea Areas University of Hull, 20–21 July 1992", 26 *Marine Pollution Bulletin* (1993), pp. 9–13.

Glowka, Lyle, "Putting Marine Scientific Research on a Sustainable Footing at Hydrothermal Vents", 27 *Marine Policy* (2003), pp. 303–312.

Gordon, J.D.M., "Environmental and Biological Aspects of Deepwater Demersal Fishes", in R. Shotton (ed.), *Deep Sea 2003: Conference on the Governance and Management of Deep-sea Fisheries: Part 1: Conference Reports. Queenstown, New Zealand, 1–5 December 2003* (Rome: FAO, 2005), pp. 70–88.

Grigg, Richard W., "Precious Coral Fisheries of Hawaii and the U.S. Pacific Islands", 55 *Marine Fisheries Review* (1993), pp. 50–60.

Grotius, Hugo, *The Freedom of the Seas, or the Right Which Belongs to the Dutch to Take Part in the East Indian Trade* (New York: Oxford University Press, 1916).

Hayashi, Moritaka, "The 1995 Agreement on the Conservation and Management of Straddling and Highly Migratory Fish Stocks: Significance for the Law of the Sea Convention", 29 *Ocean & Coastal Management* (1995), pp. 51–69.

——, "The Role of the United Nations in Managing the World's Fisheries", in Gerald H. Blake, William J. Hildesley, Martin A. Pratt, Rebecca J. Ridley and Clive H. Schofield (eds.), *The Peaceful Management of Transboundary Resources* (London: Graham & Trotman/Martinus Nijhoff, 1995), pp. 373–393.

——, "Global Governance of Deep-Sea Fisheries", 19 *International Journal of Marine and Coastal Law* (2004), pp. 289–298.

——, "Regional Fisheries Management Organisations and Non-Members", in Tafsir Malick Ndiaye and Rüdiger Wolfrum (eds.), *Law of the Sea, Environmental Law and Settlement of Disputes: Liber Amicorum Judge Thomas A. Mensah* (Leiden: Martinus Nijhoff Publishers, 2007), pp. 751–765.

Henriksen, Tore, Geir Hønneland and Are K. Sydnes, *Law and Politics in Ocean Governance: The UN Fish Stocks Agreement and Regional Fisheries Management Regimes* (Leiden: Martinus Nijhoff Publishers, 2006).

Herr, Richard, "The International Regulation of Patagonian Toothfish: CCAMLR and High Sea Fisheries Management", in Olav Schram Stokke (ed.), *Governing High Seas Fisheries: The Interplay of Global and Regional Regimes* (New York: Oxford University Press, 2001), pp. 303–328.

Hewison, Grant J., "The Precautionary Approach to Fisheries Management: An Environmental Perspective", 11 *International Journal of Marine and Coastal Law* (1996), pp. 301–332.

Hey, Ellen, *The Regime for the Exploitation of Transboundary Marine Fisheries Resources: The United Nations Law of the Sea Convention Cooperation between States* (Dordrecht: Martinus Nijhoff Publishers, 1989).

——, "The Provisions of the United Nations Law of the Sea Convention on Fisheries Resources and Current International Fisheries Management Needs", in *The Regulation of Driftnet Fishing on the High Seas: Legal Issues*, FAO Legislative Study No. 47 (Rome: FAO, 1991), pp. 1–11.

——, "Global Fisheries Regulations in the First Half of the 1990s", 11 *International Journal of Marine and Coastal Law* (1996), pp. 459–490.

——, "Reviewing Implementation of the LOS Convention and Emerging International *Public* Law", in Alex G. Oude Elferink (ed.), *Stability and Change in the Law of the Sea: The Role of the LOS Convention* (Leiden: Martinus Nijhoff Publishers, 2005), pp. 75–88.

Holmila, Erkki, "Common Heritage of Mankind in the Law of the Sea", 1 *Acta Societatis Martensis* (2005), pp. 187–205.

Howarth, William, "The Interpretation of "Precaution" in the European Community Common Fisheries Policy", 20 *Journal of Environmental Law* (2008), pp. 213–244.

Huebert, Rob, "The Law of the Sea and the Arctic: An Unfulfilled Legacy", 18 *Ocean Yearbook* (2004), pp. 193–219.

Infante, María Teresa, "The Settlement of Disputes regarding the Law of the Sea and its Bearing on the Legal Nature of the Exclusive Economic Zone", in Francisco Orrego Vicuña (ed.), *The Exclusive Economic Zone: A Latin American Perspective* (Boulder, CO: Westview Press, 1984), pp. 159–171.

Jackson, Andrew, "Developments in the Southeast Atlantic, 1997–1999: Meetings of Coastal States and Other Interested Parties on a Fisheries Management Organization for the South East Atlantic (the SEAFO Process)", in Myron H. Nordquist and John Norton Moore (eds.), *Current Fisheries Issues and the Food and Agriculture Organization of the United Nations* (The Hague: Martinus Nijhoff Publishers, 2000), pp. 55–67.

Johnston, Douglas M., *The International Law of Fisheries: A Framework for Policy-Oriented Inquiries* (New Haven: Yale University Press, 1965).

——, "The Future of the Arctic Ocean: Competing Domains of International Public Policy", 17 *Ocean Yearbook* (2003), pp. 596–624.

Joyner, Christopher C. and Peter N. De Cola, "Chile's Presential Sea Proposal: Implications for Straddling Stocks and the International Law of Fisheries", 24 *Ocean Development and International Law* (1993), pp. 99–121.

Juda, Lawrence, *International Law and Ocean Use Management* (London: Routledge, 1996).

Kanehara, Atsuko, "A New or Pathological Tendency in the International Regulation of Sovereign States: From the Perspective of the International Regulation of Whaling", 47 *Japanese Annual of International Law* (2004), pp. 34–68.

Kaye, Stuart M., *International Fisheries Management* (The Hague: Kluwer Law International, 2001).

Kimball, Lee A., "Deep-Sea Fisheries of the High Seas: The Management Impasse", 19 *International Journal of Marine and Coastal Law* (2004), pp. 259–287.

Kindt, John Warren and Charles J. Wintheiser, "The Conservation and Protection of Marine Mammals", 7 *University of Hawaii Law Review* (1985), pp. 301–375.

Kiss, Alexandre, "The Rights and Interests of Future Generations and the Precautionary Principle", in David Freestone and Ellen Hey (eds.), *The Precautionary Principle and International Law: The Challenge of Implementation* (The Hague: Kluwer Law International, 1996), pp. 19–29.

Koers, Albert W., *International Regulation of Marine Fisheries: A Study of Regional Fisheries Organizations* (West Byfleet, Surrey: Fishing News Books Ltd., 1973).

Kwiatkowska, Barbara, "The High Seas Fisheries Regime: At a Point of No Return?" 8 *International Journal of Marine and Coastal Law* (1993), pp. 327–358.

——, "The 2006 *Barbados/Trinidad and Tobago* Award: A Landmark in Compulsory Jurisdiction and Equitable Maritime Boundary Delimitation", 22 *International Journal of Marine and Coastal Law* (2007), pp. 7–60.

Lack, M., K. Short and A. Willock, *Managing Risk and Uncertainty in Deep-sea Fisheries: Lessons from Orange Roughy* (TRAFFIC Oceana and WWF Australia, 2003).

Lagoni, Rainer, "Offshore Bunkering in the Exclusive Economic Zone", in Tafsir Malick Ndiaye and Rüdiger Wolfrum (eds.), *Law of the Sea, Environmental Law and Settlement of Disputes: Liber Amicorum Judge Thomas A. Mensah* (Leiden: Martinus Nijhoff Publishers, 2007), pp. 613–627.

Leonard, L. Larry, *International Regulations of Fisheries* (Washington: Carnegie Endowment for International Peace, 1944).

Lévy, Jean-Pierre, "La Première Décennie de l'Autorité Internationale des Fonds Marins", 109 *Revue Générale de Droit International Public* (2005), pp. 101–122.

Lodge, Michael W., "Improving International Governance in the Deep Sea", 19 *International Journal of Marine and Coastal Law* (2004), pp. 299–316.

Maffei, Maria Clara, "The Protection of Endangered Species of Animals in the Mediterranean Sea", in Edward L. Miles and Tullio Treves (eds.), *The Law of the Sea: New Worlds, New Discoveries* (Honolulu: The Law of the Sea Institute, 1993), pp. 253–298.

Maguire, J.-J., M. Sissenwine, J. Csirke, R. Grainger and S. Garcia, The State of World Highly Migratory, Straddling and Other High Seas Fishery Resources and Associated Species, FAO Fisheries Technical Paper No. 495 (2006).

Mahmoudi, Said, "Common Heritage of Mankind, Common Concern of Humanity", in Jean-Pierre Beurier, Alexandre Kiss and Said Mahmoudi (eds.) *New Technologies and Law of the Marine Environment* (The Hague: Kluwer Law International, 2000), pp. 215–223.

Marashi, S.H., The Role of FAO Regional Fishery Bodies in the Conservation and Management of Fisheries, Fisheries Circular No. 916, FIPL/C916 (1996).

McConnell, Moira L., "... Darkening Confusion Mounted Upon Darkening Confusion: The Search for the Elusive Genuine Link", 16 *Journal of Maritime Law and Commerce* (1985), pp. 365–396.

McDorman, T.L., "The Role of the Commission on the Limits of the Continental Shelf: A Technical Body in a Political World", 17 *International Journal of Marine and Coastal Law* (2002), pp. 301–324.

McDougal, Myres Smith and William T. Burke, *The Public Order of the Oceans: A Contemporary International Law of the Sea* (New Haven: Yale University Press, 1962).

McNair, Arnold Duncan, *The Law of Treaties* (Oxford: Clarendon Press, 1961).

Meltzer, Evelyne, "Global Overview of Straddling and Highly Migratory Fish Stocks: The Nonsustainable Nature of High Seas Fisheries", 25 *Ocean Development and International Law* (1994), pp. 255–344.

——, "Global Overview of Straddling and Highly Migratory Fish Stocks: Maps and Charts Detailing RFMO Coverage and Implementation", 20 *International Journal of Marine and Coastal Law* (2005), pp. 571–604.

Meseguer, José Luis, "Le Régime Juridique de l'Exploitation de Stocks Communs de Poissons au-delà des 200 Milles", XXVIII *Annuaire Français de Droit International* (1982), pp. 885–899.

Meyers, Herman, *The Nationality of Ships* (The Hague: Martinus Nijhoff Publishers, 1967).

Miles, Edward L. and William T. Burke, "Pressures on the United Nations Convention on the Law of the Sea of 1982 Arising From New Fisheries Conflicts: The Problem of Straddling Stocks", 20 *Ocean Development and International Law* (1989), pp. 343–357.

Miller, D.G.M., "Management and Governance Conventions and Protocols: SEAFC, WCPFC and SADC", in R. Shotton (ed.), *Deep Sea 2003: Conference on the Governance and Management of Deep-sea Fisheries: Part 1: Conference Reports. Queenstown, New Zealand, 1–5 December 2003* (Rome: FAO, 2005), pp. 596–637.

Miller, Denzil G.M., Eugene N. Sabourenkov and David C. Ramm, "Managing Antarctic Marine Living Resources: The CCAMLR Approach", 19 *International Journal of Marine and Coastal Law* (2004), pp. 317–363.

Miovski, Lourene, "Solutions in the Convention on the Law of the Sea to the Problem of Overfishing in the Central Bering Sea: Analysis of the Convention, Highlighting the Provisions Concerning Fisheries and Enclosed and Semi-Enclosed Seas", 26 *San Diego Law Review* (1989), pp. 525–574.

Molenaar, Erik Jaap, *Coastal State Jurisdiction over Vessel-Source Pollution* (The Hague: Kluwer Law International, 1998).

——, "The Concept of Real Interest and Other Aspects of Co-operation through Regional Fisheries Management Mechanisms", 15 *International Journal of Marine and Coastal Law* (2000), pp. 475–531.

——, "CCAMLR and Southern Ocean Fisheries", 16 *International Journal of Marine and Coastal Law* (2001), pp. 465–499.

——, "The South Tasman Rise Arrangement of 2000 and Other Initiatives on Management and Conservation of Orange Roughy", 16 *International Journal of Marine and Coastal Law* (2001), pp. 77–124.

——, "Regional Fisheries Management Organizations: Issues of Participation, Allocation and Unregulated Fishing", in Alex G. Oude Elferink and Donald R. Rothwell (eds.), *Oceans Management in the 21st Century: Institutional Frameworks and Responses* (Leiden: Martinus Nijhoff Publishers, 2004), pp. 69–86.

——, "Unregulated Deep-Sea Fisheries: A Need for a Multi-Level Approach", 19 *International Journal of Marine and Coastal Law* (2004), pp. 223–246.

——, "Addressing Regulatory Gaps in High Seas Fisheries", 20 *International Journal of Marine and Coastal Law* (2005), pp. 533–570.

——, "Marine Biodiversity in Areas Beyond National Jurisdiction", 22 *International Journal of Marine and Coastal Law* (2007), pp. 89–124.

——, "Current Legal and Institutional Issues Relating to the Conservation and Management of High Seas Deep-Sea Fisheries", in *Report and Documentation of the Expert Consultation on Deep-sea Fisheries in the High Seas, Bangkok, Thailand, 21–23 November 2006*, FAO Fisheries Report, No. 838 (Rome: FAO, 2007), pp. 113–139.

——, "Arctic Fisheries and International Law: Gaps and Options to Address Them", 1/2012 *Carbon and Climate Law Review* (2012), pp. 63–77.

Mooney-Seus, Marjorie L. and Andrew A. Rosenberg, Regional Fisheries Management Organizations (RFMOs): Progress in Adopting Precautionary Approach and Ecosystem-Based Management, 10 February 2007.

Moore, Gerald, "The Food and Agriculture Organisation Compliance Agreement", 10 *International Journal of Marine and Coastal Law* (1995), pp. 412–425.

Morgera, Elisa, "Competence or Confidence? The Appropriate Forum to Address Multi-Purpose High Seas Protected Areas", 16 *Review of European Community & International Environmental Law* (2007), pp. 1–11.

Morishita, Joji, "What is the Ecosystem Approach for Fisheries Management", 32 *Marine Policy* (2008), pp. 19–26.

Mossop, Joanna, "Protecting Marine Biodiversity on the Continental Shelf Beyond 200 Nautical Miles", 38 *Ocean Development and International Law* (2007), pp. 283–304.

Mouton, M.W., *The Continental Shelf* (The Hague: Nijhoff, 1952).

Munro, G., A. Van Houtte and R. Willmann, The Conservation and Management of Shared Fish Stocks: Legal and Economic Aspects, FAO Fisheries Technical Paper No. 465 (2004).

Murombo, Tumai, "The Role of International Environmental Diplomacy in the Sustainable Use of Marine Biodiversity in Areas Beyond National Jurisdiction: Ending Deep Sea Trawling", 40 *Comparative and International Law Journal of Southern Africa* (2007), pp. 172–192.

Nandan, Satya N. and Shabtai Rosenne (eds.), *United Nations Convention on the Law of the Sea 1982: A Commentary*, vol. II (Dordrecht: Martinus Nijhoff Publishers, 1993).
——, *United Nations Convention on the Law of the Sea 1982: A Commentary*, vol. III (The Hague: Martinus Nijhoff Publishers, 1995).
Nandan, Satya N., Michael W. Lodge and Shabtai Rosenne (eds.), *United Nations Convention on the Law of the Sea 1982: A Commentary*, vol. VI (The Hague: Martinus Nijhoff Publishers, 2002).
Nelson, Dolliver, "The Development of the Legal Regime of High Seas Fisheries", in Alan Boyle and David Freestone (eds.), *International Law and Sustainable Development: Past Achievements and Future Challenges* (New York: Oxford University Press, 1999), pp. 113–134.
Nelson, L.D.M., "Declarations, Statements and 'Disguised Reservations' with Respect to the Convention on the Law of the Sea", 50 *International and Comparative Law Quarterly* (2001), pp. 767–786.
O'Connell, D.P., "Sedentary Fisheries and the Australian Continental Shelf", 49 *American Journal of International Law* (1955), pp. 185–209.
——, *The International Law of the Sea*, vol. I (Oxford Clarendon Press, 1982).
Oda, Shigeru, *International Control of Sea Resources* (Leyden: Sijthoff, 1963).
——, "Sharing of Ocean Resources: Unresolved Issues in the Law of the Sea", 3 *New York Law School Journal of International and Comparative Law* (1981), pp. 1–14.
——, "Fisheries under the United Nations Convention on the Law of the Sea", 77 *American Journal of International Law* (1983), pp. 739–755.
——, "Proposals Regarding a 12-mile Limit for the Territorial Sea by the United States in 1970 and Japan in 1971: Implications and Consequences", 22 *Ocean Development and International Law* (1991), pp. 189–197.
Oral, Nilufer, "Protection of Vulnerable Marine Ecosystems in Areas Beyond National Jurisdiction: Can International Law Meet the Challenge?" in Anastasia Strati, Maria Gavouneli and Nikos Skourtos (eds.), *Unresolved Issues and New Challenges to the Law of the Sea: Time Before and Time After* (Leiden: Martinus Nijhoff Publishers, 2006), pp. 85–108.
Örebech, Peter, Ketill Sigurjonsson and Ted L. McDorman, "The 1995 United Nations Straddling and Highly Migratory Fish Stocks Agreement: Management, Enforcement and Dispute Settlement", 13 *International Journal of Marine and Coastal Law* (1998), pp. 119–142.
Orellana, Marcos A., "The Swordfish Dispute between the EU and Chile at the ITLOS and the WTO", 71 *Nordic Journal of International Law* (2002), pp. 55–81.
Orrego Vicuña, Francisco, *The Exclusive Economic Zone: Regime and Legal Nature under International Law* (Cambridge: Cambridge University Press, 1989).
——, "The 'Presential Sea': Defining Coastal States' Special Interests in High Seas Fisheries and Other Activities", 35 *German Yearbook of International Law* (1992), pp. 264–292.
——, "Toward an Effective Management of High Seas Fisheries and the Settlement of the Pending Issues of the Law of the Sea", 24 *Ocean Development and International Law* (1993), pp. 81–92.
——, *The Changing International Law of High Seas Fisheries* (Cambridge: Cambridge University Press, 1999).
——, "The Law Governing High Seas Fisheries: In Search of New Principles", 18 *Ocean Yearbook* (2004), pp. 383–394.
——, "The Law of the Sea and the Antarctic Treaty System: New Approaches to Offshore Jurisdiction", in Christopher C. Joyner and Sudhir K. Chopra (eds.), *The Antarctic Legal Regime* (Dordrecht: Martinus Nijhoff Publishers, 1988), pp. 97–127.
Oude Elferink, Alex G., "Fisheries in the Sea of Okhotsk High Seas Enclave: The Russian Federation's Attempts at Coastal State Control", 10 *International Journal of Marine and Coastal Law* (1995), pp. 1–18.

——, "The Outer Continental Shelf in the Arctic: The Application of Article 76 of the LOSC Convention in a Regional Context", in Alex G. Oude Elferink and Donald R. Rothwell (eds.), *The Law of the Sea and Polar Maritime Delimitation and Jurisdiction* (The Hague: Martinus Nijhoff Publishers, 2001), pp. 139–156.

——, "The Sea of Okhotsk Peanut Hole: De facto Extension of Coastal State Control", in Olav Schram Stokke (ed.), *Governing High Seas Fisheries: The Interplay of Global and Regional Regimes* (New York: Oxford University Press, 2001), pp. 179–205.

——, "The Genuine Link Concept: Time for a Post Mortem?" in Ige Dekker and Harry Post (eds.), *On the Foundations and Sources of International Law* (The Hague: T.M.C. Asser Press, 2003), pp. 41–63.

——, "Reviewing the Implementation of the LOS Convention: The Role of the United Nations General Assembly and the Meeting of States Parties", in Alex G. Oude Elferink and Donald R. Rothwell (eds.), *Oceans Management in the 21st Century: Institutional Frameworks and Responses* (Leiden: Martinus Nijhoff Publishers, 2004), pp. 295–312.

——, "The Regime of the Area: Delineating the Scope of Application of the Common Heritage Principle and Freedom of the High Seas", 22 *International Journal of Marine and Coastal Law* (2007), pp. 143–176.

——, "The Continental Shelf in the Polar Regions: Cold War or Black-letter Law?", XL *Netherlands Yearbook of International Law* (2009), pp. 121–181.

Owen, Daniel, "The Application of the Wild Birds Directive Beyond the Territorial Sea", 13 *Journal of Environmental Law* (2001), pp. 39–78.

——, *The Powers of the OSPAR Commission and Coastal State Parties to the OSPAR Convention to Manage Marine Protected Areas on the Seabed Beyond 200 nm From the Baseline: A Report for WWF Germany* (Frankfurt: WWF Germany, 2006).

Oxman, Bernard H., "The Third United Nations Conference on the Law of the Sea: The Seventh Session (1978)", 73 *American Journal of International Law* (1979), pp. 1–41.

——, "The Territorial Temptation: A Siren Song at Sea", 100 *American Journal of International Law* (2006), pp. 830–851.

Prows, Peter, "Tough Love: The Dramatic Birth and Looming Demise of UNCLOS Property Law (and What is to be Done about it)", 42 *Texas International Law Journal* (2007), pp. 241–309.

Rau, Markus, "Comment: The Swordfish Case: Law of the Sea v. Trade", 62 *Zeitschrift für ausländisches öffentliches Recht und Völkerrecht* (2002), pp. 37–42.

Rayfuse, Rosemary Gail, *Non-Flag State Enforcement in High Seas Fisheries* (Leiden: Martinus Nijhoff Publishers, 2004).

——, "Countermeasures and High Seas Fisheries Enforcement", LI *Netherlands International Law Review* (2004), pp. 41–76.

——, "Melting Moments: The Future of Polar Oceans Governance in a Warming World", 16 *Review of European Community & International Environmental Law* (2007), pp. 196–216.

——, "Protecting Marine Biodiversity in Polar Areas Beyond National Jurisdiction", 17 *Review of European Community & International Environmental Law* (2008), pp. 3–13.

Rosenne, Shabtai, *The Law and Practice of the International Court, 1920–2005* (Leiden: Martinus Nijhoff Publishers, Fourth edition, 2006).

Rosenne, Shabtai and Alexander Yankov (eds.), *United Nations Convention on the Law of the Sea 1982: A Commentary*, vol. IV (Dordrecht: Martinus Nijhoff Publishers, 1991).

Rosenne, Shabtai and Louis B. Sohn (eds.), *United Nations Convention on the Law of the Sea 1982: A Commentary*, vol. V (Dordrecht: Martinus Nijhoff Publishers, 1989).

Rothwell, Donald R. and Stuart Kaye, "Law of the Sea and the Polar Regions: Reconsidering the Traditional Norms", 18 *Marine Policy* (1994), pp. 41–58.

Sand, Peter H., "'Green' Enclosure of Ocean Space – *Déjà Vu*?" 54 *Marine Pollution Bulletin* (2007), pp. 374–376.

Sands, Philippe, *Principles of International Environmental Law* (Cambridge: Cambridge University Press, 2nd edition, 2003).

Schiffman, Howard, *Marine Conservation Agreements: The Law and Policy of Reservations and Vetoes* (Leiden: Martinus Nijhoff Publishers, 2008).

Schram, Gunnar J. and André Tahindro, "Developments in Principles for the Adoption of Fisheries Conservation and Management Measures", in Ellen Hey (ed.), *Developments in International Fisheries Law* (The Hague: Kluwer Law International, 1999), pp. 251–286.

Scott, Shirley, "The LOS Convention as a Constitutional Regime for the Oceans", in Alex G. Oude Elferink (ed.), *Stability and Change in the Law of the Sea: The Role of the LOS Convention* (Leiden: Martinus Nijhoff Publishers, 2005), pp. 9–38.

Scovazzi, Tullio, "The Evolution of International Law of the Sea: New Issues, New Challenges", 286 *Recueil des Cours* (2000), pp. 39–243.

——, "Marine Protected Areas on the High Seas: Some Legal and Policy Considerations", 19 *International Journal of Marine and Coastal Law* (2004), pp. 1–17.

——, "Mining, Protection of the Environment, Scientific Research and Bioprospecting: Some Considerations on the Role of the International Sea-Bed Authority", 19 *International Journal of Marine and Coastal Law* (2004), pp. 383–409.

——, "New International Instruments for Marine Protected Areas in the Mediterranean Sea", in Anastasia Strati, Maria Gavouneli and Nikos Skourtos (eds.), *Unresolved Issues and New Challenges to the Law of the Sea: Time Before and Time After* (Leiden: Martinus Nijhoff Publishers, 2006), pp. 109–120.

Secretariat of the Convention on Biological Diversity, *Handbook of the Convention on Biological Diversity Including its Cartagena Protocol on Biosafety* (Montreal, Canada, 3rd edition, 2005).

Serdy, A., "Schrödinger's TAC: Superposition of Alternative Catch Limits from 2003 to 2006 under the South Tasman Rise Orange Roughy Arrangement between Australia and New Zealand", in R. Shotton (ed.), *Deep Sea 2003: Conference on the Governance and Management of Deep-sea Fisheries: Part 1: Conference Reports. Queenstown, New Zealand, 1–5 December 2003* (Rome: FAO, 2005), pp. 494–510.

Sohn, Louis B., "The Stockholm Declaration on the Human Environment", 14 *Harvard International Law Journal* (1973), pp. 423–515.

Soons, Alfred H.A., "Comments on the Genuine Link Concept", in Ige Dekker and Harry Post (eds.), *On the Foundations and Sources of International Law* (The Hague: T.M.C. Asser Press, 2003), pp. 65–71.

——, *Marine Scientific Research and the Law of the Sea* (Deventer: Kluwer Law and Taxation Publishers, 1982).

——, "Regulation of Marine Scientific Research by the European Community and its Member States", 23 *Ocean Development and International Law* (1992), pp. 259–277.

Standing Senate Committee on Fisheries and Oceans of the Senate of Canada, The Management of Atlantic Fish Stocks: Beyond the 200-Mile Limit, February 2007.

Staples, Dereck, "Management of Shared Fish Stocks: Australian Case Studies", Papers Presented at the Norway-FAO Expert Consultation on the Management of Shared Fish Stocks – Bergen, Norway, 7–10 October 2002, reproduced in FAO Fisheries Report No. 695 Supplement, pp. 159–179.

Stokke, Olav Schram, "The Loophole of the Barents Sea Fisheries Regime", in Olav Schram Stokke (ed.), *Governing High Seas Fisheries: The Interplay of Global and Regional Regimes* (New York: Oxford University Press, 2001), pp. 273–301.

——, "A Legal Regime for the Arctic?: Interplay with the Law of the Sea Convention", 31 *Marine Policy* (2007), pp. 402–408.

Stoll, Peter-Tobias and Silja Vöneky, "The Swordfish Case: Law of the Sea v. Trade", 62 *Zeitschrift für ausländisches öffentliches Recht und Völkerrecht* (2002), pp. 21–35.

Stone, Gregory S., Laurence P. Madin, Karen Stocks, Glenn Hovermale, Porter Hoagland, Mary Schumacher, Peter Etnoyer, Carolyn Sotka and Heather Tausig, "Seamount Biodiversity, Exploitation and Conservation", in Linda K. Glover and Sylvia A. Earle (eds.), *Defying Ocean's End: An Agenda for Action* (Washington: Island Press, 2004), pp. 43–70.

Sydnes, Are K., "New Regional Fisheries Management Regimes: Establishing the South East Atlantic Fisheries Organisation", 25 *Marine Policy* (2001), pp. 353–364.

Tache, Simon W., "The Nationality of Ships: The Definitional Controversy and Enforcement of Genuine Link", 16 *International Lawyer* (1982), pp. 301–312.

Tahindro, André, "Conservation and Management of Transboundary Fish Stocks: Comments in Light of the Adoption of the 1995 Agreement for the Conservation and Management of Straddling Fish Stocks and Highly Migratory Fish Stocks", 28 *Ocean Development and International Law* (1997), pp. 1–58.

Tanaka, Yoshifumi, "Obligation to Co-operate in Marine Scientific Research and the Conservation of Marine Living Resources", 65 *Zeitschrift für ausländisches öffentliches Recht und Völkerrecht* (2005), pp. 937–965.

Treves, Tullio, "The General Assembly and the Meeting of States Parties in the Implementation of the LOS Convention", in Alex G. Oude Elferink (ed.), *Stability and Change in the Law of the Sea: The Role of the LOS Convention* (Leiden: Martinus Nijhoff Publishers, 2005), pp. 55–74.

——, "Some International Law Aspects of the Use of Vessel Monitoring Systems for Preventing Illegal Unreported Unregulated Fishing", in Tafsir Malick Ndiaye and Rüdiger Wolfrum (eds.), *Law of the Sea, Environmental Law and Settlement of Disputes: Liber Amicorum Judge Thomas A. Mensah* (Leiden: Martinus Nijhoff Publishers, 2007), pp. 811–820.

Trouwborst, Arie, *Evolution and Status of the Precautionary Principle in International Law* (The Hague: Kluwer Law International, 2002).

——, *Precautionary Rights and Duties of States* (Leiden: Martinus Nijhoff Publishers, 2006).

——, "The Precautionary Principle in General International Law: Combating the Babylonian Confusion", 16 *Review of European Community & International Environmental Law* (2007), pp. 185–195.

Tudela, Sergi, Ecosystem Effects of Fishing in the Mediterranean: An Analysis of the Major Threats of Fishing Gear and Practices to Biodiversity and Marine Habitats, GFCM Studies and Reviews No. 74 (2004).

——, Proposal for a representative network of protected deep-sea sensitive habitats in the Mediterranean, SAC-GFCM, Sub-Committee of Marine Environment and Ecosystems, Rome, 27–30 September 2005 (2005).

UN Millennium Project, Environment and Human Well-being: A Practical Strategy, Report of the Task Force on Environmental Sustainability (2005).

Van Dyke, Jon M., "The United States and Japan in Relation to the Resources, the Environment, and the People of the Pacific Island Region", 16 *Ecology Law Quarterly* (1989), pp. 217–226.

——, "Modifying the 1982 Law of the Sea Convention: New Initiatives on Governance of High Seas Fisheries Resources: The Straddling Stocks Negotiations", 10 *International Journal of Marine and Coastal Law* (1995), pp. 219–227.

VanderZwaag, David, Rob Huebert and Stacey Ferrara, "The Arctic Environmental Protection Strategy, Arctic Council and Multilateral Environmental Initiatives: Tinkering while the Arctic Marine Environment Totters", in Alex G. Oude Elferink and Donald R. Rothwell (eds.), *The Law of the Sea and Polar Maritime Delimitation and Jurisdiction* (The Hague: Martinus Nijhoff Publishers, 2001), pp. 225–248.

Verzijl, J.H.W., "The United Nations Conference on the Law of the Sea, Geneva, 1958 II", 6 *Nederlands Tijdschrift voor Internationaal Recht* (1959), pp. 115–139.

Watanabe, Hiromoto and Seiichiro Ono, "Analytical Review of the Elaboration Process of FAO Code of Conduct for Responsible Fisheries [in Japanese]", 35 *The Report of Tokyo University of Fisheries* (2000), pp. 153–176.

Westberg, Ann-Kristin, "Governance and Management of Living Marine Resources and Fisheries on the Continental Slope and in the Deep-sea: A Legal Framework and Some Points of Departure", in R. Shotton (ed.), *Deep Sea 2003: Conference on the Governance and Management of Deep-sea Fisheries: Part 1: Conference Reports. Queenstown, New Zealand, 1–5 December 2003* (Rome: FAO, 2005), pp. 711–718.

Willing, Jane, "Arrangement between the Government of Australia and the Government of New Zealand for the Conservation and Management of Orange Roughy on the South Tasman Rise", Papers Presented at the Norway-FAO Expert Consultation on the Management of Shared Fish Stocks – Bergen, Norway, 7–10 October 2002, reproduced in FAO Fisheries Report No. 695 Supplement (2003), pp. 200–205.

Wolfrum, Rüdiger and Nele Matz, "The Interplay of the United Nations Convention on the Law of the Sea and the Convention on Biological Diversity", 4 *Max Planck Yearbook of United Nations Law* (2000), pp. 445–480.

World Commission on Environment and Development, *Our Common Future* (Oxford: Oxford University Press, 1987).

Worm, B., E.B. Barbier, N. Beaumont, J.E. Duffy, C. Folke, B.S. Halpern, J.B. Jackson, H.K. Lotze, F. Micheli, S.R. Palumbi, E. Sala, K.A. Selkoe, J.J. Stachowicz and R. Watson, "Impacts of Biodiversity Loss on Ocean Ecosystem Services", 314 *Science* (2006), pp. 787–790.

Yonezawa, Kunio, "Some Thoughts on the Straddling Stock Problem in the Pacific Ocean", in Tadao Kuribayashi and Edward L. Miles (eds.), *The Law of the Sea in the 1990s: A Framework for Further International Cooperation* (Honolulu: The Law of the Sea Institute, 1992), pp. 127–135.

Young, Tomme Rosanne, The Legal Framework for MPAs and Successes and Failures in Their Incorporation into National Legislation, in Report and documentation of the Workshop on Marine Protected Areas and Fisheries Management: Review of Issues and Considerations, Rome, 12–14 June 2006, FAO Fisheries Report No. 825 (2007), pp. 221–300.

Table of International Instruments

Convention Between Belgium, Denmark, France, Germany, Great Britain and the Netherlands, for Regulating the Police of the North Sea Fisheries, The Hague, 6 May 1882, *BFSP*, vol. 73, p. 39.
Convention between Great Britain, the United States, Japan, and Russia, respecting Measures for the Preservation and Protection of the Fur Seals in the North Pacific Ocean, Washington, 7 July 1911, entered into force 15 December 1911, *BFSP*, vol. 104, p. 175.
Convention for the Regulation of Whaling, Geneva, 24 September 1931, entered into force 16 January 1935, *LNTS*, vol. 155, p. 349.
International Agreement for the Regulation of Whaling, London, 8 June 1937, *LNTS*, vol. 190, p. 79.
Convention for the Regulation of the Meshes of Fishing Nets and the Size Limits of Fish, London, 5 April 1946, entered into force on 5 April 1953, *UNTS*, vol. 231, p. 199.
International Convention for the Regulation of Whaling, Washington, 2 December 1946, entered into force 10 November 1948, *UNTS*, vol. 161, p. 72.
Agreement for the Establishment of the General Fisheries Commission for the Mediterranean, Rome, 24 September 1949, entered into force 20 February 1952, amended lastly in November 1997, *UNTS*, vol. 126, p. 238.
International Convention for the Northwest Atlantic Fisheries, Washington, 8 February 1949, entered into force 3 July 1950, *UNTS*, vol. 157, p. 157.
Interim Convention on Conservation of North Pacific Fur Seals, Washington, 9 February 1957, entered into force 14 October 1957, *UNTS*, vol. 314, p. 105.
Convention on Fishing and Conservation of the Living Resources of the High Seas, Geneva, 29 April 1958, entered into force 20 March 1966, *UNTS*, vol. 559, p. 285.
Convention on the Continental Shelf, Geneva, 29 April 1958, entered into force 10 June 1964, *UNTS*, vol. 499, p. 311.
Convention on the High Seas, Geneva, 29 April 1958, entered into force 30 September 1962, *UNTS*, vol. 450, p. 82.
Convention on the Territorial Sea and Contiguous Zone, Geneva, 29 April 1958, entered into force 10 September 1964, *UNTS*, vol. 516, p. 205.
Antarctic Treaty, Washington, 1 December 1959, entered into force 23 June 1961, *UNTS*, vol. 402, p. 71.
Convention on the Conservation of the Living Resources of the Southeast Atlantic, Rome, 3 October 1969, entered into force 24 October 1971, *UNTS*, vol. 801, p. 101.
Vienna Convention on the Law of Treaties, Vienna, 23 May 1969, entered into force on 27 January 1980, *UNTS*, vol. 1155, p. 331.
Agreement concerning shrimp between Brazil and the United States, Brasília, 9 May 1972, entered into force 14 February 1973, *UNTS*, vol. 894, p. 34.
Convention on Future Multilateral Cooperation in the Northwest Atlantic Fisheries, Ottawa, 24 October 1978, entered into force 1 January 1979, *UNTS*, vol. 1135, p. 371.
Convention on Future Multilateral Co-operation in North-East Atlantic Fisheries, London, 18 November 1980, entered into force 17 March 1982, *UNTS*, vol. 1285, p. 129.
Convention on the Conservation of Antarctic Marine Living Resources, Canberra, 20 May 1980, entered into force 7 April 1982, *UNTS*, vol. 1329, p. 47.

United Nations Convention on the Law of the Sea, Montego Bay, 10 December 1982, entered into force 16 November 1994, *UNTS*, vol. 1833, p. 3.

Joint Statement on the Conservation of Fisheries between the Government of the Argentine Republic and the Government of the United Kingdom of Great Britain and Northern Ireland, London and Buenos Aires, 28 November 1990, *International Journal of Estuarine and Coastal Law*, vol. 6, p. 146.

Protocol on Environmental Protection to the Antarctic Treaty, Madrid, 4 October 1991, entered into force 14 January 1998, *ILM*, vol. 30 (1991), p. 1455.

Agenda 21, Rio de Janeiro, 13 August 1992.

Convention for the Protection of the Marine Environment of the North-East Atlantic, Paris, 22 September 1992, entered into force 25 March 1998, *ILM*, vol. 32 (1993), p. 1069.

Convention on Biological Diversity, Rio de Janeiro, 5 June 1992, entered into force 29 December 1993, *UNTS*, vol. 1760, p. 142.

Agreement to Promote Compliance with International Conservation and Management Measures by Fishing Vessels on the High Seas, Rome, approved by FAO Conference on 24 November 1993, entered into force 24 April 2003, *UNTS*, vol. 2221, p. 120.

Convention on the Conservation and Management of Pollock Resources in the Central Bering Sea, Washington, 16 June 1994, entered into force 8 December 1995, *ILM*, Vol. 34, p. 67.

Agreement for the Implementation of the Provisions of the United Nations Convention on the Law of the Sea of 10 December 1982 relating to the Conservation and Management of Straddling Fish Stocks and Highly Migratory Fish Stocks, New York, 4 August 1995, entered into force 11 November 2001, *UNTS*, vol. 2167, p. 3.

Code of Conduct for Responsible Fisheries adopted by the FAO Conference on 31 October 1995, available at <http://www.fao.org/DOCREP/005/v9878e/v9878e00.htm>.

Kyoto Declaration, adopted at the International Conference on the Sustainable Contribution of Fisheries to Food Security, Kyoto, Japan, 4–9 December 1995, available at <http://www.fao.org/DOCREP/006/AC442E/AC442E00.HTM>.

Declaration on the Establishment of the Arctic Council, 19 September 1996, *ILM*, vol. 35, p. 1387.

Annex V on the Protection and Conservation of the Ecosystems and Biological Diversity of the Maritime Area to the Convention for the Protection of the Marine Environment of the North-East Atlantic, Sintra, 22–23 July 1998, available at <http://www.fco.gov.uk/resources/en/pdf/pdf14/fco_ts34-01protectionmarine>.

Arrangement between the Government of Australia and the Government of New Zealand for the Conservation and Management of Orange Roughy on the South Tasman Rise, Signed on 12 January 1998 and 18 February 1998, respectively, reproduced in D. Staples, "Management of Shared Fish Stocks: Australian Case Studies", Papers Presented at the Norway-FAO Expert Consultation on the Management of Shared Fish Stocks – Bergen, Norway, 7–10 October 2002.

Agreement between the Government of Iceland, the Government of Norway and the Government of the Russian Federation concerning Certain Aspects of Cooperation in the Area of Fisheries, St. Petersburg, 15 May 1999, entered into force 15 July 1999, *LOSB*, vol. 41, p. 53.

Protocol between the Government of Iceland and the Government of the Russian Federation under the Agreement between the Government of Iceland, the Government of Norway and the Government of the Russian Federation concerning Certain Aspects of Cooperation in the Area of Fisheries, St. Petersburg, 15 May 1999, entered into force 15 July 1999, available at <http://faolex.fao.org/docs/texts/mul44149.doc>.

Protocol between the Government of Norway and the Government of Iceland under the Agreement between the Government of Norway, the Government of Iceland and the Government of the Russian Federation concerning Certain Aspects of Cooperation in the Area of Fisheries, St. Petersburg, 15 May 1999, entered into force 15 July 1999, *LOSB*, vol. 41, p. 56.

Arrangement between the Government of Australia and the Government of New Zealand for the Conservation and Management of Orange Roughy on the South Tasman Rise, Signed for New Zealand on 17 February 2000 and for Australia on 25 February 2000, reproduced in D. Staples, "Management of Shared Fish Stocks: Australian Case Studies", Papers Presented at the Norway-FAO Expert Consultation on the Management of Shared Fish Stocks – Bergen, Norway, 7–10 October 2002.

Regulations on Prospecting and Exploration for Polymetallic Nodules in the Area, adopted at the sixth session of the International Seabed Authority (20–31 March and 3–14 July 2000), available at <http://www.isa.org.jm/>.

Convention on the Conservation and Management of Highly Migratory Fish Stocks in the Western and Central Pacific Ocean, Honolulu, 5 September 2000, entered into force 19 June 2004, available at <http://www.wcpfc.int/doc/convention-conservation-and-management-highly-migratory-fish-stocks-western-and-central-pacific->.

Framework Agreement for the Conservation of Living Marine Resources on the High Seas of the South Pacific, Santiago, Chile, 14 August 2000, not yet in force, available at <http://www.cpps-int.org/spanish/tratadosyconvenios/tratadosregionales/acuerdodegalapagos.htm>.

Convention on the Conservation and Management of Fishery Resources in the South East Atlantic Ocean, Windhoek, 20 April 2001, entered into force 13 April 2003, available at <http://www.seafo.org/AUConventionText.html>.

International Plan of Action to Prevent, Deter and Eliminate Illegal, Unreported and Unregulated Fishing, approved by the FAO Committee on Fisheries on 2 March 2001, available at <http://www.fao.org/DOCREP/003/y1224e/y1224e00.HTM>.

Reykjavik Declaration on Responsible Fisheries in the Marine Ecosystem, 4 October 2001, available at <ftp://ftp.fao.org/fi/DOCUMENT/reykjavik/y2198too_dec.pdf>.

South African Development Community Protocol on Fisheries, Maputo, 14 August 2001, entered into force 8 August 2003.

Agreement establishing the Caribbean Regional Fisheries Mechanism, Belize City, Belize, 4 February 2002, entered into force 4 February 2002, available at <http://www.caricom-fisheries.com/website_content/main/agreement_establishing_the_crfm.pdf>.

Johannesburg Plan of Implementation, Johannesburg, 4 September 2002, available at <http://www.un.org/esa/sustdev/documents/WSSD_POI_PD/English/WSSD_PlanImpl.pdf>.

Ministerial Declaration at the Conference on the Governance of High Seas Fisheries and the UN Fish Agreement: Moving from Words to Action, 2 May 2005, available at <http://www.dfo-mpo.gc.ca/fgc-cgp/declaration_e.htm>.

Outcome of the Review Conference, 26 May 2006, Report of the Review Conference on the Agreement for the Implementation of the Provisions of the United Nations Convention on the Law of the Sea of 10 December 1982 relating to the Conservation and Management of Straddling Fish Stocks and Highly Migratory Fish Stocks, A/CONF.210/2006/15, Annex.

Revised Statutes of the Western Central Atlantic Fishery Commission, Rome, 20–25 November 2006, Adopted at the Hundred and Thirty-first Session of the Council of the FAO, available at <http://www.fao.org/docrep/meeting/011/j8533e.htm>.

Southern Indian Ocean Fisheries Agreement, Rome, 7 July 2006, entered into force 21 June 2012, available at <http://www.fao.org/fileadmin/user_upload/legal/docs/035t-e.pdf>.

Declaration on Deep-sea Bottom Trawling to Protect Biodiversity in the High Seas, 24–25 October 2006, available at <http://www.forumsec.org/pages.cfm/newsroom/press-statements/press-statements-2006/thirty-seventh-pacific-islands-forum.html>.

Establishment of New Mechanisms for Protection of Vulnerable Marine Ecosystems and Sustainable Management of High Seas Bottom Fisheries in the North Western Pacific Ocean, Busan, Republic of Korea, 31 January–2 February 2007, available at <http://www.fpir.noaa.gov/Library/IFD/NWPBT_InterimMeasure-1-1.pdf>.

Interim Measures adopted by Participants in Negotiations to Establish South Pacific Regional Fisheries Management Organisation, 30 April–4 May 2007, available at <http://www.southpacificrfmo.org/assets/3rd-Meeting-April-2007-Renaca/Plenary-III/SPRFMO-Interim-MeasuresFinal.doc>.

Convention on Cooperation in the Northwest Atlantic Fisheries, Lisbon, 28 September 2007, not yet in force, available at <http://www.nafo.int/publications/meetproc/2008/gc/gcsep07/annex17.html>.

Vava'u Declaration on Pacific Fisheries Resources, 16–17 October 2007, available at <http://pidp.eastwestcenter.org/pireport/special/2007_PIF_Comunique.pdf>.

Third Arrangement Implementing the Nauru Agreement Setting Forth Additional Terms and Conditions of Access to the Fisheries Zones of the Parties, Koror, Palau, 16 May 2008, amended on 11 September 2010, available at <http://www.pnatuna.com/system/files/3rd%20Implementing%20Arrangement%20(Amended%20-%2011September%202010).pdf>.

Memorandum of Understanding between the North East Atlantic Fisheries Commission (NEAFC) and the OSPAR Commission, September 2008, reproduced in Report of the Extraordinary Meeting of the North-East Atlantic Fisheries Commission, 1 and 2 July 2008, volume II – Annexes, pp. 23–25.

Convention on the Conservation and Management of High Seas Fishery Resources in the South Pacific Ocean, Auckland, New Zealand, 14 November 2009, entered into force 24 August 2012, [2012] OJ L67/3.

Agreement on Port State Measures to Prevent, Deter and Eliminate Illegal, Unreported and Unregulated Fishing, Rome, 22 November 2009, not yet in force, [2011] OJ L191/3.

Regulations on Prospecting and Exploration for Polymetallic Sulphides in the Area, adopted at the 16th session of the International Seabed Authority (16 April–7 May 2010), available at <http://www.isa.org.jm/>.

Convention on the Conservation and Management of High Seas Fisheries Resources in the North Pacific Ocean, Tokyo, 24 February 2012, not yet in force, on file with author.

The Future We Want, Rio de Janeiro, Brazil, 22 June 2012, available at <http://www.uncsd2012.org/thefuturewewant.html>.

Regulations on Prospecting and Exploration for Cobalt-rich Ferromanganese Crusts in the Area, adopted at the 18th session of the International Seabed Authority (9–27 July 2012), available at <http://www.isa.org.jm/>.

Table of Cases

Permanent Court of International Justice

The Case of the S.S. "Lotus", Judgment of 7 September 1927, *PCIJ, Recueil des Arrêts, Série A, No. 10*, p. 3.

Interpretation of the Convention of 1919 concerning Employment of Women during the Night, Advisory Opinion of 15 November 1932, *PCIJ, Série A/B No. 50*, p. 365.

International Court of Justice

North Sea Continental Shelf (Federal Republic of Germany/Denmark; Federal Republic of Germany/Netherlands), Judgment of 20 February 1969, *I.C.J. Reports 1969*, p. 3.

Fisheries Jurisdiction Case (Federal Republic of Germany v. Iceland), Judgment (Merits) of 25 July 1974, *I.C.J. Reports 1974*, p. 175.

Fisheries Jurisdiction Case (United Kingdom of Great Britain and Northern Ireland v. Iceland), Judgment (Merits) of 25 July 1974, *I.C.J. Reports 1974*, p. 3.

Nuclear Tests Case (Australia v. France), Judgment of 20 December 1974, *I.C.J. Reports 1974*, p. 253.

Nuclear Tests Case (New Zealand v. France), Judgment of 20 December 1974, *I.C.J. Reports 1974*, p. 457.

Aegean Sea Continental Shelf (Greece v. Turkey), Request for the Indication of Interim Measures of Protection, Order of 11 September 1976, *ICJ Reports 1976*, p. 3.

Aegean Sea Continental Shelf (Greece v. Turkey), Judgment (Jurisdiction) of 19 December 1978, *ICJ Reports 1978*, p. 3.

Military and Paramilitary Activities in and against Nicaragua (Nicaragua v. United States of America), Judgment (Jurisdiction of the Court and Admissibility of the Application) of 26 November 1984, *I.C.J. Reports 1984*, p. 392.

Case concerning the Continental Shelf (Libyan Arab Jamahiriya/Malta), Judgment of 3 June 1985, *I.C.J. Reports 1985*, p. 13.

Case concerning the Frontier Dispute (Burkina Faso/Republic of Mali), Judgment of 22 December 1986, *I.C.J. Reports 1986*, p. 554.

Case concerning the Gabcikovo-Nagymaros Project (Hungary/Slovakia), Judgment of 25 September 1997, *I.C.J. Reports 1997*, p. 7.

Fisheries Jurisdiction Case (Spain v. Canada), Judgment (Jurisdiction) of 4 December 1998, *I.C.J. Reports 1998*, p. 432.

Case concerning the Land and Maritime Boundary between Cameroon and Nigeria (Cameroon v. Nigeria: Equatorial Guinea Intervening), Judgment of 10 October 2002, *I.C.J. Reports 2002*, p. 303.

Sovereignty over Pedra Branca/Pulau Batu Puteh, Middle Rocks and South Ledge (Malaysia/Singapore), Judgment of 23 May 2008, *I.C.J. Reports 2008*, p. 12.

International Tribunal for the Law of the Sea

(Both cases listed below are available on the website of the International Tribunal for the Law of the Sea at <http://www.itlos.org/>.)

The M/V "Saiga" (No. 2) Case (Saint Vincent and the Grenadines v. Guinea), Judgment of 1 July 1999.

Southern Bluefin Tuna Cases (New Zealand v. Japan; Australia v. Japan) of 27 August 1999, *Order (Request for provisional measures).*

Arbitration

Bering Sea Fur-Seals Case, Award of the Tribunal of Arbitration Constituted under the Treaty Concluded at Washington, the 29th of February 1892, Between the United States of American and Her Majesty the Queen of the United Kingdom of Great Britain and Ireland of 15 August 1893, *International Environmental Law Reports (IELR)*, vol. 1 (Cairo A.R. Robb ed.), p. 67.

The North Atlantic Coast Fisheries Case between Great Britain and the United States, Award of the Tribunal of 7 September 1910, *The Hague Court Reports*, vol. I, p. 141.

Case concerning filleting within the Gulf of St. Lawrence between Canada and France, Decision of 17 July 1986, *International Law Reports*, vol. 82, p. 590.

Southern Bluefin Tuna Case, Award on Jurisdiction and Admissibility of 4 August 2000, *RIAA*, vol. XXIII, p. 1.

Arbitration Regarding the Iron Rhine ("Ijzeren Rijn") Railway between the Kingdom of Belgium and the Kingdom of the Netherlands, Award of the Arbitral Tribunal of 24 May 2005, *RIAA*, vol. XXVII, p. 35.

Arbitration Between Barbados and the Republic of Trinidad and Tobago, Award of the Arbitral Tribunal of 11 April 2006, *RIAA*, vol. XXVII, p. 147.

Index

www.ingramcontent.com/pod-product-compliance
Lightning Source LLC
LaVergne TN
LVHW020651300926
849161LV00004B/901